The PCR Technique:
Quantitative PCR

The PCR Technique:
Quantitative PCR

Edited by

James W. Larrick
Palo Alto Institute of Molecular Medicine
Mountain View, CA

BioTechniques® Books
(Division of Eaton Publishing)

James W. Larrick, MD, PhD
Palo Alto Institute of Molecular Medicine
2462 Wyandotte Street
Mountain View, CA 94043

Library of Congress Cataloging-in-Publication Data

[CATALOGING IN PROGRESS]

ISBN 1-881299-06-6

Printed in the United States of America

9 8 7 6 5 4 3 2 1

CONTENTS

CONTRIBUTORS

P. Alard
Institut de Recherches Scientifiques
 sur le Cancer
Paris, France

J. Albert
Swedish Institute for Infectious Disease
 Control
Stockhom, Sweden

M. Axelsson
Swedish Institute for Infectious Disease
 Control
Stockhom, Sweden

S.A. Berman
Louisiana State University Medical Center
Shreveport, LA

Poppo H. Boer
University of Ottawa
Ottawa, Ontario, Canada

M.J. Brisco
Flinders Medical Center
Bedford Park, South Australia

Irena Bronstein
Tropix, Inc.
Bedford, MA

Sherry Bursztajn
McLean Hospital and Harvard Medical School
Belmont, MA

L. Butler
Tropix, Inc.
Bedford, MA

C.F. Calvo
Institut de Recherches Scientifiques sur le
 Cancer
Paris, France

Thomas A. Cebula
US Food and Drug Administration
Washington, DC

B. Charpentier
Institut de Recherches Scientifiques sur le
 Cancer and Hôpital de Bicêtre and Université
 Paris-Sud
Paris, France

G. Chavanel
Institut de Recherches Scientifiques sur le
 Cancer
Paris, France

Heidi Chehadeh
Immuno AG
Vienna, Austria

J. Condon
Flinders Medical Center
Bedford Park, South Australia

Teresa Cowen
Memorial University of Newfoundland
St. John's, Newfoundland, Canada

Joseph Crivello
University of Connecticut
Storrs, CT

Enoch Satyaselan Daniel
Memorial University of Newfoundland
St. John's, Newfoundland, Canada

Eric de Kant
Academisch Ziekenhuis Utrecht
Utrecht, The Netherlands

J.L. Dieguez-Lucena
University of Málaga
Málaga, Spain

Peter Doris
Texas Tech University School of Medicine
Lubbock, TX

Melanie Ehrlich
Tulane Cancer Center
Tulane Medical School
New Orleans, LA

Cornelis Elferink
Institute of Chemical Toxicology
Wayne State University
Detroit, MI

N. Fernández-Arcás
University of Málaga
Málaga, Spain

Patricia Forgez
INSERM U339
Hôpital Saint-Antoine
Paris, France

J. García-Villanova
University of Málaga
Málaga, Spain

Biswendu B. Goswami
US Food and Drug Administration
Washington, DC

Bernard Grandchamp
INSERM U409
Faculté de Médecine Bichat
Paris, France

Amanda Hayward-Lester
Texas Tech University Health Sciences
 Center
Lubbock, TX

viii

David George Haegert
Memorial University of Newfoundland
St. John's, Newfoundland, Canada

Richard Herrmann
Freie Universität
Berlin, Germany

T. Jalava
Orion Corporation
Espoo, Finland

Kathy Jessen-Eller
University of Connecticut
Groton, CT

Sonya Kashyap
University of Ottawa Heart Institute
Ottawa, Ontario, Canada

A. Kallio
Orion Corporation
Espoo, Finland

Walter H. Koch
US Food and Drug Administration
Washington, DC

F. Kury
Labordiagnostika GesmbH "Vienna Lab"
Vienna, Austria

Olivier Lantz
INSERM U 267
Hôpital Paul-Brousse
Villejuif, France

James W. Larrick
Palo Alto Institute of Molecular Medicine
Mountain View, CA

William Lear
University of Ottawa Heart Institute
Ottawa, Ontario, Canada

Paivi Lehtovaara
Orion Corporation
Espoo, Finland

Jeffrey D. Lifson
SAIC-Frederick National Cancer Institute
Cancer Research and Development
Frederick, MD

K.-C. Luk
Genelabs
Redwood City, CA

Joakim Lundeberg
KTH, Royal Institute of Technology
Stockholm, Sweden

Marie-Claire Malinge
Faculté Xavier Bichat
Paris, France

Josef W. Mannhalter
Immuno AG
Vienna, Austria

Chris S. Martin
Tropix, Inc.
Bedford, MA

Andrew McColgan
University of Ottawa Heart Institute
Ottawa, Ontario, Canada

Michael McDonnell
University of Ottawa Heart Institute
Ottawa, Ontario, Canada

Sherissa Microys
University of Ottawa Heart Institute
Ottawa, Ontario, Canada

Alec A. Morley
Flinders Medical Center
Bedford Park, South Australia

S.H. Neoh
Flinders Medical Center
Bedford Park, South Australia

A. Ntodou-Thomé
INSERM U339
Hôpital Saint-Antoine
Paris, France

Peter Oefner
Stanford University
Stanford, CA

Michael Piatak, Jr.
Becton Dickinson Microbiology Systems
Sparks, MD

Christiane Picat
Faculté Xavier Bichat
Paris, France

Enrico Picozza
Perkin-Elmer
Wilton, CT

Catherine Porcher
Faculté Xavier Bichat
Paris, France

Jeyanthi Ramamoorthy
University of Ottawa Heart Institute
Ottawa, Ontario, Canada

M. Ranki
Orion Corporation
Espoo, Finland

Zahra Rassi
University of Ottawa Heart Institute
Ottawa, Ontario, Canada

John J. Reiners, Jr.
Wayne State University
Detroit, MI

Armando Reyes-Engel
University of Málaga
Málaga, Spain

Christoph F. Rochlitz
Freie Universität
Berlin, Germany

W. Rostène
INSERM U339
Hôpital Saint-Antoine
Paris, France

M. Ruiz-Galdón
University of Málaga
Málaga, Spain

C. Schneeberger
University of Vienna Medical School
Vienna, Austria

M. Sebagh
Institut de Recherches Scientifiques sur le
    Cancer and Hôpital Paul-Brousse and
    Université Paris-Sud
Paris, France

A. Senik
Institut de Recherches Scientifiques sur le
    Cancer
Paris, France

Paul D. Siebert
CLONTECH Laboratories
Palo Alto, CA

Hans Söderlund
VTT Biotechnology and Food Research
Technical Research Centre of Finland
Espoo, Finland

F. Souazé
INSERM U339
Hôpital Saint-Antoine
Paris, France

P. Speiser
University of Vienna Medical School
Vienna, Austria

X. Su
McLean Hospital and
    Harvard Medical School
Belmont, MA

T.F. Sullivan
McLean Hospital and
    Harvard Medical School
Belmont, MA

Pamela Sykes
Flinders Medical Center
Bedford Park, South Australia

C.Y. Tran
INSERM U339
Hôpital Saint-Antoine
Paris, France

Mathias Uhlén
KTH, Royal Institute of Technology
Stockholm, Sweden

Tatiana Vener
KTH, Royal Institute of Technology
Stockholm, Sweden

Johan Wahlberg
KTH, Royal Institute of Technology
Stockholm, Sweden

B. Williams
Applied Imaging
Santa Clara, CA

D. Weill
Institut de Recherches Scientifiques sur le
    Cancer
Paris, France

Robert Zeillinger
University of Vienna Medical School
Vienna, Austria

Gerold Zerlauth
Immuno AG
Vienna, Austria

Klaus Zimmerman
Immuno AG
Vienna, Austria

X.-Y. Zhang
Tulane Medical School
New Orleans, LA

Preface

Since the early 1980s *BioTechniques* has grown to be the most widely distributed life science laboratory techniques journal in the world. Most, if not all, bioscience laboratories on the planet use material presented in this journal. This observer has seen copies of *BioTechniques* on the lab benches of biomedical laboratories in such diverse places as Cuba, Saudia Arabia, China and Bolivia. Thus, it comes as no surprise that this forum has served to present many of the subtle refinements and improvements of various novel laboratory methods. Such is the case with quantitative PCR, one of the most useful and now widely used methods in molecular analysis.

This volume collects a number of important papers focused on various aspects of quantitative PCR which first appeared in *BioTechniques* from 1991 to 1996—a period of time that witnessed the introduction, rapid development and improvement in this laboratory method. The papers present novel methods to quantitate PCR products, improvements and modifications of previously published methods, and applications of quantitative PCR. Each chapter has been reviewed and updated by the original authors. Although advances in this methodology have appeared in other journals, the articles appearing in *BioTechniques* represent a broad range of experiences and an excellent sampling of quite useful techniques.

The book has been divided into three parts. The first section includes three papers on the general theory and analysis of PCR products. In addition to these papers, the reader is referred to the recent theoretical work of Stolovitsky and Cecchi (5) and Connolly et al. (1). To begin, the chapter by Zimmerman and Mannhalter provides a comprehensive review of the topic with an excellent set of references to the earlier literature. Souazé et al. demonstrate the reproducible quantitation of mRNA for the neurotensin receptor, and Cowan and Haegert describe a method to correct for biases introduced by differences in priming efficiencies when multiple PCR primers are being used simultaneously. Correction for PCR amplification biases is particularly important when any set of multiple primers is used to study the relative expression levels of different transcripts (see Tarnuzzer et al., Reference 6).

The middle section includes a number of methods to quantitate and then detect PCR products. Lifson and Piatak (Chapter 5) pioneered the use of quantitative competitive PCR (QC-PCR) applied to detection of HIV-1 viral load (4). Their work, along with others, initiated a fundamental paradigm change in models of AIDS pathogenesis: the virus does not enter a latent phase but rather continues to replicate at high levels throughout infection (3). This paradigm shift resulted in the development of successful combination anti-retroviral treatment strategies. References and a discussion of the impact of this methodology on our understanding and treatment of this devastating disease can be found in the update to their original chapter (p. 58-66).

Quantitation has been achieved by using internal standards of various designs, competitive PCR (such as PCR MIMICS: Chapter 6, Seibert and Larrick), limiting dilution (Chapter 7, Sykes et al.), and additive PCR (Chapter 7, Reyes-Engel et al.). Methods to measure the amplified products range from the originally used agarose gel-based methods to the more facile ELISA (see Chapter 8, Alard et al.), video image analysis (Chapter 9, Chehadeh et al.), HPLC detection (Chapter 10, Hayward-Lester et al.; Chapter 11, Zeillinger et al.) and chemiluminescence (Chapter 12, Jessen-Eller et al.; Chapter 13, Martin et al.; and Chapter 14, Su et al.). These latter assays are particularly suited for the development of high throughput assays. Lundeberg et al. (Chapter 15) describe a novel colorimetric assay, which is based on the co-amplification of target DNA with a fragment incorporating a *lac* operator sequence. This sequence incorporated in the amplified DNA is recognized by a fusion protein composed of the *E. coli LacI* repressor and beta-galactosidase. Some of these more recent methods have now been adapted to kits and large throughput quantitative analysis of PCR products. This section finishes with chapters describing quantitation using a DNA sequencer contributed by Porcher et al. (Chapter 16) and digoxigenin-labeled primers and ELISA detection using DIG-specific antibody conjugated with alkaline phosphatase (Chapter 17, Lear et al.).

The final section comprises papers that offer further applications of quantitative PCR. For example, quantitative RT-PCR is used to examine expression of the MDR1 gene (de Kant et al., Chapter 18); expression of CYP1A1 heterogeneous nuclear RNA (Elferink and Reiners, Chapter 19); hepatitis A virus (Goswami et al., Chapter 20); and hepatitis B virus (Jalava et al., Chapter 21). Vener et al. (Chapter 22) present a single-tube assay for quantitation of HIV-1 using multiple competitors. Amplified fragments are captured by streptavidin and detected by a standard automated sequencer. Finally, Zhang and Ehrlich (Chapter 23) use semi-nested PCR to detect rare cells bearing *bcl-2* translocations. Although PCR is widely used in diagnostic laboratories, routine use of quantitative PCR is not as common. As the methods become more standardized this will change. As noted above, quantitation of HIV-1 has changed the management of infected patients and is likely to have a similar effect on other viral diseases. As demonstrated by Zhang and Ehrlich, the application of quantitiative PCR to the management of oncology patients is also particularly promising.

As with any methodology, quantitative PCR is a "work in progress". Although all of the methods described in this volume utilize PCR, there are subtle differences in assay format, accuracy, and reliability of quantitation, and to date the pros and cons of the various methods have not been rigorously compared. Clearly, the next generation of PCR quantitation will be increasingly automated. One especially promising method deserves mention. Heid et al. (2) describe the development of "real-time" quantitative PCR using the TaqMan™ probe. Capillary electrophoresis and ion-pair reversed-phase

HPLC provide alternative detection systems compatible with high throughput automation. Perhaps this volume will stimulate this type of work. In any event, this volume will serve as a resource for all those interested in this important technology, and these papers will be a springboard to future applications and technology improvement.

1.**Connolly, A.R., L.G. Cleland and B.W. Kirkham.** 1995. Mathematical considerations of competitive polymerase chain reaction. J. Immunol. Methods *187*:201-211.
2.**Heid, C.A., J. Stevens, K.J. Livak and P.M. Williams.** 1996. Real time quantitative PCR. Genome Res. *6*:10.
3.**Perelson, A.S., A.U. Neumann, M. Markowitz, J.M. Leonard and D.D. Ho.** 1996. HIV-1 dynamics *in vivo*: virion clearance rate, infected cell lifespan, and viral generation time. Science *271*:1582.
4.**Piatak, M., M.S. Saag, L.C. Yang, S.J. Clark, J.C. Kappes, K.-C. Luk, B.H. Hahn, G.M. Shaw and J.D. Lifson.** 1993. High levels of HIV-1 in plasma during all stages of infection determined by competitive PCR. Science *259*:1749.
5.**Stolovitsky, G. and G. Cecchi.** 1996. Efficiency of DNA replication in the polymerase chain reaction. Proc. Nat. Acad. Sci. USA *93*:12947.
6.**Tarnuzzer, R.W., S.P. Macauley, W.G. Farmerie, S. Caballero, M.R. Ghassemifar, J.T. Anderson, C.P. Robinson, M.B. Grant, M.G. Humphreys-Beher, L. Franzen, A.B. Peck and G.S. Schultz.** 1996. Competitive RNA templates for detection and quantitation of growth factors, cytokines, extracellular matrix components and matrix metalloproteinases by RT-PCR. BioTechniques *20*:670.

James W. Larrick, MD, PhD
Palo Alto Institute of Molecular Medicine
Mountain View, CA

Section I

General Theory and Analysis

CHAPTER 1

Technical Aspects of Quantitative Competitive PCR

Klaus Zimmermann and Josef W. Mannhalter

Immuno AG, Vienna, Austria

INTRODUCTION

The polymerase chain reaction (PCR), first described by Saiki et al. (79), is a highly sensitive and specific methodology for detection of nucleic acids and a useful tool for quantitation of the amount of specific nucleic acids present in a sample. The simplest approach to quantitation of PCR and reverse transcription PCR (RT-PCR) products (reviewed by References 17,21,31 and 32) is measurement of the amount of amplification product in the exponential phase by reference to the dilution series of an external standard. However, accurate quantitation with this type of PCR is hampered by a number of variabilities that can occur during sample preparation or in the course of the reaction, and minor variations in reaction conditions are greatly magnified during the amplification process. These variabilities may partly be overcome by normalizing the amount of PCR products of the specific template with respect to an internal reference template such as the cellular gene β-globin (20) amplified in the same reaction tube.

Alternatively, limiting dilution using a nested primer methodology (57,90) can be used in combination with Poisson statistics for evaluation of the results.

The most precise quantitation of DNA and RNA can, however, be obtained by competitive PCR and competitive RT-PCR (reviewed in References 16–18,21,31,32 and 84). This assay is based on competitive co-amplification of a specific target sequence together with known concentrations of an internal standard in one reaction tube. The internal standard has to share primer recognition sites with the specific template, both specific template and internal standard must be PCR-amplified with the same efficiency and it must be possible to analyze the PCR-amplified products of specific template and internal standard separately. Quantitation is then performed by comparing the PCR signal of the specific template with the PCR signals obtained with known concentrations of the competitor (the internal standard). Ever since this method was first described (6,38,99), it has been widely used for quantitation of cellular RNA and DNA as well as viral and bacterial nucleic acids. Examples reported on include the quantitation of cytokine expression

(Reprinted from BioTechniques 21:268-279, 1996)

(6,38,52,91,99,103) and of viral nucleic acids such as hepatitis B (46), hepatitis C (9,40,53,58,75,78,106), human cytomegalovirus (35), herpes simplex virus (74) or human immunodeficiency virus type 1 (HIV-1) (3,5,36,51,62,70,71,81,88,92) and quantitation of bacteria, especially of slow-growing species such as mycobacteria (55). Furthermore, the utility of competitive PCR or RT-PCR has been demonstrated in the quantitation of mitochondrial DNA (100) or mRNA expression (29,69) and assessment of hereditary deficiencies (e.g., Reference 43) or leukemias (22,93).

Considering the hundreds of published papers on the use of competitive PCR, it is not surprising that a great variety of protocols exist. In the present article, we shall review this methodology, concentrating in particular on technical aspects of competitive PCR, such as the construction of competitive templates, different PCR strategies and modes of detecting PCR products.

CONSTRUCTION OF INTERNAL STANDARDS FOR COMPETITIVE PCR

Generation and testing of suitable internal standards and the choice of primer pairs are among the most crucial and time-consuming aspects of setting up a competitive PCR protocol. The simplest way to choose suitable primers is to use, as far as possible, primers already widely used and/or well described. If new primers have to be designed, computer programs in combination with sequence databases may be helpful tools.

General concepts of PCR primer design have been reviewed by Dieffenbach et al. (26). Since different primer pairs for the same gene can exhibit up to 1000-fold differences in sensitivity (44), special emphasis should be placed on testing the specificity and efficiency of primer pairs in advance before internal standards are generated. Once the appropriate primer pair has been selected, various strategies for the construction of internal standards may be used.

Internal standards for competitive PCR or RT-PCR are DNA or RNA fragments sharing the primer recognition sequences with the specific target yet yielding PCR products that are distinguishable from the wild-type template. The easiest way to distinguish between wild-type template and internal standard is by differences in the size of the two products. This can be achieved, for example, by constructing standards having the same sequence as the specific target but containing a deletion or an insertion. The simplest construction procedure is to use a composite primer containing two specific target sequences at a predetermined distance from each other (thus resulting in a deletion) and a second primer specific for the opposite strand (14,33,48, 61,77). The amplification of wild-type templates with such primers results in a PCR product that is shorter than the wild-type template and can thus easily be identified (e.g., after electrophoretic or HPLC separation). Alternatively, elongated internal standards can be constructed using a "looped oligo" method by amplifying cDNA with a primer containing a non-templated nucleotide insertion between template sequences (80).

When deletions or insertions in the center of the internal standard are desired, a splice overlap extension PCR can be used (39,72,82,92). First, two parts of a gene or different genes are amplified in separate PCRs. Since the downstream primer of the first template and the upstream primer of the second template are designed to contain complementary sequences, the two parts of the gene or the different genes can be linked together by subsequently amplifying the two different PCR products in one reaction tube. In similar approaches, internal standards can also be constructed by PCR amplification of a mixture of religated PCR products (1,37). Finally, if the sequence of the specific template contains one or two internal restriction sites, these sites can be used to obtain a construct with an insertion or a deletion (3,11,70,72,104).

Differences between wild type and standard can also be obtained by incorporation of restriction sites (2,6,35,38,59,68,88). Following PCR, the standards containing such restriction sites are digested with the respective restriction enzyme. Since the digested products are smaller in size, they can easily be distinguished from the wild-type template and can be analyzed separately.

Exchange of nucleotides in specific templates is another way to construct competitive internal standards (54,75,78,106). After competitive PCR, the wild-type fragment and competitive internal standards containing nucleotide exchanges can be identified following differential hybridization with probes specific for wild-type and competitor sequences.

A basically different approach to the construction of competitive internal standards is to use a nonspecific spacer gene or spacer DNA. These competitors have nucleotide sequences totally different from the wild type, except for wild-type primer sequences attached to their 5' ends. Such approaches have been used for quantitation of cytochrome P450 by using the glutathione transferase μ gene as the spacer sequence for an internal standard (96) or for quantitation of glyceraldehyde 3-phosphate dehydrogenase with an internal standard based on the v-erb B oncogene (85). Nonspecific spacer sequences are also frequently used to construct multicompetitor standards containing priming sites for various genes of interest. Such multicompetitor standards can be constructed by directly linking together several primer sites and then attaching them to a spacer sequence (4,7,52,86,99).

Finally, generation of competitive standards by amplification of multiple DNA fragments from cellular material in a PCR with low-stringency annealing conditions has been described (34,94). This procedure results in a variety of different-sized PCR products, the most suitable of which is purified and used as internal standard in a quantitative competitive PCR assay.

The competitive templates constructed in one of the ways described above are usually loaded on a gel, separated, excised and extracted. They can then be used directly in a competitive PCR assay (14,48,85), but more often, they are cloned into plasmids, preferably by the help of additionally introduced restriction sites (e.g., References 9, 61,72,104 and 110).

Cloning of competitive standards into plasmids usually involves enzymatic reactions such as digestion with specific restriction enzymes, generation of blunt ends and ligation (e.g., References 61,70,88 and 92). In addition to these conventional strategies, other simpler procedures have been developed. Should a plasmid containing the wild-type sequence be available, this plasmid can be used directly as an internal standard after introducing a deletion into the wild-type sequence as described by Zarlenga et al. (108). Furthermore, PCR products can be engineered to contain terminal sequences identical to sequences of the two ends of a linearized vector. These two DNAs can be co-transfected simply into *Escherichia coli*, which results in incorporation of the insert into the vector (49,67). Finally, PCR products and linearized plasmid can also be ligated together in a PCR by overlap extension of complementary ends (50,83). Following denaturation, heterologous re-annealing and cyclization, a recombinant plasmid is formed that is then used to transform competent *E. coli*.

It must furthermore be ensured that the concentration of the internal standard is determined accurately. It is thus advisable not only to quantitate the competitive plasmid to be used as an internal standard spectrophotometrically, but also to check its quantity and quality electrophoretically. In addition, the concentration of internal standards should be checked by PCR in combination with end-point dilution, taking into consideration the Poisson distribution of positive samples (61).

ANALYSIS OF COMPARABLE AMPLIFICATION OF WILD-TYPE AND INTERNAL STANDARD TEMPLATES

A prerequisite for the proper performance of quantitative competitive PCR is a comparable amplification efficiency of competitive internal standard and wild-type templates, and it is especially important to carefully control for this when setting up a new PCR-based quantitation procedure. The easiest mode of control is to repeatedly quantify several known amounts of specific wild-type template with the help of the newly constructed internal standard. If these analyses result in the calculated copy numbers, comparable amplification efficiencies between standard and wild-type template can be assumed, and the newly constructed internal standard can be safely used for quantitation. If the results of such quantitation experiments point to minor differences in amplification efficiency between wild type and standard, this may be compensated for by using a correction factor (60). Another frequently used procedure for the determination of amplification efficiencies is an analysis of known amounts of wild-type template and competitive internal standard in a PCR protocol that uses fluorescently or radioactively labeled primers. After various numbers of cycles, the PCR products are separated and quantitated by measuring their radioactivity or fluorescence (19,81).

When controlling for amplification efficiencies of wild-type and internal standard templates, it is also important to keep in mind that possible differences may become more pronounced when the amplification reaction is driven to the post-exponential (plateau) phase (6,60). To ensure proper quantitation, it has therefore been suggested that the amplification efficiencies of wild-type and internal standard PCR products be measured during the exponential, post-exponential and plateau phases (10,19). In general, it can be concluded that differences in sequences of wild type and standard as well as small differences in size do not substantially affect amplification efficiency. However, as an inverse exponential relationship between template size and amplification efficiency has been observed (60), it is obviously important to keep the size difference between wild type and standard as small as possible (60,61).

Another problem to be dealt with concerns the possible formation of heteroduplexes, which can occur when wild-type and internal standard templates differ in only one or a few bases (either small differences in size or small diversities in the nucleotide sequences) (6,60,70). Since heteroduplexes would interfere with subsequent quantitation procedures, great care should be taken to avoid their formation (2,24,42).

All of the above-mentioned considerations are equally important for the construction and analysis of internal standards to be used for quantitation of DNA by competitive PCR or of RNA by competitive RT-PCR. However, for the construction of RNA standards, additional precautions have to be taken. For synthesis of competitive internal standard RNA, the competitive DNA constructs are placed in plasmid vectors under the control of an RNA polymerase promoter (e.g., T7, SP6 or T3). After linearization of the vector, large amounts of RNA are transcribed (61,70). The RNA is then purified by DNase I digestion, extraction and precipitation. Subsequently, quality and copy number of the competitive RNA standard (compared to the copy number of contaminating residual DNA) should be determined both spectrophotometrically and by gel electrophoresis (61). Several researchers have described an interesting shortcut to this common methodology for synthesizing competitive internal standard RNA. They used a composite primer to construct a recombinant competitor cDNA carrying the T7 RNA polymerase recognition sequence at its 5′ end (39,77,96). After amplification with this composite primer and a second specific primer, the resulting PCR product is used directly for transcription by T7 RNA polymerase.

PCR AND RT-PCR

Extraction of Nucleic Acids

To ensure proper performance of the quantitative competitive PCR assay, the methods used for preparing the nucleic acids to be quantitated must be chosen with care and, if necessary, optimized. This is especially essential for preparation of RNA. DNA preparation, on the other hand, can follow more straightforward procedures. With respect to DNA quantitation, it has been shown that crude cell lysates can be used with the same reproducibility as purified DNA (88,110). Furthermore, the use of crude cell lysates entails less labor and avoids the loss of specific sample, which may occur during DNA purification. If, for one reason or another, the use of purified DNA for quantitation appears preferable, loss of specific sample or DNA degradation should be controlled for, e.g., by quantitating a control gene in a second competitive PCR, as described by Deng et al. (25).

Before quantitation of RNA, it is especially important to optimize RNA extraction procedures and to perform careful quality control of the extracted RNA. Genomic DNA can either be eliminated by digestion with DNase I, or, more elegantly, RT-PCR primers spanning one or more introns (38) can be designed. The simplest means of controlling RNA quality is to both measure optical density and run a denaturing RNA gel capable of detecting possible RNA degradation. It has been suggested that extraction efficiency can be estimated by performing a second extraction and determining the mean RNA loss (61). To ensure comparable extraction in a series of samples, housekeeping genes are usually quantitated in addition, and the copy number of the gene of interest is calculated with respect to the copy number of the housekeeping gene (7,86). However, it cannot be taken for granted that the so-called housekeeping genes will maintain a steady level of expression under all circumstances. Expression of housekeeping genes is often increased, especially in activated cells.

The addition of competitor RNA molecules before extraction is also a simple means of controlling extraction efficiency. When detection of RNA viruses is desired, this can most elegantly be performed by using virus mutants as competitors, provided these mutants differ from the wild type in the genome section amplified by the two primers (66). To achieve a meaningful quantitation, several known concentrations of competitive internal standard RNA must be added separately to the same concentration of wild-type RNA in different vials (70), followed by RT and PCR amplification. Before quantitation of RNA using the nucleic acid sequence-based amplification (NAS-BA™) methodology, addition of two or more known concentrations of different-sized competitive RNA standards to the same vial has been suggested (97).

Reverse Transcription

After having selected the most appropriate method of RNA preparation, great care should be taken to optimize RT of RNA. Usually, enzymes from Moloney murine leukemia virus (MMLV) or avian myeloblastosis virus (AMV) are used for RT, but the RT-PCR process can be simplified by the use of *Tth* DNA polymerase possessing both RT and polymerase activity (64). Some authors reported that *Tth* had a lower RT reactivity than RT from MMLV and therefore might not be the most suitable enzyme for detecting low copy numbers (23). However, if the conditions for *Tth* are very carefully optimized, low copy numbers may also be detected with this enzyme (105).

To control for the RT step, the sample should always be reverse-transcribed in the presence of varying amounts of competitive internal standard RNA followed by PCR amplification. In the scientific literature, several protocols do exist in which RNA is first reverse-transcribed, and the resulting cDNA is then quantitated with competitive DNA templates (e.g., Reference 4). However, this does not take into account the variability of the RT step, and such a procedure is therefore not suitable when exact quantitation is desired. Furthermore, the efficiency of amplification between single- or double-stranded DNA templates may vary (13).

Amplification

Polymerase enzymes such as the widely used *Taq* DNA polymerase or alternative enzymes such as DynaZyme™ (Finnzymes Oy, Espoo, Finland) and *Tth* usually work very well in the buffers supplied with the enzyme when used according to the manufacturers' instructions. As has been reviewed by Erlich et al. (28), concentration of enzyme, primers and nucleotides, the number of cycles and denaturation, annealing and extension times affect the specificity and sensitivity of the PCR. Since $MgCl_2$ concentrations and annealing temperature especially influence the specificity of the reaction, great care should be taken to optimize these parameters. Starting the reaction at a high temperature ("hot start") may improve performance of the PCR by eliminating undesired hybridization events (65).

Most quantitative competitive PCR protocols described suggest the use of 30–50 PCR cycles, which in most cases drives the PCR beyond the exponential phase. From time to time, publications appear (27,78,101) that recommend restricting quantitative competitive analysis to the exponential phase of product accumulation. This exponential phase would have to be determined carefully and might differ not only due to the nature of the specific template to be quantitated, but also with respect to the equipment being used; therefore, it would greatly complicate PCR-based quantitation of nucleic acids and thus be a major drawback in routine application of this technique. However, recent studies have addressed this problem (10,19,63,84) and have shown that amplification of both wild-type and competitive internal standard templates proceeded during the exponential and non-exponential phases up to the plateau phase with equal efficiency.

General Considerations for Optimal Performance of Quantitative Competitive PCR or RT-PCR

In theory, a single concentration of competitive internal standard co-amplified with the specific template may be sufficient for quantitation, especially if the internal standard/specific template ratio is compared with a calibration curve obtained by comparing constant amounts of internal standard with several known concentrations of wild-type template (46,107). However, since it has been demonstrated that quantitative measurements of PCR-amplified products are most accurate when ratios of wild-type and competitive internal standard templates are equal or similar (3,73), the performance of quantitative PCR protocols using co-amplification of one concentration of specific template with several dilutions of internal standard is preferred.

One concentration of specific template is also compared with only one concentration of standard when DNA is quantitated by performing PCR under low-stringency conditions (12). With this method, it is not necessary to construct a competitive internal standard and add it to the amplification tube. The standard is generated instead from cellular material co-amplified during the low-stringency PCR. However, as the ratio of cellular material to specific DNA cannot be changed, this method does not allow titration with different concentrations of competitive standard template and thus does not allow precise quantitation.

Accuracy of quantitative competitive RT-PCR can be potentiated by simultaneous determination of a gene of interest and a reporter gene in one reaction tube (multiplex PCR) (2,24,27,30). The expression of the titrated gene of interest is then calculated with reference to the titrated level of the reporter gene. The use of competitive templates with several tandem-arranged internal standards may be preferable for such an approach (62,76,98), as this may reduce possible errors that could occur when mixtures of specific template with several different standards have to be prepared. However, since the respective amplifications in a multiplex competitive PCR should not interfere with each other and all of the templates and primer pairs used have to be amplified with similar efficiency, not more than a few different primer-template systems can be included in a single assay.

Another important parameter that has to be considered when performing quantitative PCR analyses concerns the detection limit for specific DNA templates. With a single round of amplification, this detection limit has been reported to be 10 copies per PCR tube (70). However, problems may arise when a small number of specific copies have to be detected in a high background of nonspecific cellular DNA. To overcome this problem and raise the sensitivity and specificity of the quantitative competitive PCR assay, nested (11,41,42,106) or hemi-nested (62,95,110) PCR protocols have been developed and successfully used for quantitation of both DNA and RNA templates.

It is also important not to overestimate the accuracy of quantitative competitive PCR. With well-established tests of this type performed by experienced scientists, standard deviations between 10% and 20% (2,9,24,71,104,

110) have been reported for the analysis of replicate portions of the same sample on different occasions. Therefore, for reliable discrimination of twofold differences in copy numbers between two samples, it is preferable to calculate a mean of replicate determinations of the same sample.

DETECTION AND ANALYSIS OF PCR PRODUCTS

For detection and analysis of amplified products, a large variety of procedures are available (reviewed in Reference 47). It should generally be kept in mind that the detection method chosen should, whenever possible, avoid additional manipulations, which not only take more time, but increase the possibility of inaccurate quantitation.

PCR products of different sizes are usually separated by conventional gel electrophoresis (e.g., References 61,70,110), capillary gel electrophoresis (30,87) or HPLC (24). The easiest and most commonly used method is to use gel electrophoresis for separation of the PCR products and to visualize the DNA bands by ethidium bromide or an equivalent fluorescent dye (2,7,39,61, 70,77,86,110). The gel materials to be used are either agaroses specially designed for separating small DNA fragments or polyacrylamide gels. For quantitation of the respective DNA bands, a video image analysis system can be used. The gels can be either scanned directly on the transilluminator (61,70) or scanned after photography (2,15,77,84,89). In doing so, keep in mind that the amount of incorporated dye and the resulting fluorescent emission depends on the length of the nucleic acids. Since the determination of the equivalence point of the two different fragments is based on molar amounts, a correction factor that takes into account the diverse fluorescence emitted by different-sized fragments must be used (e.g., Reference 70) to ensure correct quantitation. The respective ratios of unknown wild-type template and known competitive standard concentrations are then used to generate a diagram that can be utilized to quantitate precisely the amount of wild-type template. The number of specific template copies in the sample is usually calculated by determining the competition equivalence point in a regression plot, as described by Piatak et al. (70). In brief, the $\log_{10}$ of the ratio of the signal intensity of the competitive internal standard over the wild-type fragment, usually multiplied by a correction factor, is plotted as a function of the $\log_{10}$ of the concentration of the competitive internal standard template. This yields a linear plot, and interpolation on the plot for a Y value of 0 gives the numbers of copies present in the test sample.

A procedure more sensitive than ethidium bromide staining, which is also frequently used for quantitation of competitive PCR products, is based on the use of radioactively or fluorescence-labeled dNTPs (27,36,38,48,104) or oligonucleotides (3,99). After electrophoretic separation, the radioactively labeled bands can be excised from gels and the cpm counted (36,81,99). However, drying of the gels and direct autoradiography and scanning (27) or radioimaging of dried gels (3,48,109) is simpler and, as this involves less manipulation, possibly also more precise. A disadvantage of direct labeling is that it also visualizes all nonspecific bands. To avoid this, PCR products are separated electrophoretically, transferred to a membrane and quantitated after hybridization with radioactively (92) or nonradioactively labeled DNA probes. Alternatively, the competitive PCR products can be hybridized with a labeled oligonucleotide before separation (liquid hybridization) and auto-radiographed after gel electrophoresis (66,88).

PCR products generated by using fluorescently labeled primers can be separated on polyacrylamide gels and elegantly quantitated using an auto-mated laser fluorescent DNA sequencer (19,62,72). This laser-induced fluo-rescence (LIF) is a highly sensitive methodology and in combination with computer software offers a means of automating post-PCR analysis. LIF can also be applied for quantitation of competitive PCR products following capil-lary electrophoresis (30,87).

Finally, HPLC has recently been used for separation and quantitative determination of competitive PCR products (24).

To achieve more automated quantitation, methods that involve capturing of PCR products on solid supports (e.g., microplates or magnetic beads) fol-lowed by quantitation of the captured material have been reported. The most frequently used procedure involves amplification of internal standard and specific template in the presence of biotinylated primers followed by captur-ing of the PCR products on microplates coated with streptavidin (46,56,91). Alternatively, 5′ aminated PCR products can be captured onto carboxylated wells of microplates (54), or PCR products are captured by probes covalently attached to organosilan surface-activated plates (8). PCR products captured in one of the ways described above are hybridized with probes specific for wild-type or internal standard sequences. These probes usually contain ra-dioactive labels or have the capacity to interact with enzymes. Quantitation is performed by measuring incorporated radioactivity (46) or digoxigenin content (9,54) or by immunoenzymatic detection (56,91,106).

Competitive RT-PCR assays have also been developed for an automated system that makes use of electrochemiluminescence (QPCR System 5000™, Perkin-Elmer, Norwalk, CT, USA) for quantitation of PCR products (45, 102). One of the primers used for amplification is biotinylated to allow cap-ture of the PCR products by magnetic beads coated with streptavidin. The PCR products can be quantitated separately by hybridization with TBR (tris(2,2′-bipyridine)ruthenium(II) chelate)-labeled probes specific for wild-type or competitive internal standard templates, or alternatively TBR-labeled primers are used for direct quantitation of the amplified products.

CONCLUSIONS

Competitive PCR and RT-PCR controlling for tube-to-tube variations in amplification efficiency by the addition of competitive internal standards are one of the most widely used approaches for quantitation of nucleic acids. If the competitive template is properly chosen and constructed, the amplification efficiency of specific target and internal standard remain equal during the entire PCR process. Therefore, competitive PCR need not be stopped during the exponential phase of amplification, which is one of the greatest advantages of this methodology. Most protocols described for precise quantitation of a specific template require several amplification reactions per sample, and attempts are currently being made to simplify quantitation by introducing several different-sized competitive templates into the same reaction tube. Another way to simplify and speed up competitive PCR involves simultaneous detection of several different specific templates in one sample by using a multiplex PCR protocol.

A great potential for competitive PCR obviously lies in the automatic analysis of separated PCR products. For research applications, easy-to-perform methodologies (such as electrophoretic separation on ethidium bromide-stained gels and measurement of the PCR products by video image analysis) not using sophisticated and expensive instruments will also continue to be widely used. The future of this procedure, however, lies in routine diagnostic and clinical applications.

REFERENCES

1. **Ali, S.A. and A. Steinkasserer.** 1995. PCR-ligation-PCR mutagenesis: a protocol for creating gene fusions and mutations. BioTechniques *18*:746-750.
2. **Apostolakos, M.J., W.H.T. Schuermann, M.W. Frampton, M.J. Utell and J.C. Willey.** 1993. Measurement of gene expression by multiplex competitive polymerase chain reaction. Anal. Biochem. *213*:277-284.
3. **Arnold, B.L., K. Itakura and J.J. Rossi.** 1992. PCR-based quantitation of low levels of HIV-1 DNA by using an external standard. Genet. Anal. Tech. Appl. *9*:113-116.
4. **Babu, J.S., S. Kanangat and B.T. Rouse.** 1993. Limitations and modifications of quantitative polymerase chain reaction. Application to measurement of multiple mRNAs present in small amounts of sample RNA. J. Immunol. Methods *165*:207-216.
5. **Bagnarelli, P., S. Menzo, A. Valenza, A. Manzin, M. Giacca, F. Ancarani, G. Scalise, P.E. Varaldo and M. Clementi.** 1992. Molecular profile of human immunodeficiency virus type 1 infection in symptomless patients and in patients with AIDS. J. Virol. *66*:7328-7335.
6. **Becker-André, M. and K. Hahlbrock.** 1989. Absolute mRNA quantification using the polymerase chain reaction (PCR). A novel approach by a PCR aided transcript titration assay (PATTY). Nucleic Acids Res. *17*:9437-9446.
7. **Benavides, G.R., B. Hubby, W.M. Grosse, R.A. McGraw and R.L. Tarleton.** 1995. Construction and use of a multi-competitor gene for quantitative RT-PCR using existing primer sets. J. Immunol. Methods *181*:145-156.
8. **Berndt, C., M. Bebenroth, K. Oehlschlegel, F. Hiepe and N. Schößler.** 1995. Quantitative polymerase chain reaction using a DNA hybridization assay based on surface-activated microplates. Anal. Biochem. *225*:252-257.
9. **Besnard, N.C. and P.M. Andre.** 1994. Automated quantitative determination of hepatitis C virus viremia by reverse transcription-PCR. J. Clin. Microbiol. *32*:1887-1893.
10. **Bouaboula, M., P. Legoux, B. Pésségué, B. Delpech, X. Dumont, M. Piechaczyk, P. Casellas and D. Shire.** 1992. Standardization of mRNA titration using a polymerase chain reaction method involving co-amplification with a multispecific internal control. J. Biol. Chem. *267*:21830-21838.
11. **Bruisten, S.M., M.H.G.M. Koppelman M.T.L. Roos, A.E. Loeliger, P. Reiss, C.A.B. Boucher and**

H.G. Huisman. 1993. Use of competitive polymerase chain reaction to determine HIV-1 levels in response to antiviral treatments. AIDS *7(suppl. 2)*:S15-S20.

12. **Caballero, O.L., L.L. Villa and A.J.G. Simpson.** 1995. Low stringency-PCR (LS-PCR) allows entirely internally standardized DNA quantitation. Nucleic Acids Res. *23*:192-193.

13. **Carding, S.R., D. Lu and K. Bottomly.** 1992. A polymerase chain reaction assay for the detection and quantitation of cytokine gene expression in small numbers of cells. J. Immunol. Methods *151*:277-287.

14. **Celi, F.S., M.E. Zenilman and A.R. Shuldiner.** 1993. A rapid and versatile method to synthesize internal standards for competitive PCR. Nucleic Acids Res. *21*:1047.

15. **Chehadeh, H.E., G. Zerlauth and J.W. Mannhalter.** 1995. Video image analysis of quantitative competitive PCR products: comparison of different evaluation methods. BioTechniques *18*:26-28.

16. **Clementi, M., P. Bagnarelli, A. Manzin and S. Menzo.** 1994. Competitive polymerase chain reaction and analysis of viral activity at the molecular level. Genet. Anal. Tech. Appl. *11*:1-6.

17. **Clementi, M., S. Menzo, P. Bagnarelli, A. Manzin, A. Valenza and P.E. Varaldo.** 1993. Quantitative PCR and RT-PCR in virology. PCR Methods Appl. *2*:191-196.

18. **Clementi, M., S. Menzo, A. Manzin and P. Bagnarelli.** 1995. Quantitative molecular methods in virology. Arch. Virol. *140*:1523-1539.

19. **Cottrez, F., C. Auriault, A. Capron and H. Groux.** 1994. Quantitative PCR: validation of the use of a multispecific internal control. Nucleic Acids Res. *22*:2712-2713.

20. **Coutlée, F., Y. He, P. Saint-Antoine, C. Olivier and A. Kessous.** 1995. Coamplification of HIV type 1 and β-globin gene DNA sequences in a nonisotopic polymerase chain reaction assay to control for amplification efficiency. AIDS Res. Hum. Retroviruses *11*:363-371.

21. **Cross, N.C.P.** 1995. Quantitative PCR techniques and applications. Br. J. Haematol. 89:693-697.

22. **Cross, N.C.P., L. Feng, A. Chase, J. Bungey, T.P. Hughes and J.M. Goldman.** 1993. Competitive polymerase chain reaction to estimate the number of BCR-ABL transcripts in chronic myeloid leukemia patients after bone marrow transplantation. Blood *82*:1929-1936.

23. **Cusi, M.G., M. Valassina and P.E. Valensin.** 1994. Comparison of M-MLV reverse transcriptase and *Tth* polymerase activity in RT-PCR of samples with low virus burden. BioTechniques *17*:1034-1036.

24. **de Kant, E., C.F. Rochlitz and R. Herrmann.** 1994. Gene expression analysis by a competitive and differential PCR with antisense competitors. BioTechniques *17*:934-942.

25. **Deng, G., M. Yu and H.S. Smith.** 1993. An improved method of competitive PCR for quantitation of gene copy number. Nucleic Acids Res. *21*:4848-4849.

26. **Dieffenbach, C.W., T.M.J. Lowe and G.S. Dveksler.** 1993. General concepts for PCR primer design. PCR Methods Appl. *3*:S30-S37.

27. **Dostal, D.E., K.N. Rothblum and K.M. Baker.** 1994. An improved method for absolute quantification of mRNA using multiplex polymerase chain reaction: determination of renin and angiotensinogen mRNA levels in various tissues. Anal. Biochem. *223*:239-250.

28. **Erlich, H.A., D. Gelfand and J.J. Sninsky.** 1991. Recent advances in the polymerase chain reaction. Science *252*:1643-1651.

29. **Fandrey, J. and H.F. Bunn.** 1993. In vivo and in vitro regulation of erythropoietin RNA: measurement by competitive polymerase chain reaction. Blood *81*:617-623.

30. **Fasco, M.J., C.P. Treanor, S. Spivack, H.L. Figge and L.S. Kaminsky.** 1995. Quantitative RNA-polymerase chain reaction-DNA analysis by capillary electrophoresis and laser-induced fluorescence. Anal. Biochem. *224*:140-147.

31. **Ferre, F.** 1992. Quantitative or semi-quantitative PCR: reality versus myth. PCR Methods Appl. *2*:1-9.

32. **Foley, K.P., M.W. Leonard and J.D. Engel.** 1993. Quantitation of RNA using the polymerase chain reaction. Trends Genet. *9*:380-385.

33. **Förster, E.** 1994. An improved general method to generate internal standards for competitive PCR. BioTechniques *16*:18-20.

34. **Förster, E.** 1994. Rapid generation of internal standards for competitive PCR by low-stringency primer annealing. BioTechniques *16*:1006-1008.

35. **Fox, J.C., P.D. Griffiths and V.C. Emery.** 1992. Quantification of human cytomegalovirus DNA using the polymerase chain reaction. J. Gen. Virol. *73*:2405-2408.

36. **Furtado, M.R., R. Murphy and S.M. Wolinsky.** 1993. Quantification of human immunodeficiency virus type 1 tat mRNA as a marker for assessing the efficacy of antiretroviral therapy. J. Infect. Dis. *167*:213-216.

37. **Galea, E. and D.L. Feinstein.** 1992. Rapid synthesis of DNA deletion constructs for mRNA quantitation: analysis of astrocyte mRNAs. PCR Methods Appl. *2*:66-69.

38. **Gilliland, G., S. Perrin, K. Blanchard and H.F. Bunn.** 1990. Analysis of cytokine mRNA and DNA: detection and quantitation by competitive polymerase chain reaction. Proc. Natl. Acad. Sci. USA *87*:2725-2729.

39. **Grassi, G., L. Zentilin, S. Tafuro, S. Diviacco, A. Ventura, A. Falaschi and M. Giacca.** 1994. A rapid procedure for the quantitation of low abundance RNAs by competitive reverse transcription-polymerase chain reaction. Nucleic Acids Res. *22*:4547-4549.

40. **Gretch, D., L. Corey, J. Wilson, C. dela Rosa, R. Willson, R. Carithers, Jr., M. Busch, J. Hart, M. Sayers and J. Han.** 1994. Assessment of hepatitis C virus RNA levels by quantitative competitive RNA polymerase chain reaction: high-titer viremia correlates with advanced stage of disease. J. Infect. Dis. *169*:1219-1225.

41. **Grünebach, F., E.-U. Griese and K. Schumacher.** 1994. Competitive nested polymerase chain reaction for quantification of human *MDR*1 gene expression. J. Cancer Res. Clin. Oncol. *120*:539-544.

42. **Hahn, M., V. Dörsam, P. Friedhoff, A. Fritz and A. Pingoud.** 1995. Quantitative polymerase chain reaction with enzyme-linked immunosorbent assay detection of selectively digested amplified sample and control DNA. Anal. Biochem. *229*:236-248.

43. **Hanspal, M., J.S. Hanspal, K.E. Sahr, E. Fibach, J. Nachman and J. Palek.** 1993. Molecular basis of spectrin deficiency in hereditary pyropoikilocytosis. Blood *82*:1652-1660.

44. **He, Q., M. Marjamäki, H. Soini, J. Mertsola and M.K. Viljanen.** 1994. Primers are decisive for sensitivity of PCR. BioTechniques *17*:82-87.

45. **Heroux, J.A. and A.M. Szczepanik.** 1995. Quantitative analysis of specific mRNA transcripts using a competitive PCR assay with electrochemiluminescent detection. PCR Methods Appl. *4*:327-330.

46. **Jalava, T., P. Lehtovaara, A. Kallio, M. Ranki and H. Söderlund.** 1993. Quantification of hepatitis B virus DNA by competitive amplification and hybridization on microplates. BioTechniques *15*:134-139.

47. **Jenkins, F.J.** 1994. Basic methods for the detection of PCR products. PCR Methods Appl. *3*:S77-S82.

48. **Jin, C.-F., M. Mata and D.J. Fink.** 1994. Rapid construction of deleted DNA fragments for use as internal standards in competitive PCR. PCR Methods Appl. *3*:252-255.

49. **Jones, D.H. and B.H. Howard.** 1991. A rapid method for recombination and site-specific mutagenesis by placing homologous ends on DNA using polymerase chain reaction. BioTechniques *10*:62-66.

50. **Jones, D.H., K. Sakamoto, R.L. Vorce and B.H. Howard.** 1990. DNA mutagenesis and recombination. Nature *344*:793-794.

51. **Jurriaans, S., J.T. Dekker and A. de Ronde.** 1992. HIV-1 viral DNA load in peripheral blood mononuclear cells from seroconverters and long-term infected individuals. AIDS *6*:635-641.

52. **Kanangat, S., A. Solomon and B.T. Rouse.** 1992. Use of quantitative polymerase chain reaction to quantitate cytokine messenger RNA molecules. Mol. Immunol. *29*:1229-1236.

53. **Kaneko, S., S. Murakami, M. Unoura and K. Kobayashi.** 1992. Quantitation of hepatitis C virus RNA by competitive polymerase chain reaction. J. Med. Virol. *37*:278-282.

54. **Kohsaka, H., A. Tanigushi, D.D. Richman and D.A. Carson.** 1993. Microtiter format gene quantification by covalent capture of competitive PCR products: application to HIV-1 detection. Nucleic Acids Res. *21*:3469-3472.

55. **Kolk, A.H.J, G.T. Noordhoek, O. de Leeuw, S. Kuijper and J.D.A van Embden.** 1994. *Mycobacterium smegmatis* strain for detection of *Mycobacterium tuberculosis* by PCR used as internal control for inhibition of amplification and for quantitation of bacteria. J. Clin. Microbiol. *32*:1354-1356.

56. **Lehtovaara, P., M. Uusi-Oukari, P. Buchert, M. Laaksonen, M. Bengtström and M. Ranki.** 1993. Quantitative PCR for hepatitis B virus with colorimetric detection. PCR Methods Appl. *3*:169-175.

57. **Luque, F., A. Caruz, J.A. Pineda, Y. Torres, B. Larder and M. Leal.** 1994. Provirus load changes in untreated and zidovudine-treated human immunodeficiency virus type 1-infected patients. J. Infect. Dis. *169*:267-273.

58. **Manzin, A., P. Bagnarelli, S. Menzo, F. Giostra, M. Brugia, R. Francesconi, F.B. Bianchi and M. Clementi.** 1994. Quantitation of hepatitis C virus genome molecules in plasma samples. J. Clin. Microbiol. *32*:1939-1944.

59. **Martino, T.A., M.J. Sole, L.Z. Penn, C.-C. Liew and P. Liu.** 1993. Quantitation of enteroviral RNA by competitive polymerase chain reaction. J. Clin. Microbiol. *31*:2634-2640.

60. **McCulloch, R.K., C.S. Choong and D.M. Hurley.** 1995. An evaluation of competitor type and size for use in the determination of mRNA by competitive PCR. PCR Methods Appl. *4*:219-226.

61. **Menzo, S., P. Bagnarelli, M. Giacca, A. Manzin, P.E. Varaldo and M. Clementi.** 1992. Absolute quantitation of viremia in human immunodeficiency virus infection by competitive reverse transcription and polymerase chain reaction. J. Clin. Microbiol. *30*:1752-1757.

62. **Michael, N.L., T. Mo, A. Merzouki, M. O'Shaughnessy, C. Oster, D.S. Burke, R.R. Redfield, D.L. Birx and S.A. Cassol.** 1995. Human immunodeficiency virus type 1 cellular RNA load and splicing patterns predict disease progression in a longitudinally studied cohort. J. Virol. *69*:1868-1877.

63. **Morrison, C. and F. Gannon.** 1994. The impact of the PCR plateau phase on quantitative PCR. Biochim. Biophys. Acta *1219*:493-498.

15

64.Mulder, J., N. McKinney, C. Christopherson, J. Sninsky, L. Greenfield and S. Kwok. 1994. Rapid and simple PCR assay for quantitation of human immunodeficiency virus type 1 RNA in plasma: application to acute retroviral infection. J. Clin. Microbiol. *32*:292-300.

65.Mullis, K.B. 1991. The polymerase chain reaction in an anemic mode: How to avoid cold oligodeoxyribonuclear fusion. PCR Methods Appl. *1*:1-4.

66.Natarajan, V., R.J. Plishka, E.W. Scott, H.C. Lane and N.P. Salzman. 1994. An internally controlled virion PCR for the measurement of HIV-1 RNA in plasma. PCR Methods Appl. *3*:346-350.

67.Oliner, J.D., K.W. Kinzler and B. Vogelstein. 1993. *In vivo* cloning of PCR products in *E. coli*. Nucleic Acids Res. *21*:5192-5197.

68.Perrin, S. and G. Gilliland. 1990. Site-specific mutagenesis using asymmetric polymerase chain reaction and a single mutant primer. Nucleic Acids Res. *18*:7433-7438.

69.Peten, E.P., L.J. Striker, M.A. Carome, S.J. Elliott, C.-W. Yang and G.E. Striker. 1992. The contribution of increased collagen synthesis to human glomerulosclerosis: a quantitative analysis of α 2IV collagen mRNA expression by competitive polymerase chain reaction. J. Exp. Med. *176*:1571-1576.

70.Piatak, M., Jr., K.-C. Luk, B. Williams and J.D. Lifson. 1993. Quantitative competitive polymerase chain reaction for accurate quantitation of HIV DNA and RNA species. BioTechniques *14*:70-81.

71.Piatak, M., Jr., M.S. Saag, L.C. Yang, S.J. Clark, J.C. Kappes, K.-C. Luk, B.H. Hahn, G.M. Shaw and J.D. Lifson. 1993. High levels of HIV-1 in plasma during all stages of infection determined by competitive PCR. Science *259*:1749-1754.

72.Porcher, C., M.-C. Malinge, C. Picat and B. Grandchamp. 1992. A simplified method for determination of specific DNA or RNA copy number using quantitative PCR and an automatic DNA sequencer. BioTechniques *13*:106-113.

73.Raeymaekers, L. 1993. Quantitative PCR: theoretical considerations with practical implications. Anal. Biochem. *214*:582-585.

74.Ramakrishnan, R., M. Levine and D.J. Fink. 1994. PCR-based analysis of herpes simplex virus type 1 latency in the rat trigeminal ganglion established with a ribonucleotide reductase-deficient mutant. J. Virol. *68*:7083-7091.

75.Ravaggi, A.Z. , C. Mazza, A. Albertini and E. Cariani. 1995. Quantification of hepatitis C virus RNA by competitive amplification of RNA from denatured serum and hybridization on microtiter plates. J. Clin. Microbiol. *33*:265-269.

76.Reiner, S.L., S. Zheng, D.B. Corry and R.M. Locksley. 1993. Constructing polycompetitor cDNAs for quantitative PCR. J. Immunol. Methods *165*:37-46; and Corrigenda, 1994, J. Immunol. Methods *173*:133 and J. Immunol. Methods *175*:275.

77.Riedy, M.C., E.A. Timm, Jr. and C.C. Stewart. 1995. Quantitative RT-PCR for measuring gene expression. BioTechniques *18*:70-76.

78.Rüster, B., S. Zeuzem and W.K. Roth. 1995. Quantification of hepatitis C virus RNA by competitive reverse transcription and polymerase chain reaction using a modified hepatitis C virus RNA transcript. Anal. Biochem. *224*:597-600.

79.Saiki, R.K., S. Scharf, F. Faloona, K.B. Mullis, G.T. Horn, H.A. Erlich and N. Arnheim. 1985. Enzymatic amplification of beta-globin genomic sequences and restriction site analysis for diagnosis of sickle cell anemia. Science *230*:1350-1354.

80.Sarkar, G. and M.E. Bolander. 1994. The "looped oligo" method for generating reference molecules for quantitative PCR. BioTechniques *17*:864-866.

81.Scadden, D.T., Z. Wang and J.E. Groopman. 1992. Quantitation of plasma human immunodeficiency virus type 1 RNA by competitive polymerase chain reaction. J. Infect. Dis. *165*:1119-1123.

82.Schanke, J.T., L.M. Quam and B.G. Van Ness. 1994. Flip PCR for DNA sequence motif inversion. BioTechniques *16*:414-416.

83.Shuldiner, A.R., K. Tanner, L.A. Scott, C.A. Moore and J. Roth. 1991. Ligase-free subcloning: a versatile method to subclone polymerase chain reaction (PCR) products in a single day. Anal. Biochem. *194*:9-15.

84.Siebert, P.D. and J.W. Larrick. 1992. Competitive PCR. Nature *359*:557-558.

85.Siebert, P.D. and J.W. Larrick. 1993. PCR MIMICS: competitive DNA fragments for use as internal standards in quantitative PCR. BioTechniques *14*:244-249.

86.Siegling, A., M. Lehmann, C. Platzer, F. Emmrich and H.-D. Volk. 1994. A novel multispecific competitor fragment for quantitative PCR analysis of cytokine gene expression in rats. J. Immunol. Methods *177*:23-28.

87.Stålbom, B.-M., A. Torvén and L.G. Lundberg. 1994. Application of capillary electrophoresis to the post-polymerase chain reaction analysis of rat mRNA for gastric H+, K+ -ATPase. Anal. Biochem. *217*:91-97.

88.Stieger, M., C. Démollière, L. Ahlborn-Laake and J. Mous. 1991. Competitive polymerase chain reaction assay for quantitation of HIV-1 DNA and RNA. J. Virol. Methods *34*:149-160.

89. **Sutherland, J.C., B.M. Sutherland, A. Emrick, D.C. Monteleone, E.A. Ribeiro, J. Trunk, M. Son, P. Serwer et al.** 1991. Quantitative electronic imaging of gel fluorescence with CCD cameras: applications in molecular biology. BioTechniques *10*:492-497.

90. **Sykes, P.J., S.H. Neoh, M.J. Brisco, E. Hughes, J. Condon and A.A. Morley.** 1992. Quantitation of targets for PCR by use of limiting dilution. BioTechniques *13*:444-449.

91. **Taniguchi, A., H. Kohsaka and D.A. Carson.** 1994. Competitive RT-PCR ELISA: a rapid, sensitive and non-radioactive method to quantitate cytokine mRNA. J. Immunol. Methods *169*:101-109.

92. **Telenti, A., P. Imboden and D. Germann.** 1992. Competititive polymerase chain reaction using an internal standard: application to the quantitation of viral DNA. J. Virol. Methods *39*:259-268.

93. **Thompson, J.D., I. Brodsky and J.J. Yunis.** 1992. Molecular quantification of residual disease in chronic myelogenous leukemia after bone marrow transplantation. Blood *79*:1629-1635.

94. **Überla, K., C. Platzer, T. Diamantstein and T. Blankenstein.** 1991. Generation of competitor DNA fragments for quantitative PCR. PCR Methods Appl. *1*:136-139.

95. **Ulrich, P.P., J.M. Romeo, L.J. Daniel and G.N. Vyas.** 1993. An improved method for the detection of hepatitis C virus RNA in plasma utilizing heminested primers and internal control RNA. PCR Methods Appl. 2:241-249.

96. **Vanden Heuvel, J.P., F.L. Tyson and D.A. Bell.** 1993. Construction of recombinant RNA templates for use as internal standards in quantitative RT-PCR. BioTechniques *14*:395-398.

97. **van Gemen, B., R. van Beuningen, A. Nabbe, D. van Strijp, S. Jurriaans, P. Lens and T. Kievits.** 1994. A one-tube quantitative HIV-1 RNA NASBA nucleic acid amplification assay using electro-chemiluminescent (ECL) labelled probes. J. Virol. Methods *49*:157-168.

98. **Virdi, A.S., S. Krishna and B.C. Sykes.** 1992. Tandem competitive polymerase chain reaction (TC-PCR): a method for determining ratios of RNA and DNA templates. Mol. Cell. Probes *6*:375-380.

99. **Wang, A.M., M.V. Doyle and D.F. Mark.** 1989. Quantitation of mRNA by the polymerase chain reaction. Proc. Natl. Acad. Sci. USA *86*:9717-9721.

100. **Wang, H., L. Fliegel, C.E. Cass, A.M.W. Penn, M. Michalak, J.H. Weiner and B.D. Lemire.** 1994. Quantification of mitochondrial DNA in heteroplasmic fibroblasts with competitive PCR. BioTechniques *17*:76-82.

101. **Wiesner, R.J., B. Beinbrech and J.C. Rüegg.** 1993. Quantitative PCR. (Letter) Nature *366*:416.

102. **Wilkinson, E.T., S. Cheifetz and S.A. De Grandis.** 1995. Development of competitive PCR and the QPCR system 5000 as a transcription-based screen. PCR Methods Appl. *4*:363-367.

103. **Wolf, S.S. and A. Cohen.** 1992. Expression of cytokines and their receptors by human thymocytes and thymic stromal cells. Immunology *77*:362-368.

104. **Xia, H.-Z., C.L. Kepley, K. Sakai, K. Chelliah, A.-M.A. Irani and L.B. Schwartz.** 1995. Quantitation of Tryptase, Chymase, FcϵRIα and FcϵRIγ mRNAs in human mast cells and basophils by competitive reverse transcription-polymerase chain reaction. J. Immunol. *154*:5472-5480.

105. **Young, K.K., R.M. Resnick and T.W. Myers.** 1993. Detection of hepatitis C virus RNA by a combined reverse transcription-polymerase chain reaction assay. J. Clin. Microbiol. *31*:882-886.

106. **Yun, Z., J. Lundeberg, B. Johansson, A. Hedrum, O. Weiland, M. Uhlén and A. Sönnerborg.** 1994. Colorimetric detection of competitive PCR products for quantification in hepatitis C viremia. J. Virol. Methods *47*:1-13.

107. **Zachar, V., R.A. Thomas and A.S. Goustin.** 1993. Absolute quantification of target DNA: a simple competitive PCR for efficient analysis of multiple samples. Nucleic Acids Res. *21*:2017-2018.

108. **Zarlenga, D.S., A. Canals and L. Gasbarre.** 1995. Method for constructing internal standards for use in competitive PCR. BioTechniques *19*:324-326.

109. **Zenilman, M.E., W. Graham, K. Tanner and A.R. Shuldiner.** 1995. Competitive reverse-transcriptase polymerase chain reaction without an artificial internal standard. Anal. Biochem. *224*:339-346.

110. **Zimmermann, K., D. Schögl and J.W. Mannhalter.** 1994. Hemi-nested quantitative competitive PCR of HIV-1. BioTechniques *17*:440-442.

Update to:

Technical Aspects of Quantitative Competitive PCR

Klaus Zimmerman and Josef W. Mannhalter

Immuno AG, Vienna, Austria

Since the publication of our review on the technical aspects of quantitative competitive PCR (QC-PCR), the use of competitive PCR for quantitation of specific nucleic acid sequences has become increasingly widespread and appears more and more to be the method of choice. With respect to technical aspects, there are only a few important new developments and trends to report on.

CONSTRUCTION OF INTERNAL STANDARDS FOR COMPETITIVE PCR

As we have already reported in the original article, a great variety of fundamentally new techniques have been published. Great efforts have been made to construct multi-sequence internal standards, as described, for example, by Tarnuzzer et al. (8) for quantitation of 52 specific messages, including growth factors, cytokines, extracellular matrix components and metalloproteinases.

ANALYSIS OF COMPARABLE AMPLIFICATION OF WILD-TYPE AND INTERNAL STANDARD TEMPLATES

The accuracy of QC-PCR is still a matter of concern. It has been suggested that the ratio of wild-type molecules should be fixed to a narrow limit (7) in order to obtain accurate results. Furthermore, Henley et al. (5) have discussed the influence of heteroduplex formation on QC-PCR performance.

PCR AND RT-PCR

We should point out that the use of an inactivated DNA polymerase (AmpliTaq® Gold; Perkin-Elmer, Norwalk, Ct, USA) that gains its full activity only after lengthy heat incubation might dramatically improve the specificity and sensitivity of the PCR.

Perhaps the most important new development in QC-PCR methodology is the addition of known amounts of two or more different sized competitors to one sample followed by co-amplification of the mixture in a single reaction tube (2,9,11). This one-tube assay reduces the amount of sample and labor required for QC-PCR.

A step towards automatization of QC-PCR is the development of a novel "real-time" quantitative PCR method, i.e., the TaqMan™ Probe (4), which does not require post-PCR sample handling.

QC-PCR is now also being used more and more for applications other than quantitation of specific genes. It can, for example, be employed for determining the amount of residual DNA in biological products (6).

DETECTION AND ANALYSIS

Very frequently, the amplified products are still separated by gel electrophoresis and analyzed either by video image analysis (11) or by an automated laser-fluorescent DNA sequencer (2). However, there is also considerable interest in automatizable detection systems for QC-PCR such as capillary electrophoresis (1,10) and ion-pair reversed-phase HPLC (3).

REFERENCES

1. **Connolly, A.R., L.G. Cleland and B.W. Kirkham.** 1995. Mathematical considerations of competitive polymerase chain reaction. J. Immunol. Methods *187*:201:211.
2. **Hämmerle, T., F.G. Falkner and F. Dorner.** 1996. A sensitive PCR assay system for the quantitation of viral genome equivalents: hepatitis C virus (HCV). Arch. Virol. *141*:2103:2114.
3. **Hayward-Lester, A., P.J. Oefner and P.A. Doris.** 1996. Rapid quantification of gene expression by competitive RT-PCR and ion-pair reversed-phase HPLC. BioTechniques *20*:250-257.
4. **Heid, C.A., J. Stevens, K.J. Livak and P.M. Williams.** 1996. Real time quantitative PCR. Genome Res. *6*:10.
5. **Henley, W.N., K.E. Schuebel and D.A. Nielsen.** 1996. Limitations imposed by heteroduplex formation on quantitative RT-PCR. Biochem. Biophys. Res. Commun. *226*:113-117.
6. **Himmelspach, M., F. Gruber, G. Antoine, F.G. Falkner, F. Dorner and T. Hämmerle.** 1996. Specific quantitation of genomic DNA in the femtogram range by amplification of repetitive sequences. Anal. Biochem. *242*:240-247.
7. **Souazé, F., A. Ntodou-Thomé, C.Y. Tran, W. Rostène and P. Forgez.** 1996. Quantitative RT-PCR: limits and accuracy. BioTechniques *21*:280-285.
8. **Tarnuzzer, R.W., S.P. Macauley, W.G. Farmerie, S. Caballero, M.R. Ghassemifar, J.T. Anderson, C.P. Robinson, M.B. Grant, M.G. Humphreys-Beher, L. Franzen, A.B. Peck and G.S. Schultz.** 1996. Competitive RNA templates for detection and quantitation of growth factors, cytokines, extracellular matrix components and matrix metalloproteinases by RT-PCR. BioTechniques *20*:670-674.
9. **Vener, T., M. Axelsson, J. Albert, M. Uhlén and J. Lundeberg.** 1996. Quantification of HIV-1 using multiple competitors in a single-tube assay. BioTechniques *21*:248-255.
10. **Williams, S.J., C. Schwer, A.S.M. Krishnarao, C. Heid, B.L. Karger and P.M. Williams.** 1996. Quantitative competitive polymerase chain reaction : analysis of amplified products of the HIV-1 *gag* gene by capillary electrophoresis with laser-induced fluorescence detection. Anal. Biochem. *236*:146-152.
11. **Zimmermann, K., D. Schögl, B. Plaimauer and J.W. Mannhalter.** 1996. Quantitative multiple competitive PCR of HIV-1 DNA in a single reaction tube. BioTechniques *21*:480-484.

CHAPTER 2

Quantitative RT-PCR: Limits and Accuracy

F. Souazé, A. Ntodou-Thomé, C.Y. Tran, W. Rostène and P. Forgez

INSERM Unit 339, Hôpital Saint-Antoine, Paris, France

ABSTRACT

In this paper we determine the limits and accuracy of quantitative reverse transcription (RT)-PCR using a modification of the original protocol. The quantification of mRNA with this procedure requires a preliminary estimation of the target molecule (TM) concentration, established from experiments with an internal control molecule (ICM). A definitive quantification is then attained from serial dilutions of the reverse transcription reaction. The success of this latter step is dependent on maintaining an equivalent number of TM and ICM in the reaction. The purpose of our study was to evaluate the influence of the deviation between the TM and the ICM on the result. We show here that we can control the accuracy of the assay by fixing the limit of the TM/ICM ratio. Indeed, when the TM/ICM ratio is between 0.66 and 1.5 (i.e., the difference between TM and ICM is 1.5-fold), the final result has an error of approximately 10%. Exceeding this limit produces errors approaching 60%, as in the case of TM/ICM = 2. When the above conditions are respected, a difference as small as 20% between two samples can be determined with an accuracy of 95%.

INTRODUCTION

The reverse transcription polymerase chain reaction (RT-PCR) is the technique of choice for analyzing extremely low abundance mRNA derived from cells or tissues. Recently, a number of reports have described different quantitative procedures for analyzing mRNA steady-state levels. PCR is now a well-established method whose sensitivity is a principal advantage over other "similar" techniques, such as Northern blots, which only provide semiquantitative results. The earliest PCR studies proceeded by comparing the amplified products from the target molecule (TM) with the amplified products from the cDNA of an abundant protein, such as actin or β-globulin. Quantification of specific mRNA molecules with RT-PCR was dependent on extrapolating the results from a control molecule. However, variations in the kinet-

(Reprinted from BioTechniques 21:280-285, 1996)

ics of the reverse transcription reaction could produce discrepancies between the calculated amount of molecules from the sample and the actual amount of molecules. For this reason, the best control molecules are internal control molecules (ICM). To ensure that the efficiency of the PCR is similar for both molecules, the ICM is created from a synthetic cRNA. This cRNA is identical to the TM except it possesses a small deletion in the amplified portion of the molecule. The resulting amplified products have different molecular weights and are distinguished easily on polyacrylamide gel electrophoresis.

Quantitative RT-PCR is based on the competitive status between both amplicons. The addition of an ICM in the RT-PCR creates competition between the TM and the ICM for the factors controlling the amplification process, such as nucleotides and primers (for review see Reference 2). The approach developed by Wang et al. (9) was based on the fact that during the exponential phase of PCR, the amount of TM could be quantified by extrapolating against the results obtained from the ICM, provided that the reaction efficiencies were identical. In order to properly set up a Wang assay, it is necessary to know the amount of the target RNA before the experiment commences and, for this reason, we adopted the modifications proposed by Nagano and Kelly (4). Consequently, a titration assay is used to estimate the quantity of target molecule. The quantitative assay is then performed using serial dilutions from an RT reaction containing equivalent amounts of TM and ICM.

The aim of our study was to evaluate how the ratio of TM to ICM affects the competition between the two amplicons and therefore the accuracy of quantitative RT-PCR. For these reasons, we devised a quantitative RT-PCR using a known number of two different cRNA molecules (one molecule was designed as the ICM and the other as the TM in nonspecific RNA). By manipulating one of the two molecules, we were able to assess the limits and the precision of quantitative RT-PCR.

MATERIALS AND METHODS

Internal Controls

Neurotensin receptor (NTR) cDNA was kindly supplied by Dr. Nakanishi. The coding region of NTR cDNA (-7 to 1301) (8) was inserted into the *SmaI-Bam*HI site of pT7/T3α18. An oligonucleotide containing a poly(dA)$_{45}$ was then inserted between the *SalI-Bam*HI sites. Deletions of 34 and 96 nucleotides were made by deleting fragments *NcoI-Nhe*I and *Hinc*II-*Nco*I, giving the plasmids named pΔ34 and pΔ96, respectively. All enzymes, unless otherwise noted, were purchased from Life Technologies (Gaithersburg, MD, USA) and the reaction conditions were those as suggested by the manufacturer. The cRNA Δ34 and the cRNA Δ96 were obtained by in vitro transcription of 8 μg each of plasmid linearized at the *Sal*I site. The transcription reaction contained 40 mM Tris-HCl pH 7.2, 10 mM dithiothreitol (DTT), 6 mM $MgCl_2$, 4 mM spermidine, 80 U RNasin® Ribonuclease Inhibitor (Promega, Madison, WI, USA), 0.5 mM dNTP and 60 U of T7 RNA polymerase. Polymerase chain reactions were carried out in a final volume of 50 µL at 37°C for 1 h. The reaction mixture was treated with 10 U of RNase-free DNase I for 15 min at 37°C. The cRNAs were purified on oligo(dT) columns (Sigma Aldrich Techware, Milwaukee, WI, USA), following the procedure described by Aviv and Leder (1). After elution, the cRNAs were precipitated with ethanol and then diluted in diethylpyrocarbonate (DEPC)-treated water containing 1 U/µL of RNasin. cRNA quality was checked by electrophoresis in formaldehyde RNA gel (6), and cRNA concentration was estimated by spectrophotometric absorbance at 260 nm. The cRNA solution was diluted to 1×10^7 molecules/µL in DEPC-treated water, put into 10-µL aliquots and stored at -80°C.

Primer Labeling

Fifty picomoles of antisense PCR primer were 5′ ^{32}P end-labeled with 20 U of polynucleotide T4 kinase in 50 µL of 10 mM $MgCl_2$, 5 mM DTT, 70 mM Tris-HCl pH 7.6 and 100 pmol [γ-^{32}P]-ATP (3000 Ci/mM) (Amersham International, Little Chalfont, Bucks, England, UK) at 37°C for 30 min. The end-labeled oligonucleotide was subsequently purified on a microcolumn. The oligonucleotide concentration was estimated by counting 1 µL of eluent on a GF/C filter (Whatman, Maidstone, Kent, England, UK) in 5 mL of dry extract scintillation fluid (Optiphase; Wallac/Pharmacia, Brussels, Belgium).

Reverse Transcription Reaction

The primer RT-NTR (5′-GCTGACGTAGAAGAG-3′) located at position 1069–1083 was used for reverse transcription of both cRNA molecules. Varying amounts of cRNA Δ34 and cRNA Δ96 were mixed with 1 µg RNA of total mRNA from Chinese hamster ovary (CHO) cells devoid of NTR mRNA, 50 pmol of RT-NTR primer in 20 mM Tris-HCl pH 8.3, 50 mM KCl, 5 mM $MgCl_2$, 10 mM DTT, 1 mM dNTP, 1 U/µL RNasin and 200 U Moloney murine leukemia virus reverse transcriptase in a final volume of 30 µL for 1 h at 37°C.

Quantitative RT-PCR Conditions

Primers used in this procedure were S-NTR (5′-CCTTCAAGGCCAAGA-CCCTC-3′) and AS-NTR (5′-CAGCCAGCAGACCACAAAGG-3′) at positions 521–540 and 947–966, respectively, giving a PCR product of 411 nucleotides for cRNA Δ34 and 349 nucleotides for cRNA Δ96. Primers were synthesized and purified using polyacrylamide gel electrophoresis. The PCR amplification was performed on 1:5 (vol:vol) of the RT reaction in a mixture containing 16 mM Tris-HCl pH 8.3, 40 mM KCl, 1.5 mM $MgCl_2$, 0.2 mM concentration of each dNTP, 25 pmol of each primer (S-NTR and AS-NTR),

10^6 cpm of a 5′ end-labeled [γ-^{32}P]ATP AS-NTR and 1 U of *Taq* DNA Polymerase (Perkin-Elmer, Norwalk, CT, USA) in a final volume of 50 μL. The amplification profile was divided into denaturation at 94°C for 30 s, annealing at 55°C for 1 min and extension at 72°C for 1 min 30 s. The 26 cycles were preceded by a denaturation at 95°C for 5 min and immediately followed by a final extension at 72°C for 10 min. The amplification was performed in a 500-μL GeneAmp® tube, containing 50 μL of Nujol mineral oil, in a DNA Thermal Cycler 480 (Perkin-Elmer).

The assay system consisted of two steps. Titration assays were performed with 10^6 molecules of cRNA Δ96 with increasing quantities (10^4 to 10^8 molecules) of cRNA D34. For precise RNA quantitative studies, a known number of cRNA Δ34 and cRNA Δ96 molecules were subjected to reverse transcription, followed by a series of 1:3 (vol:vol) dilutions and amplification by PCR. For the negative controls, the same procedure was applied to a RT reaction containing no RNA or cRNA. This control was made for each group of samples. Gel pieces from the negative control were excised, counted and used for background estimates. The experiment was rejected if the negative control contained visible bands.

Electrophoresis

Twenty microliters of each PCR was immediately loaded on 5% polyacrylamide gels in 90 mM Tris-borate and 2 mM EDTA. Electrophoresis was performed at 250 V for 3 h at room temperature in an ADJ2 apparatus (Owl Scientific, Woburn, MA, USA). After migration, the bands were stained in ethidium bromide and cut out before counting in 3 mL of scintillation fluid in a β-scintillation counter (Model LS 6000 SC; Beckman Instruments, Fullerton, CA, USA).

Data Analysis

After counting, the log of the cpm was plotted against the log of the ICM or the log of the RNA. A linear regression for each curve was calculated. The data generated from these curves were used to extrapolate the number of molecules in each sample. The experiments in Table 1 were performed five times and the results expressed as the average. The statistical significance of the results were analyzed using the Student's *t* test.

RESULTS AND DISCUSSION

The quantitative RT-PCR assay developed by Wang et al. (9) and modified by Nagano and Kelly (4) is the most sensitive technique available for the measurement of RNA. Proper implementation of this technique is dependent on completing two rounds of RT-PCR, consisting of titration and quantification assays. We performed here a series of tests to illuminate the critical steps and parameters controlling the accuracy of quantitative RT-PCR.

In a typical RT-PCR, there is an unknown concentration of TM and a known concentration of ICM. The concentration of the TM is determined by extrapolating the data based on the ICM. For our experiments, the TM was known in advance, which permitted us to determine if the expected concentration resulting from the RT-PCR was equal to its known concentration. In our system, cRNA Δ34 was considered the target molecule (TM), contained in 1 μg of CHO RNA, and cRNA Δ96 was the ICM. The known number of starting molecules is referred to as "Molecules IN" and the number of molecules calculated as a result of the RT-PCR is referred to as "Molecules OUT".

The titration assay is a preliminary step, consisting of a competitive RT-PCR that determines the departure point for the final quantification of the mRNA. An increasing amount of ICM is added to the RNA sample, followed by an RT reaction and PCR. If the initial TM and ICM concentrations were

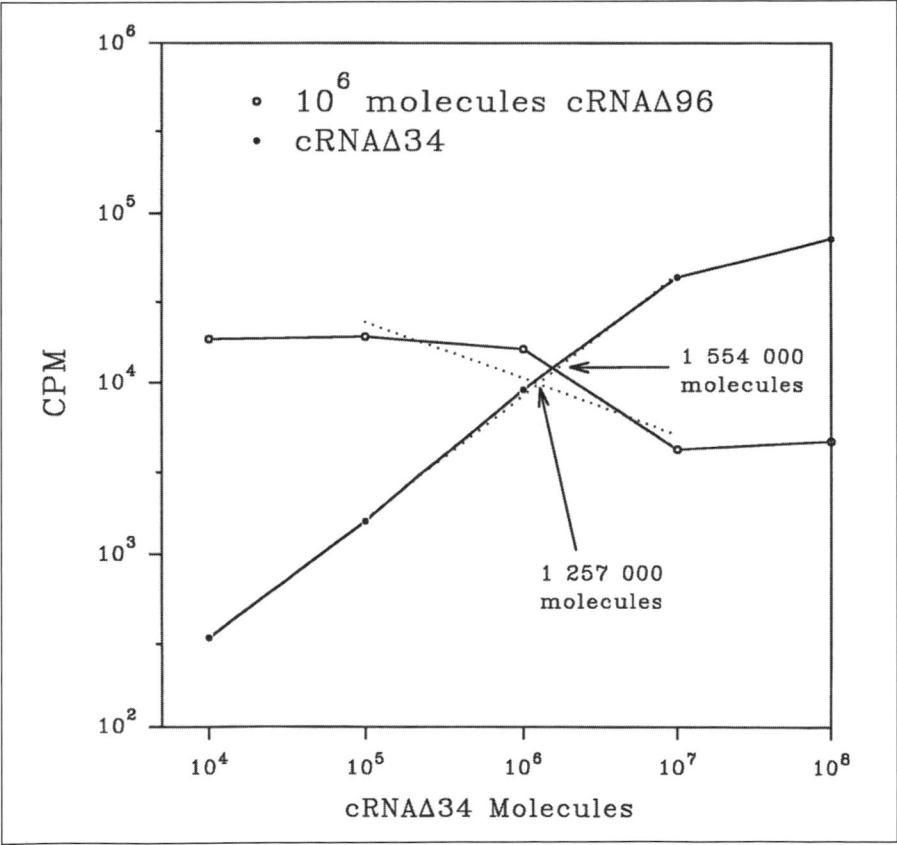

Figure 1. Titration assay. Each point containing 10^6 molecules of cRNA Δ96 and various numbers of molecules of cRNA Δ34 were reverse-transcribed and amplified as described in Materials and Methods. The amplification products were separated using polyacrylamide gel electrophoresis and detected by ethidium bromide staining. The cpm from the bands were counted after they were excised from the gel.

25

Table 1. Summary of the Results Obtained from Titration (A) and Quantitative (B) Assays, and a Comparison of Two Samples in Which Quantity Differs by 20% (C)

	A Titration Assay	B Quantitative Assay					C Sensitivity	
Molecules IN cRNA Δ34	1 000 000	1 000 000	750 000	1 500 000	2 000 000	500 000	1 000 000	800 000
Molecules OUT (mean)	1 179 825	1 091 817	831 764	1 654 190	3 083 262	306 359	1 091 817	755 681
S.D.	345 505	130 920	62 059	132 370	534 665	65 239	130 920	136 790
S.D. as % Molecules OUT	29.3	12	7.4	8	17	21.3	12	18
% IN/OUT	15.25	8.4	9.8	9.3	35.1	63.2	8.4	5.9
Student's t Test							$P<0.05$	

A varying number of cRNA Δ34 molecules is assayed with 10^6 molecules of cRNA Δ96.

identical, the resulting PCR products should be produced in equal amounts (5,6).

We evaluated the accuracy of the titration assay by determining the effect of increasing the number of cRNA Δ34 molecules on a constant quantity of ICM, i.e., 10^6 molecules of cRNA Δ96. The results were plotted using the formula Log (cpm) = f(Log ICM), and the estimated quantity of TM was calculated at the point where the two curves intersected. As shown in Figure 1, the results obtained from this procedure were inexact. The total number of molecules differed by approximately 25% (1.554×10^6 vs. 1.257×10^6), depending upon whether the calculation was made at the point were the curves crossed (solid lines) or at the point where the regressions crossed (dotted lines). In addition, while the average quantity of molecules obtained from the five experiments corresponded to what was expected, the value for the standard deviation was 30% of the final result (Table 1A). Finally, we also noticed that among these five experiments, the extreme values of the results varied up to 50% (data not shown). The titration assay, based on competitive PCR, was first described by Gilliland et al. (3). While this method is very accurate for DNA measurements, the authors noted that several pitfalls should be considered when applying this technique to the quantification of low abundance mRNAs. To eliminate the effect caused by differences in RT efficiency, which can vary up to 50% (2), a homologous cRNA internal control must be added to the RT reaction. In the case of the titration assay, where calculations are based on individual RT reactions, it is likely that these inconsistent results are caused by efficiency differences between the various RT reactions. Nevertheless, in many cases this procedure can be used when the experiment requires an accuracy of only 50%.

As can be seen from the results shown above, a second step (quantitative assay) is required to attain precise data with RT-PCR. By using the results obtained from the initial titration assay, a known quantity of RNA sample (TM) is mixed with the internal control so that the quantities of both molecules are equal. This mixture is then reverse-transcribed and a PCR is performed on samples that have been (1:3) serially diluted.

In order for RT-PCR to be quantitative, it is necessary to have nonbiased competition between the TM and the ICM during the exponential phase of the PCR. Unfortunately, nonbiased competition can be assumed only when the ratio of TM to ICM is 1:1. We designed experiments that purposely distorted the ratio of TM to ICM to determine how variations in this ratio influence the final quantification.

One way to visualize the effect of varying the TM to ICM ratio was to plot the cpm from ICM and from TM against the number of internal control molecules. The distances between the two curves illustrate the divergence between the amounts of TM and ICM. In Figure 2A, 10^6 molecules of cRNA $\Delta96$ (ICM) was assayed with 10^6 molecules of cRNA $\Delta34$ (TM). As expected, the curves were perfectly superimposed because TM and ICM were in nonbiased competition. However, variations in their respective con-

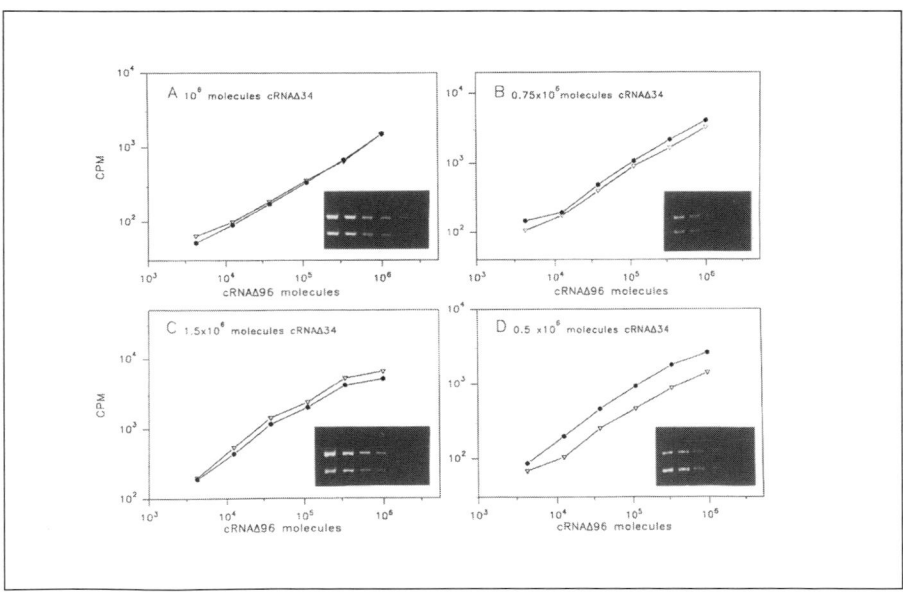

Figure 2. Visualization of the discrepancy between the amounts of TM and ICM. Varying quantities of cRNA $\Delta34$ molecules (∇), as indicated in Panels A, B, C and D, and 10^6 molecules of cRNA $\Delta96$ ($\bullet$) were reverse-transcribed, followed by serial dilutions and amplification as described in Materials and Methods. The amplification products were separated using polyacrylamide gel electrophoresis and detected by ethidium bromide staining (see insert in each panel). The bands were excised from the gel and counted. The cpm from cRNA $\Delta34$ and cRNA $\Delta96$ were plotted against the quantity of cRNA $\Delta96$ molecule contained in each dilution.

centrations induced biased competition effects. For example, when the amount of cRNA Δ34 was altered by 25% or 50% (i.e., by adding 0.75×10^6 or 1.5×10^6 molecules), the curves became slightly separated (Figure 2, B and C). The curves separated completely when the TM and ICM had a 2-fold concentration difference, as shown in Figure 2D.

The results from these same experiments were also quantified, as shown in Table 1B. A difference up to 1.5-fold between the TM and the ICM resulted in a deviation of approximately 10%, as determined from the known amount of "Molecules IN" and the calculated amount of "Molecules OUT". However, a divergence of 2-fold between the TM and the ICM caused the amount of Molecules OUT to be altered by 35%–60%, when compared to the number of Molecules IN (Table 1B).

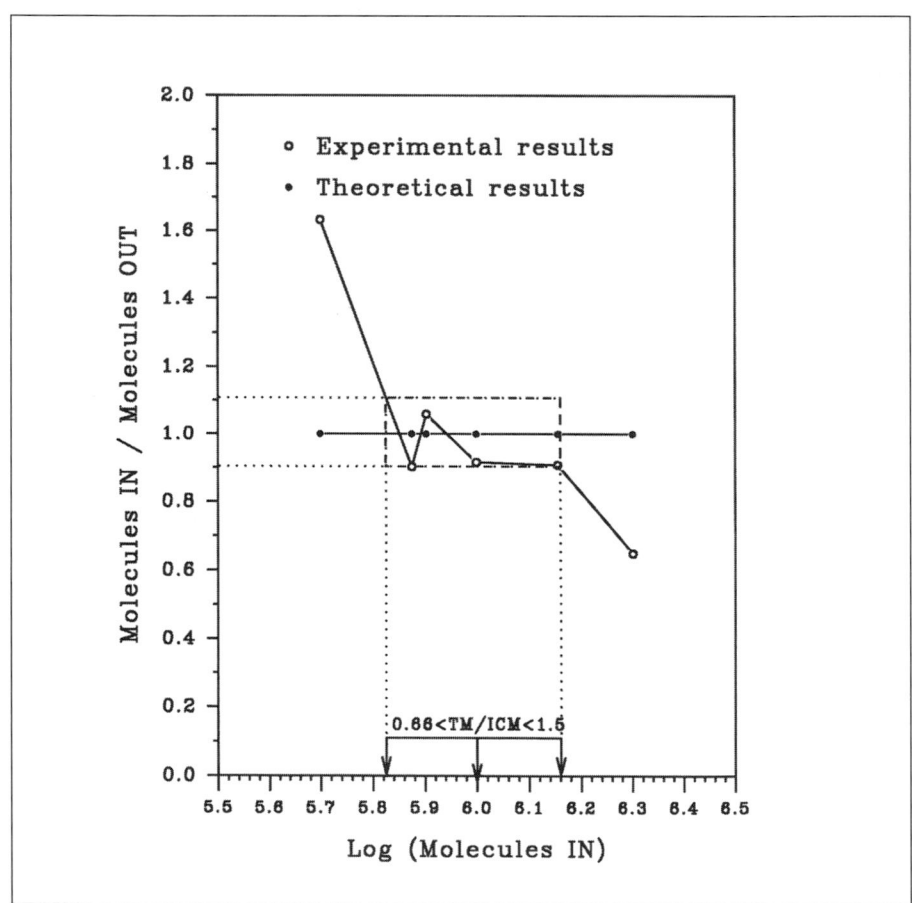

Figure 3. Accuracy of quantitative RT-PCR. A comparison between the experimental results and the theoretical results. This graph shows the variation between the number of molecules added at the beginning of the experiment and the number of calculated molecules (Molecules IN/Molecules OUT), against the number of molecules added at the beginning of the experiment (Molecules IN).

To access to the accuracy of the assay, we represented the variation of the ratio of the Molecules IN/Molecules OUT as a function of the Molecules IN. The closer the Molecules IN/Molecules OUT ratio is to 1, the more precise is the assay. In Figure 3 we visualized the discrepancy caused by distorting the ICM to TM ratio. The practical limits of quantitative RT-PCR is depicted by a box that defines the situation where the ratio of Molecules IN/OUT remains linear with a variation of 10%. These conditions are satisfied when the TM/ICM ratio is between 0.66 and 1.5 (or the difference between TM and ICM is 1.5-fold). Outside these limits, the IN/OUT ratio diverged abruptly and the fidelity of the results subsequently diminished.

To test the sensitivity of quantitative RT-PCR, we analyzed the results from two different experiments where the ratio of TM and ICM was within the above-determined limits. In this analysis, we wanted to determine if minor differences between samples could be accurately quantified. As shown in Table 1C, two experiments were performed where the quantity of starting cRNA $\Delta 34$ differed by only 20%. A Student's t test was applied to the results of this quantification and were shown to be accurate at the level of 95% confidence.

From these experiments we determined some of the fundamental aspects controlling the accuracy and sensitivity of quantitative RT-PCR. The titration assay was a preliminary step employed to set up the quantitative assay. The quantitative assay was, however, necessary for very accurate quantification. By manipulating the ratio of TM and ICM, we were able to define the practical limits of quantitative RT-PCR. As a general rule, results are accurate to 90% if the difference between TM and ICM remains equal or under 1.5-fold. Lastly, we showed that this technique was sufficiently sensitive for measuring concentration differences as small as 20%. We demonstrated that under defined limits, quantitative RT-PCR is a very useful method to precisely quantify low abundance mRNA.

ACKNOWLEDGMENT

The authors wish to express their many thanks to Dr. Neil Insdorf for his invaluable help in the writing of the manuscript and for helpful discussions. The authors also thank Drs. Kelly and Nagano for introducing them to this method. This work was supported by a fellowship to F. Souazé from the Ministère de l'enseignement supérieur et de la recherche.

REFERENCES

1. **Aviv, H. and P. Leder.** 1972. Purification of biological active globin messenger RNA by chromatography on oligothymidylic acid-cellulose. Proc. Natl. Acad. Sci. USA 69:1408-1412.
2. **Ferre, F., A. Marchese, P. Pezzoli, S. Griffin, E. Buxton and V. Boyer.** 1994. Quantitative PCR—an

overview, p. 67-88. Polymerase Chain Reaction. Birkhauser Boston, Cambridge, MA.

3. **Gilliland, G., S. Perrin, K. Blanchard and H.F. Bunn.** 1990. Analysis of cytokine mRNA and DNA: detection and quantification by competitive polymerase chain reaction. Proc. Natl. Acad. Sci. USA *87*:2725-2729.

4. **Nagano, M. and P.A. Kelly.** 1994. Tissue distribution and regulation of rat prolactin receptor gene expression. Quantitative analysis by polymerase chain reaction. J. Biol. Chem. *269*:13337-13345.

5. **Piatak Jr., M., L. Ka-Cheung, B. Williams and J.D. Lifson.** 1993. Quantitative competitive polymerase chain reaction for accurate quantitation of HIV DNA and RNA species. BioTechniques *14*:70-81.

6. **Sambrook, J., E.F. Fritsch and T. Maniatis.** 1989. Molecular Cloning: A Laboratory Manual. Cold Spring Harbor Laboratory Press, Cold Spring Harbor, NY.

7. **Stieger, M., C. Démollière, K. Ahlborn-laake and J. Mous.** 1991. Competitive polymerase chain reaction assay for quantification of HIV-1 DNA and RNA. J. Virol. Methods *34*:149-160.

8. **Tanaka, K., M. Masu and S. Nakanishi.** 1990. Structure and functional expression of the cloned rat neurotensin receptor. Neuron *4*:847-854.

9. **Wang, A.M., M.V. Doyle and D.F. Mark.** 1989. Quantitation of mRNA by the polymerase chain reaction. Proc. Natl. Acad. Sci. USA *86*:9717-9721.

Update to:

Quantitative RT-PCR: Limits and Accuracy

F. Souazé, M. Najimi, M. Mèndez-Ubach, D. Pélaprat, W. Rostène and P. Forgez

INSERM Unit 339, Hôpital Saint-Antoine, Paris, France

Quantitative PCR is an excellent method for measuring the exact number of mRNA molecules, in part because it exceeds the sensitivity limits of classical techniques such as Northern blot, hybridization assay or in situ hybridization. We have previously shown that mRNA variations as small as 20% can be quantified with an accuracy reaching 90%. This technique can be applied under various physiological conditions and pharmacological treatments. In this supplement, we investigated the use of this technique in primary cell cultures, in in vivo situations, and for the quantitative measurements of mRNA half-life.

In Table 1, quantitative RT-PCR was applied under different conditions. One of the most significant aspects of this technique was the ability to measure quantities as small as 10 000 molecules/µg of total RNA. Even under stringent conditions, this technique was highly reproducible between samples. In Table 1A, neurotensin receptor (NTR) mRNA was assayed in rat hypothalamic neurons from primary cell culture. This experiment represents conditions where the assay was performed on limited quantities of a heterogeneous material. Indeed, in this cell population, only a few cells expressed NTR. Nonetheless, we were able to detect a 40% decrease in NTR mRNA after 48 h treatment with a neurotensin agonist.

Using this technique we also estimated NTR mRNA half-life in the human adenocarcinoma cells HT-29. NTR mRNA level was measured every 30 min after actinomycin D treatment. As shown in Table 1B, even though the variations were small between time points, the accurate calculation of NTR mRNA half-life was obtained. With this technique we demonstrated that after a 72-h neurotensin agonist treatment, the NTR mRNA half-life was reduced by 60% in HT-29 cells.

NTR mRNA was also extracted from in vivo sources such as brain regions and other tissues. The variation of NTR mRNA levels in rat was studied following a 2-week treatment with an antagonist of the NTR. Shown in Table 1C are various examples where both small significant variations (duodenum) and no variations (cerebellum) were detected in NTR mRNA levels.

Finally, we recently applied our method of quantitative RT-PCR on other genes like tyrosine hydroxylase (TH). We measured TH mRNA levels in the

Table 1. Examples of Different Applications for Quantitative RT-PCR

	Sample	Mean $\times 10^6$ molecules/μg of total RNA $\pm$SEM	Number of experiments
A. Effect of agonist on NTR mRNA level in rat hypothalamic neurons in primary culture.	Control	0.061 ± 0.0046	3
	48-h agonist treatment	0.037 ± 0.0027	3
B. Study of NTR mRNA half-life. Measurement of NTR mRNA after 30 and 60 min of actinomycin D treatment.	Control $C_{0\ min}$ $C_{30\ min}$ $C_{60\ min}$	6.9 ± 0.63 4.8 ± 0.32 3.0 ± 0.7	3 3 3
	72-h agonist treatment		
	$72_{0\ min}$ $72_{30\ min}$ $72_{60\ min}$	2.06 ± 0.5 0.93 ± 0.17 0.46 ± 0.06	3 3 3
C. Measurement of NTR mRNA levels in rat tissues after NTR antagonist treatment.	Duodenum control 15-day antagonist treatment	1.7 ± 0.07 1.0 ± 0.03 *	5 5
	Cerebellum control 15-day antagonist treatment	1.1 ± 0.04 0.97 ± 0.07	5 5
D. Effect of NT agonist on TH mRNA level in neuroblastoma cells.	Control	1.8 ± 0.3	3
	5-h agonist treatment	3.0 ± 0.5	3
	72-h agonist treatment	3.5 ± 0.8	3

* $P<0.01$

human neuroblastoma cell line, CHP212, in control conditions or under different exposure times with the neurotensin agonist. Table 1D shows that TH mRNA level increases, and subsequently remains constant with the time of agonist exposure.

In conclusion, using quantitative RT-PCR as described above, is a very powerful tool to estimate mRNA variations. Applied to different models, quantitative RT-PCR is an excellent technique for quantifying small mRNA variations even when the RNA is difficult to obtain.

CHAPTER 3

Method to Identify Biases in PCR Amplification of T Cell Receptor Variable Region Genes

Enoch Satyaselan Daniel and David George Haegert

Memorial University of Newfoundland, St. John's NF, Canada

Quantitative polymerase chain reaction (PCR) methods have been used to establish the relative frequencies of expression of T cell receptor (TCR) beta (B) and alpha (A) variable (V) gene families in vitro and in various tissues in vivo. This quantitative approach is difficult as it is influenced by factors inherent in the PCR, and quantitation must be performed during the exponential phase of DNA amplification (8). An additional potential problem is that variation in the amplification efficiencies of different TCR segments is possible with several of the commonly used PCR methods (8). If these amplification biases are not corrected for, they could lead to spurious conclusions as to relative TCR segment expression levels in blood or in another tissue from an individual. Some investigators (1,5) have assumed that amplification biases do not occur and that different primer pairs amplify different TCRV gene segments with equal efficiencies. Others, aware of the potential problem, have compared the relative levels of expression of particular TCR segments between individuals or between tissues in an individual (3). In a variation of this latter approach, Lovebridge et al. (6) controlled for the amounts of starting TCRB constant (C) region templates and then compared TCRBV gene expression between different individuals. No quantitative PCR method reliably establishes the relative expression levels of different TCR segments in a tissue from a particular individual. We report a simple method to do this. The method assesses the relative efficiencies of priming of TCRBV (BV) segments when using a set of sequence-specific oligonucleotide (SSO) BV primers, and we describe a method to correct for amplification biases.

Total RNA was isolated from phytohemagglutin-stimulated peripheral blood lymphocytes using RNAzol™ B (Cinna/Biotecx; currently available from Tel-Test B, Friendswood, TX, USA) (2), and 1–2 µg was used for cDNA synthesis using the Pharmacia Biotech First-Strand cDNA Synthesis Kit (Piscataway, NJ, USA). PCR amplification of TCRBV segments was done in two rounds. For the first round, 1-µL aliquots of cDNA were amplified in separate reaction tubes with a 3′ TCRBC (BC) primer and one of a set of 23 different 5′ BV primers reported to be specific for different BV fami-

(Reprinted from BioTechniques 20:600-602, 1996)

lies or BV family members (1,10). One microliter of cDNA was amplified separately with 5′ and 3′ TCRAC (AC) primers (1). All reagents except the BV primers were prepared in a master mixture to keep concentrations constant. Each 50-µL first-round PCR contained 2.5 units *Taq* DNA polymerase (Promega, Madison, WI, USA), 240 µM deoxyribonucleoside triphosphates (dNTPs), 1.5 mM MgCl$_2$ and 0.165 µM of each member of the relevant primer pairs. PCR amplification was performed on a Perkin-Elmer Model 480 Thermal Cycler (Norwalk, CT, USA) for 30 cycles, consisting of 95°C for 1 min, 55°C for 1 min and 72°C for 1 min; preliminary

Table 1. Correlations (r Values) Between BV Priming Efficiencies and BV Repertoires of Total T Cells Before and After Correction for Amplification Biases[a]

Donor	Initial	After Correction
DB2	0.474	0.024
SC1	0.553	0.253
SC2	0.534	0.130
SN	0.419	-0.128
WG	0.106	-0.209
LH	0.417	0.154

[a]The critical value of $r = 0.444$ for $P < 0.05$

experiments with cDNA from the TCRBV8-expressing Jurkat cell line indicated 30 cycles is well within the exponential phase of the PCR (unpublished observation). Only the anticipated TCR products and no additional amplificants were detected by ethidium-bromide staining and a 3-day X-ray film exposure of 2% agarose gels.

For the second-round PCR, all BV and AC products were chloroform-extracted, then quantitated on a spectrophotometer. First-round BV products (0.07 µM) were aliquoted into 23 separate tubes, and all other reagents were added from a master mixture. The 50-µL second-round PCR components were exactly as described for the first round except for the use of first-round BV products and the following: 0.07 µM AC first-round product, 0.15 µM 3′ AC primer, 0.015 µM ^{32}P-labeled 3′ AC primer (10[6] cpm), 0.15 µM 3′ BC primer and 0.015 µM ^{32}P-labeled BC primer (10[6] cpm). The second-round PCR amplification was for 30 cycles, as described above. Other experiments showed that PCR amplification of 0.07 µM of BV6 and 12 products from a first-round PCR was in the exponential phase at 30 cycles. Since the amount of starting template was the same for all BVs, the known independent variable was the ability of various BV primers to anneal to the corresponding BV sequences. Three trials were performed on cDNA from one individual, and the mean BV cpm/AC cpm was calculated for each BV.

Figure 1 shows, first, the extent to which the BV cpm/AC cpm is reproducible for each BV. Second, the figure shows there is marked variation in the efficiency of priming of different BV segments. BVs 3, 10 and 17 primed least efficiently, and BVs 6 and 15 most efficiently, with a sixfold difference in priming efficiency between the least and the most efficiently primed segments. A one-way ANOVA ($P < 0.001$) confirmed that the mean BV/AC

34

ratios differed significantly among the 23 different BVs, i.e., there is an amplification bias for different BV segments when the BV target templates are at a constant concentration.

The 0.07-μM BV target templates used in the second-round PCR represented approximately 0.02% of the first-round PCR products. It seemed unlikely, therefore, that the second-round PCR results were influenced by carryover of primers, enzyme, etc. However, it was necessary to completely exclude this possibility and the possibility that accumulation of the TCRAC reference product in the same reactions as the BVs had affected the priming efficiencies of some or all BV segments (4,7). Also, it was essential to confirm the initial findings of an amplification bias of BV segments. This was done by repeating the experiment on cDNA from a second, genetically different individual. Double-stranded DNA from a first-round PCR of the 23 BVs was purified using the Wizard™ PCR Preps DNA Purification System (Promega), which excludes primers, buffers and dNTPs. After quantitation, 0.07 μM of each BV was PCR-amplified in duplicate for 28 cycles in separate tubes and without the AC reference template; 28 cycles were used to exclude completely the possibility that the PCR had reached the plateau phase. The results were obtained as cpm for each BV. Correlation analysis of the mean BV/AC ratio from the first individual and the BV profile from the second individual gave a Pearson correlation (r value) of 0.882; i.e., the high

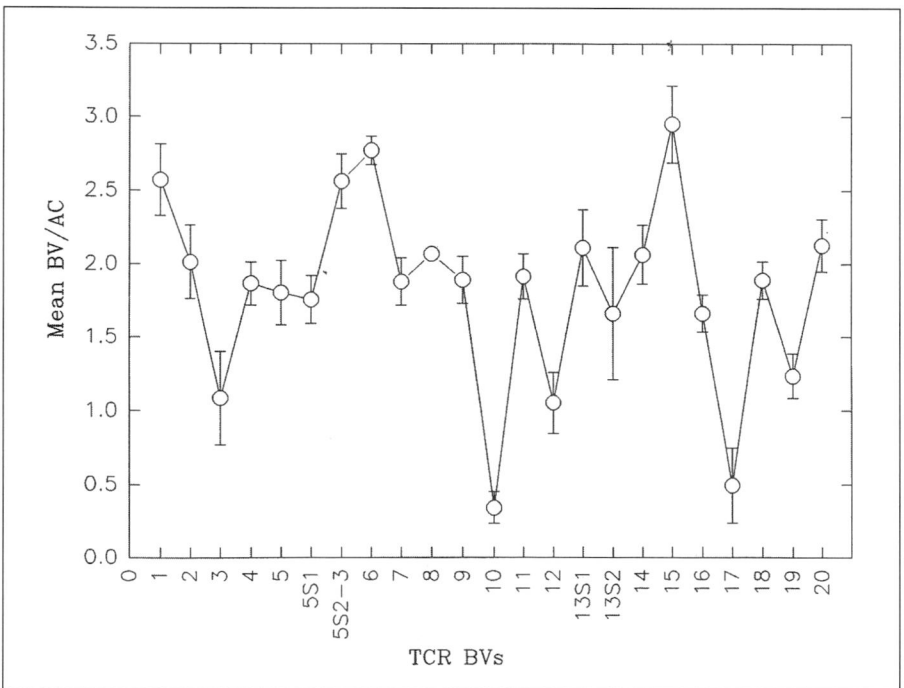

Figure 1. Relative efficiencies of 23 TCRBV primers shown as mean BV/AC ratios ± SD of 3 trials.

correlation indicates the BV profiles are remarkably similar using the different methods and confirms that there is a significant amplification bias for different BV segments—the least efficiently primed and the most efficiently primed BV segments were as in Figure 1 (unpublished observation).

The results indicate that it is not possible to compare the relative expression levels of different BV segments in a tissue or from a particular individual using PCR data obtained with SSO primers without correcting for BV segment amplification biases. The approach we used to correct for these biases, and based on data in Figure 1, is as follows: A correction factor (CF) was obtained by dividing an arbitrary integer by the mean BV/AC ratio for each BV; the results are the same using any integer other than 1. The % of each BV was multiplied by the corresponding CF, and the modified relative BV scores were calculated: % BV modified = corrected BV score / $\sum$ corrected BV scores × 100. Standard PCR analysis of the total T cell BV repertoires showed a significant Pearson correlation (r value) between priming efficiencies and observed BV repertoires in 27 of 48 individuals. Subsequent correlations after correction for amplification biases showed that there was no longer this significant correlation. Table 1 shows an example of the results of normalization, i.e., correction for amplification biases, for total T cells.

We observed amplification biases with primers from two distinct sources (1,10). Therefore, correction for amplification biases seems likely to be needed for any set of SSO BV primers. The approach we have used here could be extended to studies of the TCRBV repertoire in disease if pooled cDNA from a population-based group of individuals were used to arrive at mean BV/AC ratios. The method has obvious wider relevance, not only to analysis of the TCRAV repertoire, but also to other quantitative PCR approaches, e.g., the use of a single SSO primer pair to amplify a target molecule and an internal PCR standard that have different intervening sequences (9).

ACKNOWLEDGMENT

D.G. Haegert is supported by the Multiple Sclerosis Society of Canada.

REFERENCES

1. **Choi, Y., B. Kotzin, L. Herron, J. Callahan, P. Marrack and J. Kappler.** 1989. Interaction of *Staphylococcus aureus* toxin "superantigens" with human T cells. Proc. Natl. Acad. Sci. USA *86*:8941-8945.
2. **Chomczynski, P. and N. Sacchi.** 1987. Single-step method of RNA isolation by acid guanidinium thiocyanate-phenol-chloroform extraction. Anal. Biochem. *162*:156-159.
3. **Davey, M.P., M.M. Meyer and A.C. Bakke.** 1994. T cell receptor Vβ gene expression in monozygotic twins: discordance in CD8 subset and in disease states. J. Immunol. *152*:315-321.
4. **Genevée, C., A. Diu, J. Nierat, A. Caignard, P.-Y. Dietrich, L. Ferradini, S. Roman-Roman, F. Triebel and T. Hercend.** 1992. An experimentally validated panel of subfamily-specific oligonucleotide primers (Vα1-w29/ Vβ1-w24) for the study of human T cell receptor variable V gene segment

usage by polymerase chain reaction. Eur. J. Immunol. *22*:1261-1269.

5. **Gulwani-Akolkar, B., B. Shi, P.N. Akolkar, K. Ito, W.B. Bias and J. Silver.** 1995. Do HLA genes play a prominent role in determining T cell receptor Vα segment usage in humans? J. Immunol. *154*:3843-3851.

6. **Lovebridge, J.A., W.M.C. Rosenberg, T.B.L. Kirkwood and J.I. Bell.** 1991. The genetic contribution to human T-cell receptor repertoire. Immunology *74*:246-250.

7. **Munson, J.L., E. van Twuyver, R.J.D. Mooijart, E. Roux, I.J.M. ten Berge and L.P. de Waal.** 1995. Missing T-cell receptor Vβ families following blood transfusion. The role of HLA in development of immunization and tolerance. Hum. Immunol. *42*:43-53.

8. **Panzara, M.A., J.R. Oksenberg and L. Steinman.** 1992. The polymerase chain reaction for detection of T-cell antigen receptor expression. Curr. Opin. Immunol. *4*:205-210.

9. **Wang, A.M. and D.F. Mark.** 1990. Quantitative PCR, p. 70-75. *In* M.A. Innis, D.H. Gelfand, J.J. Sninsky and T.J. White (Eds.), PCR Protocols: A Guide to Methods and Applications. Academic Press, San Diego.

10. **Wucherpfennig, K.W., K. Ota, N. Endo, J.G. Seidman, A. Rosenzweig, H.L. Weiner and D.A. Hafler.** 1990. Shared human T cell receptor Vβ usage to immunodominant regions of myelin basic protein. Science *248*:1016-1019.

Update to:

Method to Identify Biases in PCR Amplification of T Cell Receptor Variable Region Genes

Teresa Cowan and David George Haegert

Memorial University of Newfoundland, St. John's NF, Canada

INFLUENCES OF PCR AMPLIFICATION BIASES ON THE HUMAN T CELL RECEPTOR J BETA (BJ) REPERTOIRE

In an earlier study of the T cell receptor (TCR) V beta (BV) repertoire we described a method to test the relative priming efficiencies of a set of 23 different BV PCR primers. The various primers amplified the TCR target sequences with significantly different efficiencies, and we reported a method that corrects for these amplification biases (2). This methodology has obvious implications for studies using multiple PCR primers, and we further

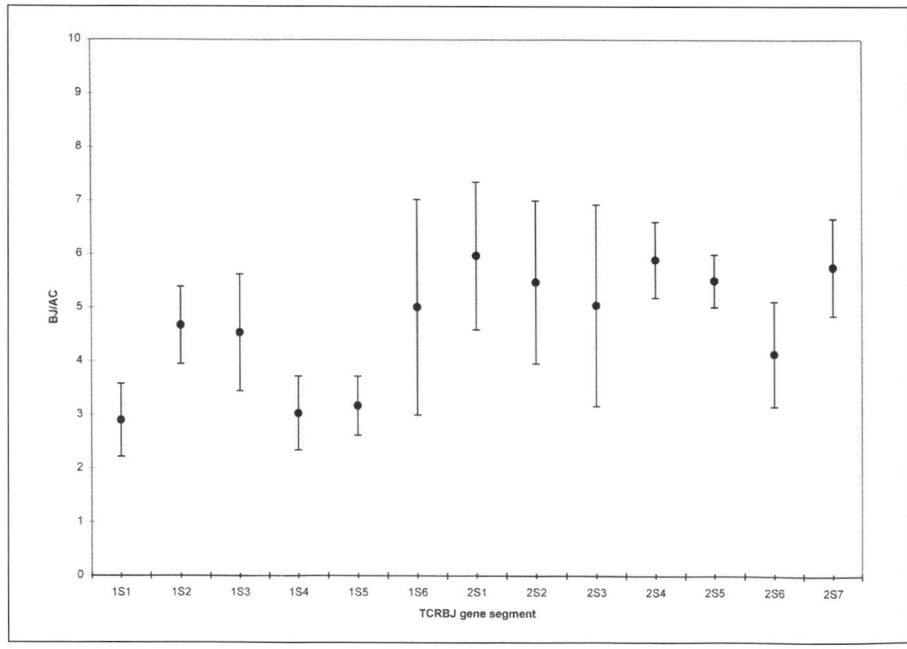

Figure 1. Relative efficiencies of 13 TCRBJ primers shown as mean BJ/AC ratios ± SD of 5 trials.

38

illustrate the importance of correcting for differences in priming efficiencies in a study of human TCR J beta (BJ) segments by quantitative PCR.

Human peripheral blood T cell receptor DNA was PCR-amplified in separate reaction tubes with 5' and 3' TCR alpha (A) constant (C) primers (1,5) and with 13 different 5' BJ primers (4) and a single 3' TCRBC primer (2). Each PCR amplification was performed in two rounds as described previously for the BV segment primer efficiency study (2) with the following exceptions. For both PCR rounds, 14 cycles were used; with the TCRBJ1S5 and 2S2 primers, 14 cycles were well within the exponential phase for both the TCR cDNA used in the first round and for the TCR first-round products used as the template in the second round. Also, BJ and AC templates in the second round were at 0.01 μM. Five trials were performed on DNA from a single individual and the mean BJ cpm/AC cpm was calculated for each BJ segment.

Figure 1 shows there is marked variability in priming efficiency of the different BJ segments. A one-way ANOVA ($P<0.005$) confirms this impression and indicates that the mean BJ/AC ratios of the various BJs differ markedly. The most efficient primers under the present test conditions were TCRBJ2S1, 2S4 and 2S7, and the least efficient primers were TCRBJ1S1, 1S4 and 1S5; the relative primer efficiencies could, of course, change under different test conditions.

The importance of testing priming efficiency under particular PCR conditions and then correcting for differences among primers was demonstrated by a study of the BJ repertoire in CD4+CD45RA+ (naive) T cells from 27 healthy individuals (Figure 2). Initially, i.e., before correcting for differences in priming efficiencies, a one-way ANOVA ($P<0.005$) indicated that the mean expression levels of individual BJ segments are significantly different

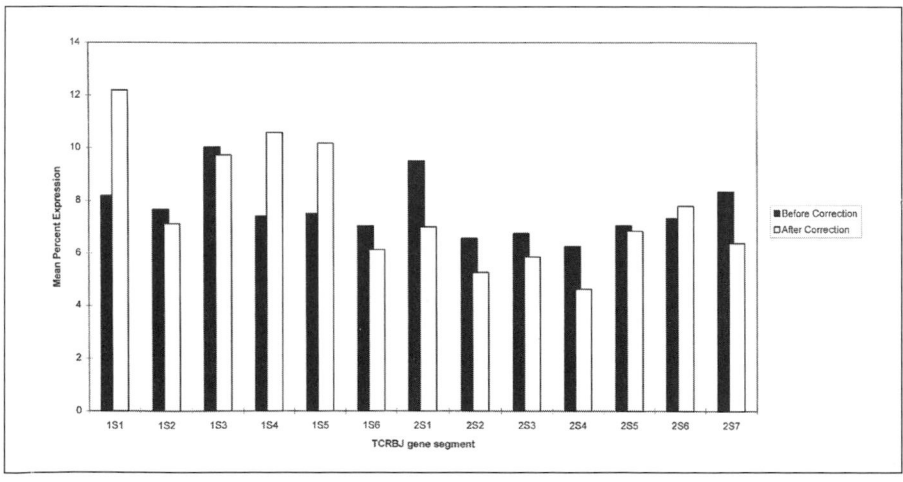

Figure 2. Mean percent expression of 13 TCRBJ segments from 27 individuals before and after correction for PCR amplification biases.

and TCRBJ1S3, 2S1 and 2S7 segments appeared to be expressed at significantly higher levels than other BJ segments. After correcting for differences in priming efficiency (Figure 2) as described (2), correlation analysis was used to test the similarity of the two BJ repertoires. The correlation coefficient (r value) of 0.44 indicates that there is no significant correlation (P=0.12) between the BJ repertoires obtained before and after correction. Moreover, after correction, TCRBJ1S1, 1S3, 1S4 and 1S5 were found to be expressed at higher levels than other BJ segments. These results indicate, therefore, that differences in TCRBJ segment priming efficiency can easily lead to erroneous conclusions as to dominant BJ segment usage. Together with our earlier report (2), the present data call into question various findings on the frequency of usage of various TCR segments in a group of individuals (see, for example, References 3 and 6). Also, our two reports suggest that the use of any set of multiple primers to study relative expression levels of different transcripts in a tissue may require testing of and then correction for PCR amplification biases.

REFERENCES

1. **Choi, Y., B. Kotzin, L. Herron, J. Callahan, P. Marrack and J. Kappler.** 1989. Interaction of *Staphylococcus aureus* toxin "superantigens" with human T cells. Proc. Natl. Acad. Sci. USA *86*:8941-8945.
2. **Daniel, E.S. and D.G. Haegert.** 1996. Method to identify biases in PCR amplification of T-cell receptor variable region genes. BioTechniques *20*:600-602.
3. **Gulwani-Akolkar, B., B. Shi, P.N. Akolkar, K. Ito, W.B. Bias and J. Silver.** 1995. Do HLA genes play a prominent role in determining T cell receptor Vα segment usage in humans? J. Immunol. *154*:3843-3851.
4. **Pannetier, C., M. Cochet, S. Darche, A. Casrouge, M. Zöller and P. Kourilsky.** 1993. The sizes of the CDR3 hypervariable regions of the murine T-cell receptor β chains vary as a function of the recombined germ-line segments. Proc. Natl. Acad. Sci. USA *90*:4319-4323.
5. **Wucherpfennig, K.W., K. Ota, N. Endo, J.G. Seidman, A. Rosenzweig, H.L. Weiner and D.A. Hafler.** 1990. Shared human T cell receptor Vβ usage to immunodominant regions of myelin basic protein. Science *248*:1016-1019.
6. **Usuku, K., N. Joshi, C.J. Hatem Jr., C.A. Alper, D.A. Schoenfeld and S.L. Hauser.** 1993. The human T-cell receptor β-chain repertoire: longitudinal fluctuations and assessment in MHC matched populations. Immunogenetics *38*:193-198.

Section II

Specific Quantitative Methods and Detection of PCR Products

Quantitative Competitive Polymerase Chain Reaction for Accurate Quantitation of HIV DNA and RNA Species

M. Piatak, Jr., K.-C. Luk, B. Williams[1] and J.D. Lifson

Genelabs, Redwood City, CA and [1]Applied Imaging, Santa Clara, CA, USA

ABSTRACT

Inherent features of the PCR make this procedure suboptimal for quantitative applications. For typical clinical specimens, the absolute amount of product derived from PCR does not always bear a consistent relationship to the amount of target sequence present at the start of the reaction. Competitive PCR approaches to quantitation of nucleic acid sequences overcome the limitations of basic PCR methods for quantitation. We have developed a competitive PCR method known as quantitative competitive PCR for quantitation of HIV DNA and RNA sequences. Key features of the technique include quantitation, based on the relative amounts of products produced from the target sequence and a competitive template introduced into test specimens, and stringent internal control of all reactions, including the reverse transcription step in RNA PCR. The method is suitable for analysis of clinical specimens and may be particularly valuable for accurate quantitation of viral load in patients undergoing treatment with experimental therapies.

INTRODUCTION

The polymerase chain reaction (PCR) provides unprecedented sensitivity in diagnostic applications, including situations where there was no previously available test procedure, or where the existing procedures had serious intrinsic limitations (21,22). However, inherent features of the PCR may constrain its use in diagnostic settings where accurate quantitation is required, as opposed to a simple positive/negative determination. While the theoretical relationship between amount of starting target sequence and amount of PCR product can be demonstrated under ideal conditions (18), this does not apply for most typical clinical specimens. Specifically, because of differential

(Reprinted from BioTechniques 14:70-81, 1993)

efficiencies and kinetics of the PCR, depending on the abundance of the target sequence in the specimen of interest (8,20), along with the variable presence of various inhibitors of the PCR in typical clinical specimens and exponential compounding of minor variations in even replicate specimens, the amount of product obtained from a given specimen following a particular number of PCR cycles cannot be assumed to be an accurate reflection of the amount of target sequence present in the starting specimen. In spite of the development of a variety of approaches for accurately quantitating the products of PCRs (7), the extremely desirable goal of truly quantitative PCR has proved elusive.

One conceptually attractive approach to controlling the variability inherent in PCR involves the use of a competitive template that is largely identical to, but somehow discriminable from, the intended target sequence to be quantitated (2,12,13). Titered, known quantities of this competitive template can then be introduced into replicate PCRs containing identical aliquots of a test sample containing an unknown quantity of target sequence. The introduced template serves as a competitor at all steps involved in amplifying out the desired target sequence. When the amounts of target sequence and competitor template are equivalent, equal amounts of the respective products will accumulate. The products can be quantitated by a variety of different techniques, and the amount of target sequence present in the original test sample can be determined by either direct or interpolated assessment of the equivalence point (where the amount of target sequence product obtained from the test sample is equal to the amount of competitive template-derived product) (2,12,13). This approach has been shown to have excellent sensitivity and accuracy of quantitation in several different systems and has the added advantage, due to the competitive principle on which it is based, of being subject to stringent internal control of all steps. As modified in this manuscript, the internal control of competitive PCR is extended to the reverse transcription step through the use of RNA-competitive templates for RNA PCR. Since in this internally controlled, competitive approach it is the relative, not absolute, amounts of target and competitive products that matter, accurate quantitation can still be obtained even if there is substantial variability in the efficiency of the PCRs performed in different wells or tubes in a titration series.

Quantitation of viral load in specimens from HIV-infected patients would appear to be an ideal application for this methodology. While it is generally believed that the level of viral replication in HIV-infected patients should correlate with disease stage and progression, or alternatively with response to therapeutic intervention, this postulate remains largely unconfirmed because there have not been appropriate procedures available for determining the level of virus present in clinical samples. Available techniques, including measurement of viral antigen levels circulating in plasma (17) and the use of various viral culture approaches (5,14), are relatively insensitive, may not correlate reliably with disease status or clinical responses to therapy and/or are prohibitively labor-intensive and expensive for widespread use. Conventional PCR-based approaches may suffer from the shortcomings noted above when applied to typical clinical specimens. We have developed a procedure for sensitive and accurate quantitation of HIV RNA and DNA species that is applicable to clinical specimens. The approach may be useful for assessing viral load in patients, within the context of overall assessment of HIV-infected patients and in the evaluation of responses to therapeutic intervention, including treatment with experimental agents.

MATERIALS AND METHODS

Production of Wild-Type Positive Control Plasmid and Competitive Template

A region in the HIV-1 *gag* gene, highly conserved among divergent viral isolates (19), was selected as our target sequence. A plasmid containing this sequence (pQP1) and a plasmid identical to pQP1, except for an internal 80-bp deletion (pQP1Δ80), were generated from HXB2c (10) for use as a positive control target sequence and competitive template, respectively (Figure 1). An approximately 1420-bp *Sac*I to *Bgl*II fragment of the *gag* gene was subcloned into pBluescript™ KS (Stratagene, La Jolla, CA, USA); the HIV sequences in pQP1 and pQP1Δ80 were placed under control of a T7 promoter to allow generation of the corresponding in vitro transcripts for use as a positive control and competitive template in RNA PCR.

PCR Primers

Primers GAG04 and GAG06 were designed to amplify an internal fragment of either 260 bp (from wild-type HIV-1 target sequences) or 180 bp (from the pQP1Δ80 competitive template). Highly conserved sequences were selected for primer sites (19). The primers were designed with consideration for length (28 and 27 bp, respectively), non-clustered distribution of G+C residues and non-divergent sequence match to the target sequence at their 3′ ends. To maintain equivalent priming efficiency with divergent sequences, primers incorporated inosine residues at the few positions where divergence from the conserved consensus sequence had been reported (3,19; Figure 1).

DNA Preparation

Total cellular DNA was extracted using the Elu-Quik™ Kit (Schleicher and Schuell, Keene, NH, USA); the extraction protocol supplied with the kit was adapted to facilitate the handling of multiple samples. For quantitative competitive (QC)-PCR analysis of cellular DNA, cell pellets ($0.5–2.0 \times 10^7$ cells) were lysed in approximately 350 µL guanidinium thiocyanate buffer, supplemented to contain 0.5% Sarkosyl™. DNA was adsorbed to glass beads, washed and eluted using 800 µL binding buffer, 250 µL glass beads, 3 × 1 mL wash buffer, 2 × 1 mL salt reduction buffer and 3 × 100 µL sterile water. The eluted samples were adjusted to 0.15 M potassium acetate (KOAc), pH 5.5, and ethanol precipitated. Pelleted DNA was then rinsed with 70% ethanol, partially dried and redissolved in 60 µL sterile water. Typical DNA yields were 5–8 µg/10^6 unstimulated peripheral blood mononuclear cells (PBMC).

RNA Preparation

Positive control and competing RNA transcripts were prepared as run-off products of the corresponding DNA templates, linearized at an *Eco*RI site located just 3′ to the HIV-1 sequence insert. In vitro transcription was performed using T7 polymerase and a commercially available kit (Stratagene). RNase-free DNase (Promega, Madison, WI, USA) was then added to a concentration of 1 U/µg DNA and incubation at 37°C continued for 20 min. The preparation was then extracted twice with phenol:chloroform:isoamyl alcohol (24:24:1) and once with chloroform:isoamyl alcohol (24:1) and then fractionated on a Select D(RF) column (5 Prime→3 Prime, Boulder, CO, USA), as per the manufacturer's protocol. The RNA-containing fractions were then treated with 1 mg/mL proteinase K at 37°C for 30 min, extracted three times

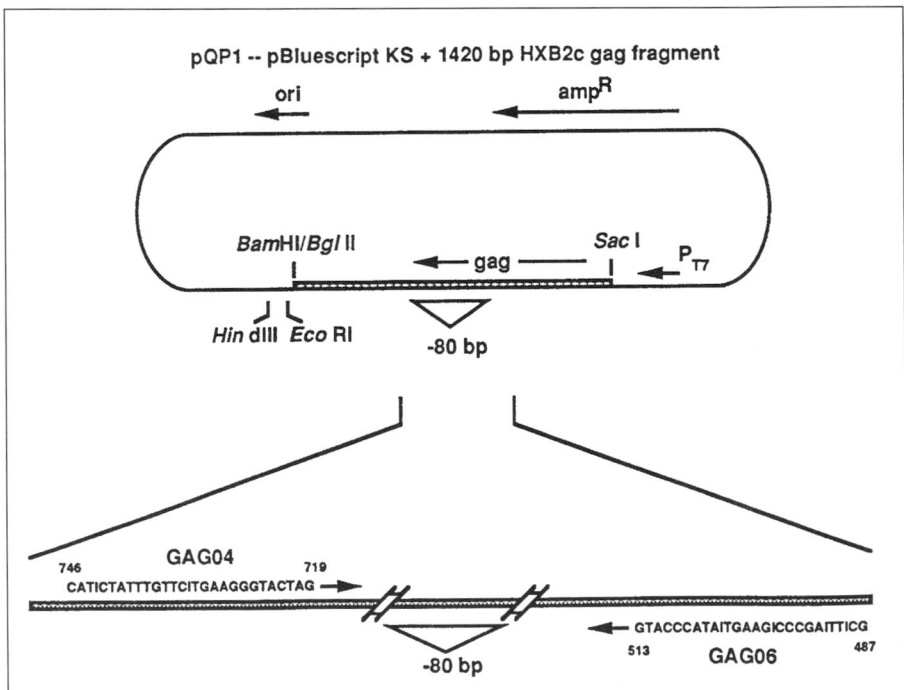

Figure 1. Schematic diagram illustrating plasmids for wild-type HIV-1 *gag* target sequence (pQP1) and deletion-containing competitive template (pQPΔ80) and relative positions of PCR primers.

A. QC-PCR: PRINCIPLE AND SCHEMATIC ILLUSTRATION

Wild type target sequence template (260 bp product) added, 10 copies constant

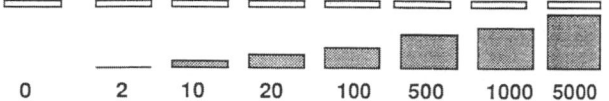

0 2 10 20 100 500 1000 5000

Competing deleted sequence template (180 bp product) added, 0-5,000 copies

B. EtBr STAINED GEL OF QC-PCR ANALYSIS: 10 COPIES HIV

1 2 3 4 5 6 7 8 9 10

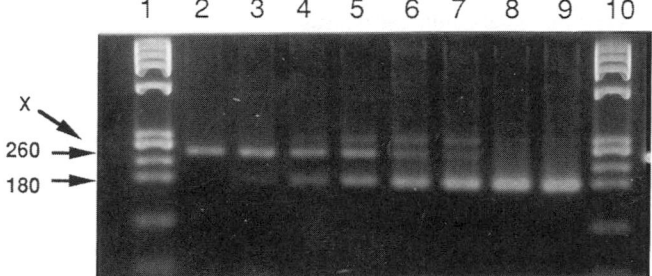

X
260 →
180 →

C. QUANTITATIVE FLUORESCENCE PLOT: 10 COPIES HIV

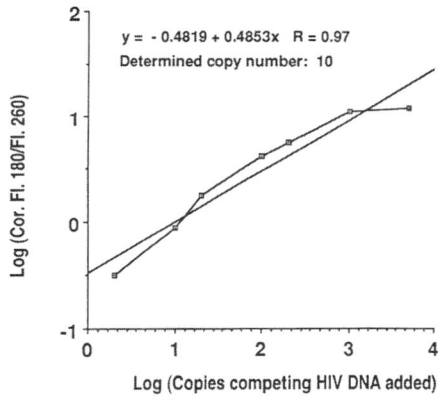

y = - 0.4819 + 0.4853x R = 0.97
Determined copy number: 10

Log (Cor. Fl. 180/Fl. 260)

Log (Copies competing HIV DNA added)

Figure 2. Schematic diagram illustrating application of QC-PCR to quantitation of HIV-1 target sequences. A. Diagram demonstrates underlying principle and expected results for QC-PCR analysis of 10 copies/reaction of wild-type HIV-1 target sequence. **B.** Ethidium bromide-stained gel for QC-PCR analysis of 10 input copies of wild-type HIV-1 *gag* DNA (pQP1). Note progressive competition between fixed amount (10 copies) of wild-type (pQP1) per reaction (260-bp band; lanes 2–9) and from 0 to 5000 copies of competitive template (pQP1Δ80) per reaction (180-bp band; lanes 2–9). Equivalence point (after correction for relative masses of product bands) is in lane 4, to which 10 copies of pQPΔ80 were added as competitor. "X" indicates putative heteroduplex formed from the pQP1 and pQPΔ80 derived products in reactions occurring near the equivalence point. Lanes 1 and 10 contain size markers (*Hae*III-cut φX174 DNA). **C.** Plot of quantitative fluorescence data obtained by computer-based video image analysis of the gel shown in B, using procedures described in Materials and Methods. Determined value of 10 copies input wild-type sequence is in exact agreement with actual input copy number.

with phenol:chloroform:isoamyl alcohol and once with chloroform:isoamyl alcohol as above and ethanol precipitated overnight at -20°C. RNA pellets were recovered by centrifugation, dissolved in RNase-free water and quantitated by measuring absorbance at 260 nm. Aliquots were stored at -70°C until needed.

For extraction of virion-associated RNA present in patient-derived plasma samples or cell culture supernatants, specimens were first subjected to ultra centrifugation (Type 70.1 rotor; Beckman Instruments, Palo Alto, CA, USA), 40 000 rpm for 1 h, room temperature) to pellet virions. The supernatant was then aspirated and the pellet suspended in 300 μL 20 mM Tris-HCl, pH 7.5, 150 mM NaCl, 2 mM EDTA containing 1 mg/mL proteinase K and 0.1% sodium dodecyl sulfate (SDS). Specimens were then incubated at 37°C for 1 h, with occasional vortex mixing, then transferred to 1.5-mL centrifuge tubes, with addition of a 100-μL buffer rinse to ensure quantitative recovery and transfer. Samples were then extracted three times with phenol: chloroform: isoamyl alcohol (24:24:1) and once with chloroform:isoamyl alcohol (24:1), and the RNA precipitated in the presence of 40 μg/mL glycogen as carrier. Precipitated RNA was recovered by ultracentrifugation (TLS 55 rotor, [Beckman Instruments], 50 000 rpm, 40 min, 5°C). The RNA pellet was partially dried, then dissolved in 40 μL sterile, RNase-free water. Samples were stored at -70°C until subsequent analysis.

PCR Setup and PCR

DNA QC-PCR: Prior to setting up PCR, test DNA samples for analysis were heated to 100°C for 5 min to ensure complete denaturation of high molecular weight DNA in the initial PCR cycles. The pQP1Δ80 competing DNA was linearized with *Eco*RI and also heated to 100°C for 3–5 min to avoid the possibility of differential denaturation kinetics. Reaction conditions were based on the recommendations of the commercial reagent supplier (Perkin- Elmer, Norwalk, CT, USA), except

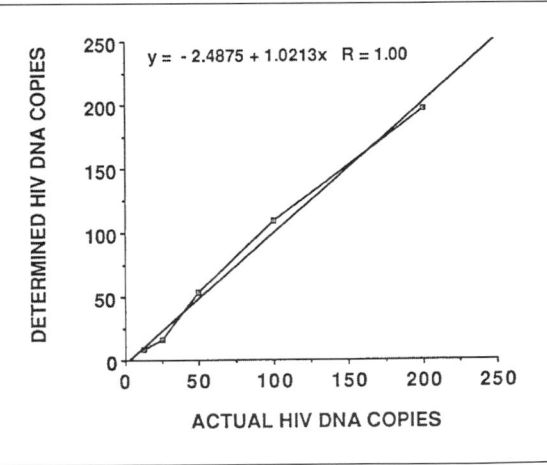

Figure 3. Detection and quantitation of known copy numbers of HIV-1 proviral sequences in a dominant background of DNA from uninfected PBMC. Determined copy numbers were calculated from quantitative fluorescence results of QC-PCR. Plot of determined HIV DNA copies vs. actual HIV DNA copies illustrates sensitivity to 12.5 copies per reaction, discrimination of twofold differences of copy number and excellent linearity over the copy number range tested.

that each reaction was conducted in a total volume of 50 µL, using 0.5 U/reaction of *Taq* polymerase. A typical reaction contained 4 µL of appropriate competing template preparation (or water), 6 µL test sample (corresponding to 10% of the total amount of test sample) and 40 µL of reaction cocktail consisting of pooled buffer, deoxyribonucleoside triphosphates (dNTPs) and enzyme to give the appropriate final concentrations. Reactions were conducted in 96-well microplates using a programmable thermal controller (Model PTC-100; MJ Research, Watertown, MA) and the following PCR cycle program:

3 cycles: 97°C for 1 min; 55°C for 2 min; 72°C for 1 min;

37 cycles: 94°C for 1 min; 55°C for 2 min; 72°C for 1 min, followed by 5°C for 5 min.

RNA QC-PCR: Commercial reagents (Perkin-Elmer) and manufacturer's suggested reaction conditions were employed with only minor modifications. The initial reaction for reverse transcription consisted of only 30 µL and typically contained 4 µL for test RNA sample (10% of total specimen) and 2 µL of competing RNA template preparation (or water) in a 96-well format. A typical competing template series consisted of 0, 10, 50, 100, 500, 1000 or 5000 copies. The reverse transcription was conducted for 45 min at 42°C. Presence of the competing template in RNA form during the reverse transcription reaction provided an internal control for this step, as opposed to performing DNA-competitive PCR on cDNA as a means of RNA quantitation (12,13). For QC-PCR analysis following reverse transcription, the initial reaction was adjusted to contain primers and additional buffer in a total of 60 µL. The PCR cycle program used was:

40 cycles: 94°C for 1 min; 50°C for 2 min; 72°C for 1 min, followed by incubation at 5°C for 5 min.

Sample analysis and quantitation: Approximately 20% of each reaction mixture was run on a 4% NuSieve®:1% Agarose (FMC BioProducts, Rockland, ME, USA) gel in 40 mM Tris-acetate, pH 7.8, 1 mM EDTA. Gels were stained with ethidium bromide for visualization

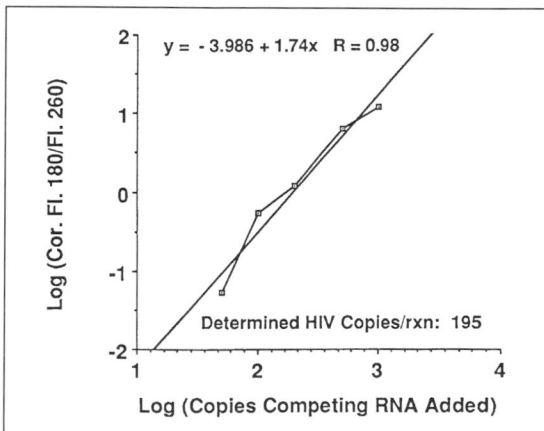

Figure 4. Quantitation of 200 copies of HIV-1 RNA. RNA QC-PCR was performed as described in Materials and Methods, using varying copy numbers of competitive template-derived (pQPΔ80) RNA per reaction to compete aginst a fixed amount (200 copies) of wild-type derived (pQP1) RNA per reaction. Determined input copy number of 195 copies wild-type sequence is in excellent agreement with actual input of 200 copies of pQP1.

under UV light. Quantitation was performed on a Lynx 4000 Molecular Biology Workstation, using matched custom software (Applied Imaging, Santa Clara, CA, USA). The system includes a UV transilluminator light box, incorporating a view-ported darkroom hood and camera stand. A high-speed video camera is used to acquire a video image of the gel, with output directly to a computer-based system for data analysis and storage, with an attached video printer for hard-copy output. Ethidium fluorescence associated with DNA bands in the gel is proportional to the amount of DNA present. Bands of interest are selected on the video screen for analysis and the integrated fluorescence intensity associated with each band determined by computer.

The PCR product band corresponding to the wild-type HIV-1 target sequence is 260 bp, while the PCR product derived from the deleted competitive HIV-1 *gag* template (pQP1Δ80) is 180 bp. Since the comparisons and equivalence point determination in the QC-PCR technique are based on molar amounts, the fluorescence associated with the 180-bp product was corrected by multiplication by a factor of 260/180 to enable direct comparison of corrected fluorescence (Cor. Fl.) intensity of the 180-bp band with measured fluorescence intensity of the 260-bp band. For determination of the competition equivalence point, it was convenient to plot the $\log_{10}$ of the ratio of the corrected fluorescence intensity of the 180-bp band over the fluorescence intensity of the 260-bp band [$\log_{10}$ (Cor. Fl. 180/Fl. 260)], as a function of the $\log_{10}$ of the number of copies of competing template added. Graphically, this transformation yields a linear plot. At the equivalence point, the corrected fluorescence of the 180-bp band should equal the measured fluorescence for the 260-bp band, and their ratio should equal 1. Interpolation on the plot for a Y value of 0 ($\log_{10}$ 1=0) gives the number of copies corresponding to the equivalence point and, hence, the number of copies present in the test sample, typically corresponding to 10% of the starting sample per reaction.

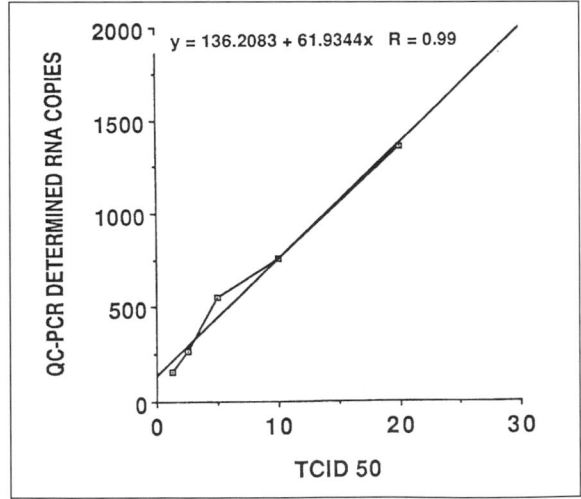

Figure 5. RNA QC-PCR determined copy number for serial twofold dilutions of infectious HIV-1 stock spiked into HIV-seronegative plasma. After processing as described in Materials and Methods, determined HIV-1 RNA copy numbers were calculated from quantitative fluorescence results of QC-PCR. Plot of determined HIV RNA copies vs. actual HIV RNA copies illustrates discrimination of twofold differences of copy number and excellent linearity over the copy number range tested.

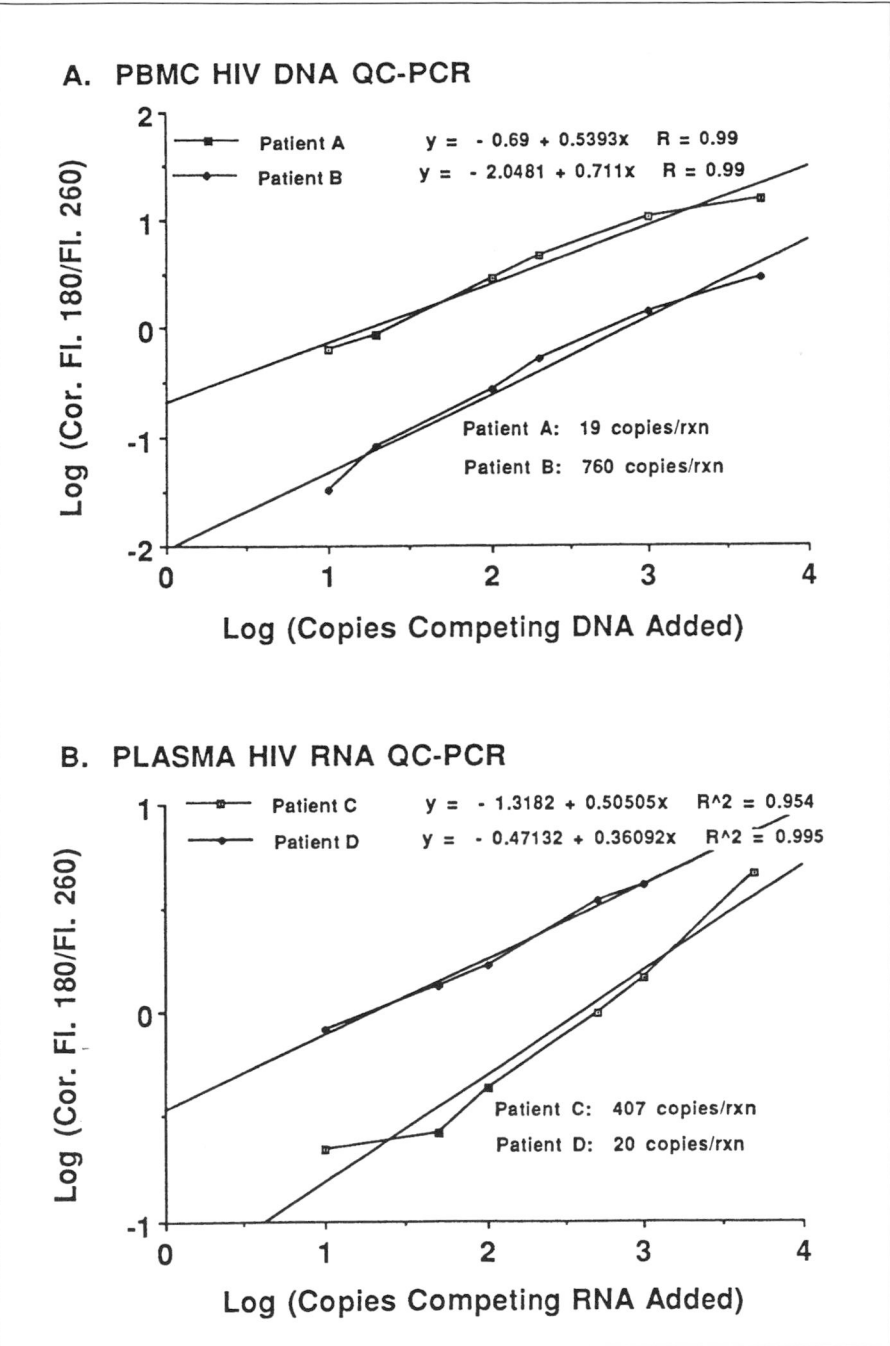

Figure 6. Application of QC-PCR to determination of HIV-1 DNA and RNA copy numbers in peripheral blood from HIV-infected patients. A. DNA QC-PCR performed on PBMC from HIV-infected patients. B. RNA QC-PCR performed on plasma from HIV-infected patients.

For DNA, final results can be expressed as "n HIV-1 DNA copies per 10^6 cells" or per "μg of PBMC DNA." Alternatively, because of potential variability in cell viability and in cell counts performed by different operators, as well as potential variability in efficiency of extraction of DNA from different specimens, it may be more accurate to report results expressed relative to diploid genome DNA content equivalents, as determined by parallel QC-PCR quantitation of a single copy cellular genomic sequence "n HIV-1 DNA copies per diploid genome DNA equivalent" (Piatak et al., unpublished). For plasma-derived virion RNA, final results can be expressed as "n HIV-1 RNA copies per mL plasma."

Infected Cells and Virus Preparations for QC-PCR Analysis

Cell culture and isolation: For model studies of assay sensitivity and calibration of quantitation of HIV DNA, the 8E5-LAV subclone of the CEM cell line was used. This chronically infected cell line has been shown to contain a single integrated copy of HIV-1_{LAV} per cell (11). 8E5-LAV cells were grown in RPMI-1640 (GIBCO BRL/Life Technologies, Gaithersburg, MD, USA) supplemented with 10% (vol/vol) heat-inactivated fetal calf serum. PBMC from healthy volunteer blood donors at the Stanford University Blood Bank were isolated by density centrifugation with Ficoll-Paque® (Pharmacia Biotech, Piscataway, NJ, USA). For sensitivity and calibration studies, infected 8E5-LAV cells were mixed at various ratios with 5×10^6 unstimulated uninfected PBMC and pelleted for DNA extraction as described above.

Preparation and analysis of infectious HIV-1 stock: For model studies of assay sensitivity and calibration of quantitation of virion-associated HIV RNA, an infectious stock of the virus isolate HIV-1_{DV} (6) was prepared by acute infection of the VB cell line (16). The stock contained 4×10^5 TCID$_{50}$ (50% tissue culture infectious dose) U/mL. Serial dilutions of this viral stock were spiked into replicate samples of plasma from a healthy HIV seronegative donor. Virions were pelleted and RNA was extracted as described above.

Preparation and analysis of blood specimens from HIV-infected subjects: Acid-citrate-dextrose anticoagulated peripheral blood specimens were obtained from HIV-infected individuals with informed consent approved by the Committee on Human Subjects of the University of California, San Francisco. After centrifugation at approximately $1000 \times g$ for 10 min, plasma was collected and the packed cells resuspended in Ca/Mg-free phosphate-buffered saline (PBS); then PBMC were isolated by density centrifugation on Ficoll-Paque at approximately $800 \times g$ for 30 min. PBMC were recovered, washed with PBS, counted and pelleted for further DNA extraction as described above. Isolated plasma samples were centrifuged at $1600 \times g$ for 10 min to remove any residual cells and either processed directly or cryopreserved (-70°C) for up to 90 days prior to being thawed and processed. Freezing prior to processing did not decrease yield from plasma specimens (Piatak et al., unpublished).

RESULTS

Figure 2 shows representative results that illustrate the basic principle of quantitation through competitive PCR. Addition of increasing amounts, from 0–5000 molecules, of the competitive template pQP1Δ80 to replicate reactions containing identical amounts (in this instance 10 molecules) of the wild-type template pQP1 results in a progressive increase in the intensity of fluorescence associated with the product band for the competitive template (180 bp) and a corresponding progressive decrease in the intensity of fluorescence associated with the product band for the wild-type template (260 bp). The point at which the intensity of the product band obtained from a known amount of added competitive template corresponds to the intensity of the product band obtained for an unknown amount of wild-type target sequence indicates the amount of the wild-type target present in the original sample (after correction for the respective formula weights of the PCR products). As shown in Figure 2, the data corrected for molar equivalence can be expressed as the ratio of the two products and plotted as a function of increasing amounts of competitive template added to the replicate reactions. On the log/log plot, the x-intercept corresponding to the point where the ratio of corrected fluorescence is 1 indicates the amount of wild-type target sequence present in the initial sample. In this example, the determined amount of 10 molecules agrees exactly with the actual input of pQP1 per reaction. In other plasmid reconstruction experiments, accurate quantitation of between 10 and 2000 molecules of wild-type template per reaction was demonstrated (data not shown). Interestingly, for reactions at or near the equivalence point, a third band was often observed on gels, migrating more slowly than, and clearly resolved from, both the pQP1Δ80 band (180 bp) and the wild-type target sequence band (260 bp). This migration pattern and the prominence of the band only in reactions conducted near the equivalence point of target and competitor suggest that the band may consist of a heteroduplex of the two expected products. Of note, due to the fact that quantitation in QC-PCR is based on relative, not absolute, quantitation of product bands for target and competitive template, the presence of such a heteroduplex band should not interfere with accurate quantitation, so long as both target and competitive products are comparably affected by the formation of heteroduplex because by definition they must be.

Figure 3 shows results obtained from DNA QC-PCR analysis of different cell preparations in which various amounts of the 8E5.LAV cell line, containing a single integrated proviral copy, were spiked into a dominant background of uninfected PBMC from a healthy donor. Twofold serial dilutions of the 8E5-LAV cell line were made corresponding to expected target copy numbers of 200, 100, 50, 25 and 12.5 copies per reaction. Excellent agreement of expected-to-determined copy number and linearity of signal were observed. The results demonstrate that the procedure is suitable for use on total DNA preparations derived from infected cells and is capable of detecting and quantitating 10 molecules of HIV-1 DNA per reaction, corresponding to a single proviral genome in a background of 5×10^4 cells.

Detection of HIV DNA copy numbers may be a useful method for monitoring clinical status or response to therapeutic intervention (1,4,9), but assessment of amounts of viral RNA, particularly virion-associated RNA, may provide important complementary information about the level of active viral replication. To demonstrate RNA QC-PCR, in vitro transcripts corresponding to the wild-type HIV-1 sequence and the sequence of the competitive template were generated. For RNA QC-PCR analysis, specified amounts of the RNA transcript for the competitive template were added to replicate aliquots of the test samples prior to the reverse transcription step. This provides for stringent internal control in every reaction for both the reverse transcription step and subsequent PCR.

Shown in Figure 4 are the results of an RNA QC-PCR analysis demonstrating accurate quantitation of 200 input molecules of wild-type transcript per reaction. Because reverse transcription of RNA into cDNA is relatively inefficient, the sensitivity of RNA QC-PCR is expected to be inherently less than DNA QC-PCR, perhaps with greater variability. From continuing optimization of the RNA QC-PCR procedure, we estimate the ultimate, reliable sensitivity of the method to be between 10 and 100 molecules per reaction. To evaluate the RNA QC-PCR procedure on samples more biologically relevant than control in vitro RNA transcripts, serial dilutions from an infectious HIV-1 stock, spiked into plasma from an HIV seronegative donor, were processed and quantitated by RNA QC-PCR analysis. As shown in Figure 5, fewer than 200 HIV-1 RNA copies could be detected, and twofold differences in viral copy number could be differentiated.

Having demonstrated that the procedure could be used to reproducibly and accurately measure HIV DNA and RNA copy numbers in various reconstructed model systems, we tested whether the approach could be used to quantitate the amount of HIV nucleic acids present in typical clinical specimens obtained from infected individuals. DNA from PBMC isolated from infected patients was prepared and analyzed by QC-PCR. Figure 6A shows representative DNA results for two different HIV-1 infected patients, while Figure 6B shows representative results for plasma virion associated HIV-1 RNA determined by RNA QC-PCR. Results from additional patients are in general agreement with estimates of HIV DNA and RNA copy numbers determined by more conventional PCR approaches.

DISCUSSION

PCR provides extraordinary sensitivity in the detection of rare copy nucleic acid sequences. However, inherent features of the PCR constrain its use for reliable quantitation of target sequences in typical specimens of interest.

Utilizing the principle of competitive PCR for quantitation, we have developed an approach for the accurate quantitation of HIV DNA and HIV RNA. Perhaps the most important feature of the competitive template approach to quantitative PCR we have used is in the fact that the competitive nature of the procedure provides stringent internal control at all steps in the process. In addition, the fact that the procedure is based on quantitation of the relative, not absolute, amounts of PCR product derived from target sequence and competitor means that accurate quantitation can be obtained even in the face of numerous factors known to alter the kinetics and efficiency of PCRs.

In reconstituted model system experiments, the procedure is capable of reliably detecting as few as 10 copies of DNA; ongoing optimization studies suggest that reproducible accurate direct quantitation of 100 molecules of RNA should be feasible. In addition, extrapolation from results above this lower limit for reliable quantitation may allow good estimation of RNA amounts below 100 copies. The relatively lower sensitivity of the RNA procedure likely reflects the less than 100% efficiency of reverse transcription and losses associated with processing of specimens containing very limited numbers of copies of target RNA.

The QC-PCR procedure is capable of reliably discriminating between twofold differences in either DNA or RNA copy number. The approach is suitable for quantitative analysis of HIV-1 DNA and HIV-1 RNA in clinical specimens and has been used to quantitate levels of PBMC-associated HIV-1 DNA and plasma HIV-1 RNA content in peripheral blood specimens from HIV-1-infected patients. We believe the internal control inherent in the procedure makes it the most rigorous approach available to PCR-based quantitation of nucleic acid sequences. This method holds much promise for monitoring in various clinical settings, including determination of viral load in the overall assessment of clinical status of HIV-1-infected patients and in evaluation of responses to treatment, including testing of experimental therapeutics.

Although in the current studies we have utilized primers and competitive templates based on the HIV-1 *gag* gene, other primer/competitive template pairs targeted to other regions of the HIV-1 genome are under development and should allow accurate differential quantitation of expression of different HIV-1 transcripts. This may be of interest in studies of control of viral expression, breaking of latency and evaluation of antiviral agents, both in vitro and in vivo. The QC-PCR methods described here may also be adaptable to other virus systems, both in vitro and in vivo, characterized by relatively low copy numbers of viral nucleic acids.

ACKNOWLEDGMENTS

Supported in part by Grant AI25922 from the National Institutes of Health. The 8E5.LAV cell line was provided by the AIDS Research and Reference Reagent Program of the NIAID. The authors thank Nai Ping Shen for technical assistance, Dr. Michael McGrath for providing blood specimens, Drs. Tom Merigan and Mitch Mitsuya for helpful discussions and sharing pre-publication data.

Note Added in Proof: Since acceptance of this manuscript, other accounts have appeared describing the use of similar competitive PCR methods for quantitation of HIV-1 DNA (**Jurrians, S., J.T. Dekker and A. de Ronde.** 1992. HIV-1 viral DNA load in peripheral blood mononuclear cells from se-roconverters and long term infected individuals. AIDS 6:635-641), HIV-1 RNA (**Scadden, D.T., Z. Wang and J.F. Groopman.** 1992. Quantitation of plasma human immunodeficiency virus type 1 RNA by competitive poly-merase chain reaction. J. Infect. Dis. *165*:1119-1123.) or both (**Menzo, S., P. Bagnarelli, M. Giacca, A. Manzin, P.F. Voraldo and M. Clementi.** 1992. Absolute quantitation of viremia in human immunodeficiency virus infection by competitive reverse transcription and polymerase chain reaction. J. Clin. Microbiol. *30*:1752-1757.).

REFERENCES

1. **Aoki, S., R. Yarchoan, R.V. Thomas, J.M. Pluda, K. Marczyk, S. Broder and H. Mitsuya.** 1990. Quantitative analysis of HIV-1 proviral DNA in peripheral blood mononuclear cells from patients with AIDS and ARC: decrease of proviral DNA content following treatment with 2′,3′-dideoxyinosine (ddI). AIDS Res. Hum. Retroviruses 6:1331-1339.
2. **Becker-André, M. and K. Hahlbrock.** 1989. Absolute mRNA quantification using the polymerase chain reaction. A novel approach by a PCR aided transcript titration assay (PATTY). Nucleic Acids Res. *17*:9437.
3. **Cassol, S., T. Salas, N. Lapointe, M. Arella, J. Rudnik and M. O'Shaughnessy.** 1991. Improved de-tection of HIV-1 envelope sequences using optimized PCR and inosine substituted primers. Mol. Cell. Probes 5:157-160.
4. **Conway, B., K.E. Adler, L.J. Bechtel, J.C. Kaplan and M.S. Hirsch.** 1990. Detection of HIV-1 DNA in crude cell lysates of peripheral blood mononuclear cells by the polymerase chain reaction and non-radioactive oligonucleotide probes. J. Acquir. Immune Defic. Syndr. *3*:1059-1064.
5. **Coombs, R.W., A.C. Collier, J.P. Allain, B. Nikora, M. Leuther, F.G. Gherset and L. Corey.** 1989. Plasma viremia in human immunodeficiency virus infection. N. Engl. J. Med. *321*:1626-1631.
6. **Crowe, S., J. Mills and M.S. McGrath.** 1987. Quantitative immunocytofluorographic analysis of CD4 surface antigen expression and HIV infection of human peripheral blood monocyte/macrophages. AIDS Res. Hum. Retroviruses *3*:135-145.
7. **Dahlen, P.O., A.J. Iitia, G. Skagius, A. Frostell, M.R. Nunn and S. Kwiatkoski.** 1991. Detection of human immunodeficiency virus type 1 by using the polymerase chain reaction and a time resolved flu-orescence based hybridization assay. J. Clin. Microbiol. *3*:798-804.
8. **Dickover, R.E., R.M. Donovan, E. Goldstein, S. Dandekar, C.E. Bush and J.R. Carlson.** 1990. Quantitation of human immunodeficiency virus DNA by using the polymerase chain reaction. J. Clin. Microbiol. *18*:2130-2134.
9. **Donovan, R.M., R.E. Dickover, E. Goldstein, R.G. Huth and J.R. Carlson.** 1991. HIV-1 proviral copy number in blood mononuclear cells from AIDS patients on zidovudine therapy. AIDS 5:766-772.
10. **Fisher, A.G., E. Collalti, L. Ratner, R.C. Gallo and F. Wong-Staal.** 1985. A molecular clone of HTLV-III with biological activity. Nature *316*:262-265.
11. **Folks, T.M., D. Powell, M. Lightfoote, S. Koenig, A.S. Fauci, S. Benn, A. Rabson, D. Daugherty, H.E. Gendelman, M.D. Hoggan, S. Venkatesan and M.A. Martin.** 1986. Biological and biochemi-cal characterization of a cloned Leu3- cell surviving infection with the acquired immune deficiency syndrome retrovirus. J. Exp. Med. *164*:280-287.
12. **Gilliland, G., S. Perrin, K. Blanchard and H.F. Bunn.** 1990. Analysis of cytokine mRNA and DNA: Detection and quantitation by competitive polymerase chain reaction. Proc. Natl. Acad. Sci. USA 87:2725-2729.

13. Gilliland, G., S. Perrin and H.R. Bunn. 1990. Competitive PCR for quantitation of mRNA, p. 60-69. *In* M.A. Innis, D.H. Gelfand, J.J. Sninsky and T.J. White (Eds.), PCR Protocols: A Guide to Methods and Applications. Academic Press, San Diego.

14. Ho, D.D., T. Moudgil and M. Alam. 1989. Quantitation of human immunodeficiency virus type 1 in the blood of infected persons. N. Engl. J. Med. *321*:1621-1625.

15. Holodniy, M., D.A. Katzenstein, D.M. Israelski and T.C. Merigan. 1991. Reduction in human immunodeficiency virus ribonucleic acid after dideoxynucleoside therapy as determined by the polymerase chain reaction. J. Clin. Invest. *88*:L1755-1759.

16. Lifson, J.D., G.R. Reyes, M.S. McGrath, B.S. Stein and E.G. Engleman. 1986. AIDS retrovirus induced cytopathology: Giant cell formation and involvement of CD4 antigen. Science *232*:1123-1127.

17. Merigan, T.C., G. Skowron, S.A. Bozzette, D. Richman, R. Uttamchandani, M.A. Fischl, R. Schooley, M. Hirsch, W. Soo, C. Petinelli, H. Schaumburg and the ddC Study Group of the AIDS Clinical Trials Group. 1989. Circulating p24 antigen levels and responses to dideoxycytidine in human immunodeficiency virus (HIV) infections, a phase I and II study. Ann. Intern. Med. *110*:189-194.

18. Mullis, K.G. and F.A. Faloona. 1987. Specific synthesis of DNA in vitro via a polymerase catalyzed chain reaction. Methods Enzymol. *155*:263-273.

19. Myers, G., B. Korber, J.A. Berzofsky, R.F. Smith and G.N. Pavlakis (Eds.). 1991. Human Retroviruses and AIDS. A Compilation and Analysis of Nucleic Acid and Amino Acid Sequences. Los Alamos National Laboratory, Los Alamos, NM.

20. Noonan, K.E., C. Beck, T.A. Holymayer, J.E. Chin, J.S. Wunder, I.L. Andrulis, A.F. Gadzar, C.L. Willman, B. Griffity, D.D. Von Hoff and I.B. Roninson. 1990. Quantitative evaluation of MDR1 (multidrug resistance) gene expression in human tumors by polymerase chain reaction. Proc. Natl. Acad. Sci. USA *87*:7160-7164.

21. Rogers, M.F., C.Y. Ou, M. Rayfield, P.A. Thomas, E.E. Schoenbaum, E. Abrams, et al. 1989. Use of the polymerase chain reaction for early detection of the proviral sequences of human immunodeficiency virus in infants born to seropositive mothers. N. Engl. J. Med. *320*:1649-1654.

22. Scarlatti, G., V. Lombardi, A. Plebani, N. Principi, C. Vegni, G. Ferraris, A. Bucceri, E.M. Fenyo, H. Wigzell, P. Rossi and J. Algbert. 1991. Polymerase chain reaction virus isolation and antigen assay in HIV-1 antibody positive mothers and their children. AIDS *5*:1173-1178.

Update to:

Quantitative Competitive Polymerase Chain Reaction for Accurate Quantitation of HIV DNA and RNA Species

Jeffrey D. Lifson and Michael Piatak, Jr.[1]

Laboratory of Retroviral Pathogenesis, SAIC Frederick, National Cancer Institute-Frederick Cancer Research and Development Center, Frederick; [1]Becton Dickinson Microbiology Systems, Sparks, MD, USA

INTERNALLY CONTROLLED PCR APPROACHES FOR QUANTIFICATION OF RETROVIRAL REPLICATION AND OTHER APPLICATIONS

Our original report described the use of internally controlled PCR and RT-PCR procedures to quantify HIV-1-derived nucleic acids. The essential feature of the method was the use of matched internal control templates, spiked at known copy numbers in a titration series into replicate aliquots of a test sample containing an unknown copy number of target template. The same primers were used to amplify with comparable efficiency both target template and internal control template. After differential quantification of the resulting products, the unknown pre-amplification copy number of the target template was determined at the titration equivalence point by interpolation onto a plot of the ratio of amplified target template and internal control template derived products vs. the known pre-amplification input copy numbers of the internal control template. The use of quantification based on determination of the amount of target template-derived product obtained relative to the amount of internal control template-derived product obtained, rather than striving for quantification based solely on the absolute amount of PCR product, controlled for a number of the inherent features of the PCR that make standard end-point PCR analysis less than optimal for quantitative applications, particularly when dealing with unknown specimens spanning a broad dynamic range (27).

Since initial publication of our work describing the use of internally controlled PCR and RT-PCR (quantitative competitive polymerase chain reaction or "QC-PCR") approaches to the quantification of HIV-1 DNA and RNA, respectively, there has been tremendous progress in several related areas of research. These include the use of nucleic acid-based quantitative assays of HIV-1 replication in basic studies of HIV-1 pathogenesis and, increasingly, application of these methods to clinical situations, including the evaluation of experimental treatments and routine patient monitoring. In addition, the interval since our original publication has also seen the establishment of internally controlled PCR and RT-PCR approaches as standard laboratory methods, with widespread applications to quantification of infectious pathogens other than HIV-1 as well as quantification of a host of other target templates far too numerous to list. Increasing sophistication has characterized the application of these methods, with incorporation of a variety of different emerging technologies for quantitative analysis of amplified products. In this Update, we will review some of the interim developments and provide a perspective on the present and future role of internally controlled PCR approaches.

HIV-1 VIRAL LOAD TESTING: FROM PREMISE TO PARADIGM SHIFT

The methods described in our original paper were developed with the intent of monitoring virologic responses in HIV-1-infected patients treated with experimental antiretroviral treatments in a clinical trial setting. At the time the work was initiated, available approaches for virologic monitoring, including both various viral culture procedures and capture immunoassays for viral antigens, did not possess the requisite sensitivity, reproducibility, and dynamic range to be usefully applied to evaluate the effects of treatment in the majority of HIV-1-infected patients (11,25,26). Interestingly, in keeping with the prevailing models of AIDS pathogenesis at the time, this was construed more as evidence in support of virologic latency during the early stages of HIV-1 infection than as indicative of technical limitations in the assay procedures. After validation of the sensitivity and reproducibility of our QC-PCR and RT-QC-PCR assays in model systems and a pilot demonstration of the feasibility of applying these methods to typical clinical specimens, we undertook a more extensive validation on well-characterized clinical specimens, with emphasis on comparison of measurements of plasma virion-associated RNA with other parameters of plasma viral load, including limiting dilution culture and antigen capture assays (11,25,26).

These studies revealed a picture of viral replication in HIV-1 infection that was very different from previous perspectives. We demonstrated unequivocally that there was extensive ongoing viral replication in all stages of HIV-1 infection, even the early asymptomatic phases of infection in which it was difficult or impossible to detect infectious virus or viral antigens circulating in the plasma (25,26). In cross-sectional studies, we established a correlation between higher levels of circulating plasma virus and more advanced clinical stages of HIV-1 disease (25). Furthermore, our analysis of longitudinal specimens showed that in primary HIV-1 infection, although with resolution of the initial phase of plasma viremia, there was a decrease in levels of circulating virus of between one and three logs, plasma virus levels remained at readily quantifiable levels, reflecting extensive ongoing viral replication (25,26). These observations, along with other studies confirming the presence of ongoing viral replication in all stages of disease and studies identifying the lymphoreticular tissues as the primary sites of viral replication in vivo (1,3,7,17,22,23) refuted the notion of virologic latency as the basis for the prolonged asymptomatic phase of HIV-1 infection and led to a fundamental revision of models of AIDS pathogenesis.

The manifest importance and commercial potential of being able to quantify viral replication in HIV-1-infected patients, reinforced by the observations described above, led to concerted efforts to develop quantitative amplification assays for viral RNA, eventually yielding commercially available assays utilizing a variety of different formats (3,18,30). The application of QC-PCR methods and these assays, along with the availability of potent new antiretroviral agents, facilitated the next significant development in improving our understanding of the pathogenesis of AIDS. These new antiretroviral drugs were used as probes to block new rounds of infection, perturbing the quasi-steady state that obtains over the short term in HIV-1-infected patients. Based on the model assumption that the drugs completely blocked new rounds of infection, it was possible to track the decay of plasma virus and model the clearance of virions and productively infected cells (4,9,24,33). The findings obtained in this manner demonstrated a prolific level of ongoing viral replication, with up to an estimated 10^{10} particles being created and cleared every day, extending the fundamental recasting of the underlying paradigm of AIDS pathogenesis.

A related observation was that the quasi-steady state level of plasma viremia that is established following acute HIV-1 infection is a reflection of the overall level of ongoing viral replication in the host. This level, or viral load "set point", has been shown to be a major prognostic determinant, with subjects having higher plasma viral load levels being at significantly higher risk for a more rapidly progressive disease course (8,15,16,20,32). In aggregate, these observations have led to a new understanding of the dynamic nature of viral replication throughout the course of HIV-1 infection, with profound impact on strategies for treatment. Current experimental clinical studies are exploring intensive combination antiretroviral treatment approaches implemented early in the course of infection to aggressively minimize viral replication and spread, and to explore the feasibility of potential eradication of virus from the infected subject. Plasma viral load testing is an essential element of these protocols and is also increasingly being integrated into the standard clinical management of HIV-1-infected patients.

NON-HIV-1 TARGETS

In the interval since original publication of our paper, there has been extensive utilization of QC-PCR methods for quantification of both HIV-1 nucleic acid species and other target templates. The recent availability of multiple different commercially available assays for quantification of HIV-1 RNA levels using formats with greater throughput has reduced the use of RT-QC-PCR for this application. However, QC-PCR approaches are still widely employed, both within and outside of AIDS research for quantitative analysis of target templates for which insufficient economic incentives exist to justify the development of commercial assay formats. For example, we have developed QC-PCR/RT-QC-PCR methods for quantification of DNA and RNA of simian immunodeficiency virus (SIV) (8). These methods have been central to further characterization of the SIV-infected macaque as one of the preferred animal models of HIV-1 infection and AIDS. In particular, QC-PCR approaches have allowed establishment of the relationship between patterns of viral replication and in vivo pathogenesis in SIV-infected macaques. This has helped to demonstrate the relevance and enhance the value of the SIV-infected macaque model in the contexts of studies of basic pathogenesis and evaluation of both therapeutic and vaccine approaches (6,8,29,32). Similar approaches applied to quantification of HIV-2 replication in infected macaques have also been valuable in evaluating therapeutic efficacy of anti-retroviral interventions and understanding pathogenesis (29). QC-PCR methods have also been developed for quantification of a variety of non-retroviral infectious targets, including clinically significant herpes viruses (5,19).

The common denominator in these studies in which QC-PCR methods have made critical contributions is perhaps the existence of experimental situations that require the development and application of a sensitive and precise, yet robust, quantitative assay for specific DNA or RNA species, in circumstances that do not require extremely high specimen throughput rates, and for which the projected demand precludes availability of a commercial assay. In this setting, QC-PCR type methods represent perhaps the best option for typical laboratories with some modest degree of molecular biology expertise, affording a conceptually valid approach that has been empirically established in numerous settings. When performed correctly, internally controlled PCR approaches provide a comforting intrinsic assurance of the accuracy of the results obtained. It is likely these features that account for the popularity and widespread application of internally controlled PCR and RT-PCR procedures for quantification of diverse target templates in a myriad of experimental contexts, ranging from botany to the neurosciences.

METHODOLOGIC CONSIDERATIONS IN INTERNALLY CONTROLLED PCR AND RT PCR

In developing internally controlled PCR methods for any particular application, there are a variety of different options with regard to specific technical aspects of the assay. The widespread successful application of the basic method in different variant forms constitutes de facto evidence of the robustness of the underlying approach. We have recently summarized our perspective on pertinent considerations in the design and application of RT-QC-PCR methods (27), and a full treatment of this topic would be far beyond the scope of the current update. However, three fundamental principles bear reiteration. First, to fulfill the conceptual principles of internally controlled PCR, it is important that the target template and internal control template be amplified using the same primers. In cases where reasonably well-defined sequence heterogeneity can be anticipated in test specimens (such as for sequence diverse clinical isolates of some viral pathogens), incorporation of neutral base substitutions at positions of anticipated sequence variation may avoid biasing of amplification efficiency in the event of mismatch of primer and target template (27). Secondly, it is important that target template and internal control be amplified with comparable efficiency. Finally, and most importantly, assay performance should be validated empirically. Regardless of the conceptual underpinnings or theoretical elegance of the design of any particular primer set or internal control template, internally controlled PCR assays should be empirically validated in the actual intended assay format, before they should be assumed to provide accurate results. So long as these three principles are followed, numerous variations on the basic theme can be pursued with respect to different aspects of assay configuration, with satisfactory results.

One of the main choice points in the development of alternative versions of internally controlled PCR assays involves design and construction of the internal control template. While there is general agreement that the primer binding sites should be identical for target template and internal control template, there are marked differences of opinion with respect to the internal sequence, with advocates both for internal sequences that are matched (2,14, 21) and mismatched (13,28,31) between target and internal control templates. As there are examples of successful assays using both approaches, the important consideration is again empirical validation of whatever format is selected for a given application.

Design of the internal control template is also constrained by the strategy to be used for differential quantification of products derived from input target and internal control templates (see Reference 27 for discussion; also see References 10 and 12 for reviews). Perhaps the most basic and most common approach is the introduction of a size differential between target and internal control templates through introduction of an insertion or deletion into the internal control template. This generally facilitates resolution of amplified products by some form of gel electrophoresis with differential quantification by computer-assisted video image analysis of fluorescent stained gels. Alternatively, size differentials between target and internal control template derived products can be resolved and quantified using instrumented formats such as HPLC or capillary electrophoresis, with or without the use of laser-induced fluorescence (34).

Sequence differences between target and internal control template-derived products can also be exploited by differential hybridization capture and detection methods, using modified oligonucleotides and enzyme conjugates to provide either colorimetric or chemiluminsecent readouts for quantification of products. Isotopic approaches can also be employed, with quantification of resolved amplified products by phosphorimager technology. Finally, introduction of a unique restriction site into the internal control template, or use of a natural site present in the target template but not in the internal control template, can also allow for resolution and differential quantification of amplified products, essential to the basic approach. Once again, in view of successful examples of assays based on each of the alternative assay and internal control template formats described above, empirical validation in the setting of the actual intended application is more important than the theoretical or design considerations of any particular assay.

Depending on the application, providing internal control for aspects of the assay in addition to the PCR amplification may be important. Thus, for quantification of RNA species by internally controlled RT-PCR, use of the internal control template in RNA form (typically through use of a purified in vitro transcript) provides an internal control for the efficiency of reverse transcription, as well as internal control of the PCR amplification of the resulting cDNA. In this context, the efficiency of reverse transcription of a particular internal control template may sometimes be found to be different from that of the corresponding target template. Thus, DNA PCR using the DNA version of the internal control template may provide accurate quantification, while due to differences in RT efficiency, the RNA version of the same internal control template may result in skewed quantification. Most often, however, any such differences in RT efficiency tend to be consistent, allowing the user to obtain accurate relative quantification of target template by employing a correction factor. Once again, this underscores the importance of empirical validation of the particular assay system to be used, under actual conditions of intended use.

Use of an RNA template spanning an intron in the target sequence, along with use of an appropriate matched internal control template, may obviate the need to DNase-treat cell-derived specimens prior to RT-PCR. Depending on the source specimen, endogenous DNA or RNA sequences shown to be present at predictable copy numbers can be used to provide internal control for the efficiency of nucleic acid recovery, an important source of experimental variability in some applications, allowing expression of results in a more meaningful normalized form. Exogenous internal control templates comprised of homologous or heterologous sequences can also be spiked into specimen lysates to provide an internal control for nucleic acid recovery in cell-free specimens or specimens lacking a suitable endogenous internal control template (30).

FINAL COMMENTS

In summary, internally controlled PCR methods have enabled key observations that have helped to significantly refine concepts of AIDS pathogenesis, leading to modifications of our understanding with practical impact on approaches to treatment and prevention of HIV-1 infection. Such methods are extremely versatile and have been employed in other applications too numerous to mention. Multiple variations on the basic approach of internally controlled PCR are feasible, and specific assay configurations can be developed to suit particular experimental applications. Regardless of the assay configuration employed, empirical validation of the assay under conditions of intended use is more important than the particulars of the assay format. While conceptually and technically straightforward and robust, internally controlled PCR methods are comparatively labor-intensive and are not well suited to high throughput applications. Nevertheless, for a laboratory with a need for sensitive, reproducible quantification of specific DNA or RNA species in a modest number of specimens, and possessing a modicum of expertise in standard molecular biology techniques and a desire to avoid a significant investment in technology development or dedicated instrumentation, internally controlled PCR methods may represent the preferred analytical approach. Based on these attributes, internally controlled PCR methods will likely continue to flourish in a variety of niche applications, no doubt undergoing continued technical refinement in the coming years.

ACKNOWLEDGMENTS

The authors thank the individuals who have worked with us over the past several years, establishing basic methods and improvements, and developing the body of experience that forms the basis of much of the commentary in this Update, including: Ka-Cheung Luk, Naiping Shen, Gabriela Vasquez, John Wages, Theresa Wiltrout, Limei Yang and Andrew Yun. Sponsored in part by the National Cancer Institute, Department of Health and Human Services (DHHS). The contents of this publication do not necessarily reflect the views or policies of the DHHS, nor does mention of trade names, commercial products, or organizations imply endorsement by the US Government.

REFERENCES

1. Bagnarelli, P., A. Valenza, S. Menzo, A. Manzin, A.G. Scalise, P.E. Varaldo and M. Clementi. 1994. Dynamics of molecular parameters of human immunodeficiency virus type 1 activity in vivo. J. Virol. 68:2495-2502.
2. Becker-André, M. and K. Hahlbrock. 1989. Absolute mRNA quantification using the polymerase chain reaction. A novel approach by PCR aided transcript titration assay (PATTY). Nucleic Acids Res. 17:9437-9446.
3. Cao, Y., D.D. Ho, J. Todd, R. Kokka, M. Urdea, J.D. Lifson, M. Piatak, S. Chen, B.H. Hahn, M.S.

Saag and G.M. Shaw. 1995. Clinical evaluation of branched DNA (bDNA) signal amplification for quantifying HIV-1 in human plasma. AIDS Res. Hum. Retroviruses *11*:353-361.

4. **Coffin, J.M.** 1995. HIV population dynamics in vivo: implications for genetic variation, pathogenesis and therapy. Science *267*:483-489.

5. **Fox, J.C., P.D. Griffiths and V.C. Emery.** 1992. Quantification of human cytomegalovirus DNA using the polymerase chain reaction. J. Gen. Virol. *73*:2405-2408.

6. **Grant, R.M., M. Kaur, H.A. McClure, A. Rosenthal, C.S. Horton, P. Carroll, R.P. Johnson, M.B. Feinberg and S.I. Staprans.** Plasma SIV dynamics during primary viremia. 14th Annual Symposium on Nonhuman Primate Models for AIDS, 23-26 October 1996, Portland, OR [abstract 46].

7. **Haase, A.T., K. Henry, M. Zupancic, G. Sedgewick, R.A. Faust, H. Melroe, W. Cavert, K. Gebhard, K. Staskus, A.-Q. Zhang, P.J. Dailey, H.H. Balfour, A. Erice and A.S. Perelson.** 1996. Quantitative image analysis of HIV-1 infection in lymphoid tissues. Science *274*:985-989.

8. **Hirsch, V.M., T.R. Fuerst, G. Sutter, M.W. Carroll, L.C. Yang, S. Goldstein, M. Piatak, W.R. Elkins, D.C. Montefiori, B. Moss and J.D. Lifson.** 1996. Patterns of viral replication correlate with outcome in simian immunodeficiency virus (SIV)-infected macaques: effect of prior immunization with a trivalent SIV vaccine in modified vaccinia virus Ankara. J. Virol. *70*:3741-3752.

9. **Ho, D.D., A.U. Neumann, A.S. Perelson, W. Chen, J.M. Leonard and M. Markowitz.** 1995. Rapid turnover of plasma virions and CD4 lymphocytes in HIV-1 infection. Nature *373*:123-126.

10. **Jenkins, F.J.** 1994, Basic methods for the detection of PCR products. PCR Methods Appl. *3*:S77-S82.

11. **Kappes, J.C., M.S. Saag, G.M. Shaw, B.H. Hahn, P. Chopra, S. Chen, E.A. Emini, R. McFarland, L.C. Yang, M. Piatak and J.D. Lifson.** 1995. Assessment of antiretroviral therapy by plasma viral load testing: standard and ICD HIV-1 p24 antigen and viral RNA (QC-PCR) assays compared. J. Acquir. Immune Defic. Syndr. *10*:139-149.

12. **Lazar, J.G.** 1994, Advanced methods for the detection of PCR products. PCR Methods Appl. *4*:S1-S14.

13. **Li, B., P. Sehajapal, A. Khanna, H. Vlassara, A. Cerami, K.H. Stenzel and M. Suthanthiran.** 1991. Differential regulation of transforming growth factor beta and interleukin-2 genes in human T cells: demonstration by usage of novel competitor DNA constructs in quantitative polymerase chain reaction. Gene *93*:125-128.

14. **McCulloch, R.K., C.S. Choong and D.M. Hurley.** 1995. An evaluation of competitor type and size for use in determination of mRNA by competitive PCR. PCR Methods Appl. *4*:219-226.

15. **Mellors, J.W., L.A. Kingsley, C.R. Rinaldo, Jr., J.A. Todd, B.S. Hoo, R.P. Kokka and P. Gupta.** 1995. Quantitation of HIV-1 RNA in plasma predicts outcome after seroconversion. Ann. Intern. Med. *8*:573-579.

16. **Mellors, J.W., C.R. Rinaldo, Jr., P. Gupta, R.M. White, J.A. Todd and L.A. Kingsley.** 1996. Prognosis in HIV-1 infection predicted by the quantity of virus in plasma. Science *272*:1167-1170.

17. **Michael, N.L., M. Vahey, D.S. Burke and R.R. Redfield.** 1992. Viral DNA and mRNA expression correlate with stage of human immunodeficiency virus (HIV) type 1 infection in humans: evidence for viral replication in all stages of HIV disease. J. Virol. *66*:310-316.

18. **Mulder, J., N. McKinney, C. Chrostopherson, J. Sninsky, L. Greenfield and S. Kwok.** 1994. Rapid and simple PCR assay for quantitation of human immunodeficiency virus type 1 RNA in plasma: application to acute retroviral infection. J. Clin. Microbiol. *32*:292-300.

19. **Nash, K.A., J.S. Klein and C.B. Inderlied.** 1995. Internal controls as performance monitors and quantitative standards in the detection by polymerase chain reaction of herpes simplex virus and cytomegalovirus in clinical specimens. Mol. Cell. Probes *9*:347-356.

20. **O'Brien, T., W.A. Blattner, D. Waters, E. Eyster, M.W. Hilgartner, A.R. Cohen, N. Luban, A. Hatzakis, L.M. Aledort, P.S. Rosenberg et al.** 1996. Serum HIV-1 RNA levels and time to development of AIDS in the multicenter hemophilia cohort study. JAMA *276*:105-110.

21. **Pannetier, C., S. Delassus, S. Darche, C. Saucier and P. Kourilsky.** 1993. Quantitative titration of nucleic acids by enzymatic amplification reactions run to saturation. Nucleic Acids Res. *21*:577-583.

22. **Pantaleo, G., C. Graziosi, J.F. Demarest, L. Butini, P.A. Pizzo, S.M. Schnittman, D.P. Kotler and A.S. Fauci.** 1991. Lymphoid organs function as major reservoirs for human immunodeficiency virus. Proc. Natl. Acad. Sci. USA *88*:9838-9842.

23. **Pantaleo, G., C. Graziosi, J.F. Demarest, L. Butini, M. Montroni, C.H. Fox, J.M. Orenstein, D.P. Kotler and A.S. Fauci.** 1993. HIV infection is active and progressive in lymphoid tissue during the clinically latent stage of disease. Nature *362*:355-358.

24. **Perelson, A.S., A.U. Neumann, M. Markowitz, J.M. Leonard and D.D. Ho.** 1996. HIV-1 dynamics in vivo: virion clearance rate, infected cell life span, and viral generation time. Science *271*:1582-1586.

25. **Piatak, M., M.S. Saag, L.C. Yang, S.J. Clark, J.C. Kappes, K.-C. Luk, B.H. Hahn, G.M. Shaw and J.D. Lifson.** 1993. High levels of HIV-1 in plasma during all stages of infection determined by

competitive PCR. Science *259*:1749-1754.

26.**Piatak, M., M.S. Saag, L.C. Yang, S.J. Clark, J.C. Kappes, K.-C. Luk, B.H. Hahn, G.M. Shaw and J.D. Lifson.** 1993. Determination of plasma viral load in HIV-1 infection by quantitative competititve polymerase chain reaction (QC-PCR). AIDS Suppl. *7*:S65-S71.

27.**Piatak, M., J. Wages, K.-C. Luk and J.D. Lifson.** 1996. Competitive RT PCR for quantification of RNA: theoretical considerations, practical advice, p. 191-221. *In* P.A. Krieg (Ed.), A Laboratory Guide to RNA: Isolation, Analysis, Synthesis. Wiley-Liss, New York.

28.**Siebert, P.D. and J.W. Larrick.** 1993. PCR mimics: competitive DNA fragments for use as internal standards in quantitative PCR. BioTechniques *14*:244-249.

29.**Travis, B.M., A. Watson, J. Ranchalis, N.L. Haigwood and S.L. Hu.** Comparative QC-PCR analysis of viral load in macaques infected by primate lentiviruses. Plasma SIV dynamics during primary viremia. 13th Annual Symposium on Nonhuman Primate Models for AIDS, 5-8 November 1996, Monterey, CA [abstract 119].

30.**Van Gemen, B., P. van der Wiel, R. van Beuningen, P. Sillekens, S. Jurriaans, C. Dries, R. Schoones and T. Kievits.** 1995. The one tube quantitative HIV-1 RNA NASBA: precision, accuracy, and application. PCR Methods Appl. *4*:S177-S184.

31.**Wang, A.M., M.V. Doyle and D.F. Mark.** 1989. Quantitation of mRNA by the polymerase chain reaction. Proc. Natl. Acad. Sci. USA *86*:9717-9721.

32.**Watson, A., J. Ranchalis, B. Travis, J. McClure, W. Sutton, P.R. Johnson, S.-L. Hu and N.L. Haigwood.** 1997. Plasma viremia in macaques infected with simian immunodeficiency virus: plasma viral load in early in infection predicts survival. J. Virol. *71*:284-290.

33.**Wei, X., S.K. Ghosh, M.E. Taylor, V.A. Johnson, E.A. Emini, P. Deutsch, J.D. Lifson, S. Bonhoeffer, M.A. Nowak, B.H. Hahn, M.S. Saag and G.M. Shaw.** 1995. Viral dynamics in human immunodeficiency virus type 1 infection. Nature *373*:117-122.

34.**Williams, S.J., C. Schwer, A.S. Krishnarao, C. Heid, B.L. Karge and P.M. Williams.** 1996. Quantitative competitive polymerase chain reaction: analysis of amplified products of the HIV-1 gag gene by capillary electrophoresis with laser induced fluorescence detection. Anal. Biochem. *236*:146-152.

CHAPTER 5

PCR MIMICS: Competitive DNA Fragments for Use as Internal Standards in Quantitative PCR

Paul D. Siebert and James W. Larrick[1]

CLONTECH Laboratories and [1]Palo Alto Institute for Molecular Medicine, Palo Alto, CA, USA

ABSTRACT

A rapid and reliable method is described for preparing competitive DNA fragments for quantitative PCR. Synthetic DNAs complementary to previously established PCR primers are ligated together with the primers to both ends of a generic DNA fragment whose length differs from the natural target gene PCR product. After a short ligation step, the properly constructed ligation products (i.e., those that have the correct primer templates on opposite sides of the generic DNA fragment) are preferentially amplified by PCR. The generation of competitive PCR fragments, MIMICS, can be completed in a single day. To perform quantitative PCR, known quantities of PCR MIMICS are spiked into PCR amplification reactions containing the experimental cDNA samples. A visual or radioactive comparison of the PCR products can then be used to determine the initial quantity of target gene. We show that competitive PCR MIMICS can be used to accurately measure small changes in mRNA levels.

INTRODUCTION

The use of the polymerase chain reaction (PCR) to examine levels of gene transcripts, often referred to as reverse transcription PCR (RT-PCR) or Message Amplification Phenotyping (MAPPing), has become very popular because it is exquisitely sensitive and rapid (see Reference 5 for a review). However, quantitation of mRNA levels or changes in mRNA levels can be problematic due to the exponential nature of PCR, where small variations in amplification efficiency can lead to dramatic changes in product yields. This has the effect of obscuring differences in levels of the target mRNA during amplification.

Several methods have been described to address this issue. They all involve the use of some form of internal standard to compare the efficiency of

(Reprinted from BioTechniques 14:244-249, 1993)

the PCR in different reactions. The internal standard can be a homologous DNA fragment that has the same primer templates as the target and thus will compete with the target DNA for the same primers. The competitor DNA is designed to generate a PCR product of a different size than the target DNA. The competitor DNA can be obtained by site-directed mutagenesis of the target DNA so that a restriction site is either added or deleted (4), or by generation of a genomic DNA PCR product that contains a short intron (4). In the former case, the PCR products are digested with a restriction enzyme so that the target and competitive PCR products can be distinguished by gel electrophoresis.

Alternatively, a nonhomologous DNA fragment of the desired size can be engineered to contain primer templates (6,10) or sometimes can be obtained by PCR amplification using the target primers and cross-species genomic DNA under reduced annealing stringency (9). Another form of the internal standard can be the use of an additional set of primers to a gene product invariant in the experiment such as a "housekeeping" gene (3,7). This is sometimes referred to as "multiplex PCR" (3).

We describe here a rapid and reliable method for generating non-homologous competitive PCR fragments, MIMICS, and show how MIMICS can be used as internal PCR standards to measure changes in levels of a specific mRNA. We also discuss the advantages of competitive PCR over the use of sets of control primers in multiplex PCR.

MATERIALS AND METHODS

Generation of the Competitive PCR Fragments (MIMICS)

Synthetic DNAs were constructed on a MilliGen Biosearch 8700 DNA synthesizer (Millipore, Bedford, MA, USA). The sequences of the synthetic DNAs are provided in the figure legends. In some cases the 5′-end of the synthetic DNAs had phosphate groups that were attached during synthesis. Competitive PCR MIMICS were constructed as follows: 100 pmol of PCR primer, and each synthetic DNA complementary to the primers were ligated simultaneously to 100 ng (ca. 0.25 pmol) of a 0.60-kb *Eco*RI-*Bam*HI v-*erb* B fragment (11) (commercially available from CLONTECH, Palo Alto, CA, USA) with 20 U of T4 DNA ligase (New England Biolabs, Beverly, MA, USA) in a volume of 20 μL. Ligation was conducted for 2 h at room temperature. Conditions for ligation final concentrations were 50 mM Tris-HCl, pH 7.6, 10 mM MgCl$_2$, 1 mM ATP and 5% (wt/vol) polyethylene glycol-8000. Twenty micrograms of glycogen (nucleic acid grade, Boehringer Mannheim, Indianapolis, IN, USA) were added, and the reaction was then diluted to 50 μL with H$_2$O. The mixture was then passed through a pre-spun spin chromatography column (CLONTECH Chroma Spin-400) at 4000 rpm in a fixed angle microcentrifuge for 2 min. One-tenth of the eluent was added to a 50 µL PCR containing 50 mM Tris-HCl, pH 8.3, 50 mM KCl, 1.5 mM MgCl$_2$, 0.2 mM each deoxyribonucleoside triphosphate (dNTP), 0.01% (wt/vol) gelatin, 0.4 µM each primer and 2 U of AmpliTaq® DNA Polymerase (Perkin-Elmer, Norwalk, CT, USA). PCR was conducted using a Perkin-Elmer DNA Thermal Cycler with the following cycle parameters: denature, 94°C for 45 s; anneal, 60°C for 45 s; extend, 72°C for 2 min. Amplification was monitored by gel electrophoresis. As soon as PCR products were clearly visible on the gel (about 25 cycles), the amplification was terminated. The amplified DNA was then passed through another spin column to remove primers and reaction components. The yield of competitive PCR MIMIC can be determined by measuring absorbance at 260 nm or estimated by comparison to dilutions of known quantities of a similarly sized DNA. In the experiments described, the 0.6-kb *Hae*III fragment of φX174 was used for this purpose. A portion of the final purified PCR MIMIC was diluted to 100 amol/µL in 50 µg/mL glycogen and used as a stock solution for all MIMIC dilutions used in subsequent PCR experiments. In this way, any inaccuracies in determining the concentrations of the MIMICS would not affect the calculations of relative changes in levels of target gene. MIMIC stocks and dilutions were made in 50 µg/mL glycogen and stored at -20°C.

Reverse Transcriptase PCR

Total RNA was prepared by the method of Chomczynski and Sacchi (2) and stored in the form of an EtOH precipitate at -20°C. First-strand cDNA was synthesized by Moloney murine leukemia virus (M-MLV)-reverse transcriptase using oligo(dT) as a primer in a 20-µL reaction using a commercially available kit (CLONTECH, 1st-Strand cDNA Synthesis Kit). One-microliter aliquots were then added to PCRs using the same conditions described above. When PCR products were to be radioactively labeled, 4 µCi of ^{32}P-dCTP (3000 Ci/mmol; Du Pont NEN, Boston, MA, USA) were added to the PCR.

RESULTS

Generation of a Competitive PCR MIMIC

A gene fragment was chosen for use as a competitive PCR MIMIC that differed in size from the target gene PCR product and possessed asymmetric restriction sites leaving 5′-overhangs. In the experiments described here, we used a 0.60-kb *Eco*RI-*Bam*HI fragment of the v-*erb* B oncogene (11). We constructed two synthetic DNAs complementary to each of the primer sequences. One complementary DNA had the addition of a 5′-AATT (*Eco*RI extension). The other had the addition of a 5′-GATC (*Bam*HI extension). The relationships between the various DNAs are shown in Figure 1A. The process for generating competitive PCR MIMICS is shown in Figure 1B. The primers and complementary synthetic DNAs were ligated simultaneously to the generic v-*erb* B fragment in the presence of T4 DNA ligase for 2 h at room temperature. Non-ligated synthetic DNA was removed by passage of the reaction through a spin column. A 10% portion of the ligation products was amplified by PCR using the target gene primers to generate a PCR MIMIC. The amplification step in MIMIC generation serves two important functions: first, it selects properly constructed competitor DNA, as only those will amplify exponentially; second, it generates large quantities of the MIMIC. The amplified MIMIC was then purified by passage through another spin column.

We first prepared a PCR MIMIC for the ubiquitously expressed housekeeping gene, glyceraldehyde 3-phosphate dehydrogenase (G3PDH) (8). Primer templates were chosen so they spanned several introns. In this way, amplification products generated by any contaminating genomic DNA would

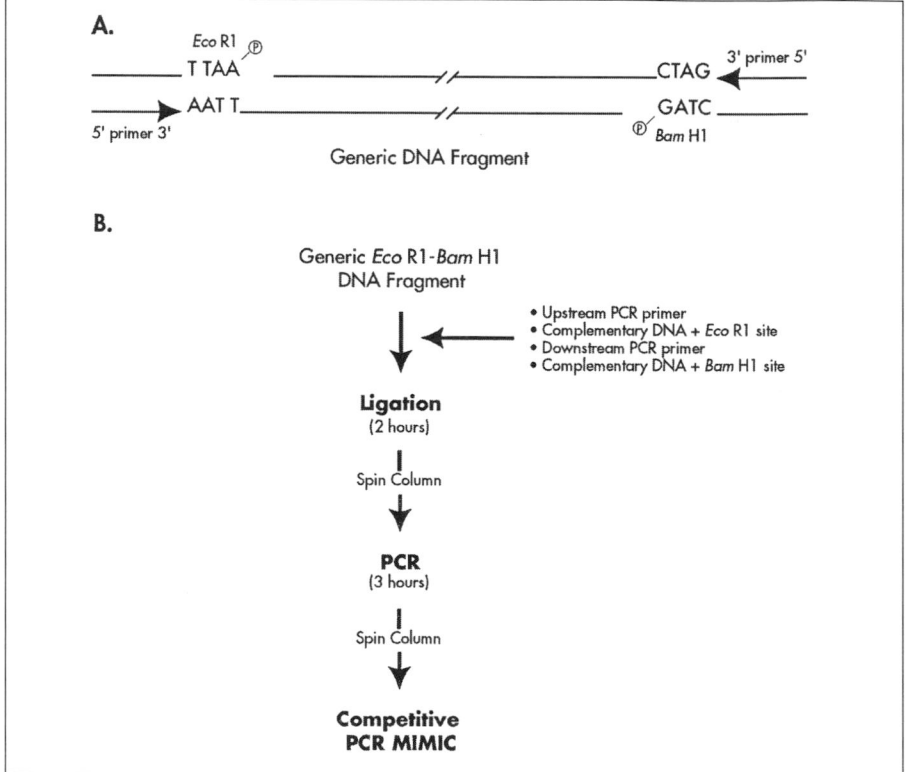

Figure 1. Diagram and flowchart describing the generation of competitive PCR MIMICS. A: Structure of the DNAs. The generic DNA fragment used was a 0.6 kb *Eco*RI-*Bam*HI fragment of the v-*erb* B oncogene (11). It contains asymmetrical restriction sites that yield 5′ overhangs. On either side of the v-*erb* B fragment are shown the synthetic DNAs. The arrows denote the upstream and downstream PCR primers (5′ to 3′). Opposite the primers are shown the complimentary DNAs having either a 4-nucleotide *Eco*RI or *Bam*HI extension sequence. The 5′-phosphate was added to the synthetic DNAs during synthesis. **B:** Flowchart illustrating the steps used to generate the PCR MIMICS. Experimental details are provided in the text.

be significantly larger than from cDNA.

Although the G3PDH target and PCR MIMIC share the same primer template, the intervening DNA sequences differ making it possible for the MIMIC and target to amplify with different efficiencies due to differences in denaturation characteristics or other factors. It was therefore necessary to show that their amplification efficiencies were similar. This was achieved by a kinetic analysis shown in Figure 2. Approximately equal molar quantities of a G3PDH cDNA and G3PDH MIMIC were added to a single PCR along with a small amount of ^{32}P-dCTP. Starting when PCR products were first visualized on an agarose gel, aliquots were removed after each cycle for a total of 7 cycles. The EtdBr agarose profile is shown in Figure 2A. The bands corresponding to the target and MIMIC were then excised from the gel and the amount of radioactivity quantitated by counting in a scintillation counter. The results, shown in Figure 2B, indicate that the G3PDH target and MIMIC shared very similar amplification kinetics.

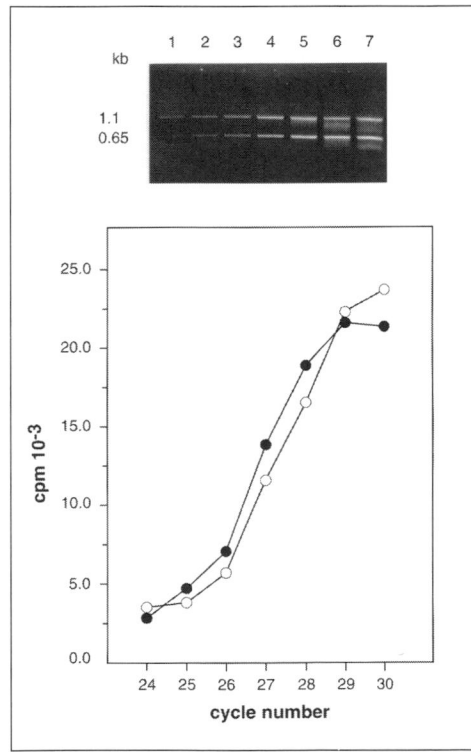

We have compared the amplification kinetics of target and PCR MIMIC pairs for a number of human genes including IL-1β (see below), IL-1α, IL-2, IL-4, IFNγ and β-actin. In all cases, the target and MIMIC pairs amplified with similar kinetics.

Competitive PCR

An example of a competitive PCR experiment is shown in Figure 3. Tenfold serial dilutions of the G3PDH MIMIC were co-amplified with a constant amount of the cloned G3PDH cDNA fragment. Again the reactions contained a small amount of ^{32}P-dCTP. A small portion of the products was then resolved on a 1.6% EtdBr-agarose gel. The EtdBr-agarose gel profile is shown in Figure 3A. A plot of the data

Figure 2. Kinetics of amplification of the G3PDH target cDNA and PCR MIMIC. 0.1 amol each of G3PDH cDNA and MIMIC were added together with ^{32}P-dCTP. Starting at 24 cycles and after each of six additional cycles, 5% portions of the reactions were removed and resolved on a 1.6% EtdBr-agarose gel. A: EtdBr-agarose gel profiles. B: Following gel electrophoresis, the bands corresponding to the target (1.1 kb) and MIMIC (0.65 kb) were excised from the gels and the amount of radioactivity determined by scintillation counting. The data are plotted was a function of the cpm vs. cycle number. Closed circles, G3PDH target. Open circles, G3PDH MIMIC.

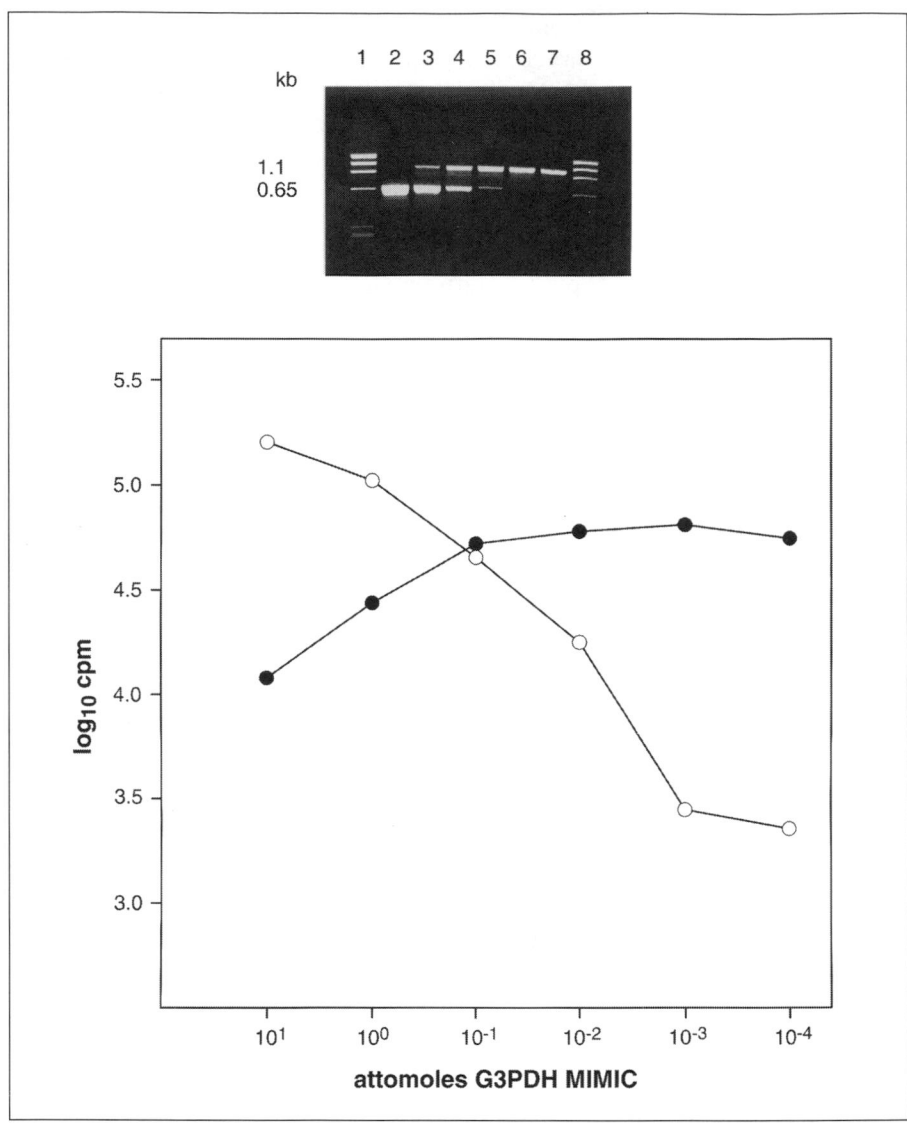

Figure 3. Titration of a G3PDH PCR MIMIC with a constant amount of a cloned G3PDH cDNA. A 0.1-amol amount of a G3PDH cDNA was added to a PCR containing 10-fold serial dilutions of a G3PDH MIMIC. After 30 cycles of amplification, 10% portions of the products were resolved on a 1.6% EtdBr-agarose gel (A). Lanes 2–7: 10^1, 10^0, 10^{-1}, 10^{-2}, 10^{-3} and 10^{-4} amol of G3PDH MIMIC, respectively. Lanes 1 and 8 contain a HaeIII-digest of φX174 as size markers. The positions of the 1.1-kb G3PDH target and 0.65-kb G3PDH MIMIC PCR products are indicated. The sequences of the synthetic DNAs were as follows: upstream G3PDH primer, 5′ TGAAGGTCGGAGTCAACGGATTTGGT 3′ (8); downstream G3PDH primer, 5′ CATGTGGGCCATGAGGTCCACCAC 3′; downstream G3PDH complementary DNA-5′ phosphate GATCACCAAATCCGTTGACTCCGACCTTCA 3′; downstream G3PDH complementary DNA-5′ phosphate AATTGTGGTGGACCTCATGGCCCACATG 3′. (B) Plot of the data shown in (A). Following electrophoresis, the bands corresponding to the G3PDH target and MIMIC were excised from the gel and the radioactivity determined by scintillation counting. The data is plotted as a function of the log cpm vs. amol of G3PDH MIMIC. Closed circles, G3PDH target. Open circles, G3PDH MIMIC.

obtained by radioactive counting of the excised bands is shown in Figure 3B.

Under competitive conditions, the absolute amount of MIMIC added to the reaction is equal to the amount of target when the molar ratio of products becomes equal for target and MIMIC pairs that are similar in size. Molar equivalence occurs at approximately the point where equivalent amounts of radioactivity are incorporated into the target and MIMIC. For precise quantitation, the differences in size between the MIMIC and target must be taken into consideration. In the experiment shown in Figure 4, this point was reached at approximately 0.1 amol, which is in very close agreement with the amount of target DNA added to the PCR (0.1 amol).

Quantitative Analysis of Changes in IL-1β mRNA

We next examined the ability of competitive PCR to accurately measure relatively small changes in the levels of a specific mRNA. For this purpose, we prepared a PCR MIMIC for IL-1β. As with the G3PDH PCR primers, the IL-1β primers were designed to span several introns, making it possible to differentiate between PCR products derived from cDNA and genomic DNA. To imitate a defined induction in IL-1β mRNA, we synthesized cDNA from both 0.5 μg and 2 μg of human lung total RNA, then performed competitive PCR. Four identical experiments were performed, each starting with the reverse transcriptase reaction, to determine the reproducibility of the entire method.

A kinetics study similar to the one shown in Figure 2 was performed which indicated that the IL-1β target and MIMIC amplified with similar efficiencies (data not shown).

To determine the appropriate amount of IL-1β MIMIC to use in the PCR amplification, we performed a preliminary titration experiment in which IL-1β from cDNA derived from 2 μg of total RNA was amplified in the presence of 10-fold serial dilutions of the IL-1β MIMIC. The results of this experiment (not shown) enabled us to provide a fine-tuned, 2-fold dilution series of the IL-1β MIMIC. Competitive PCRs were then performed with cDNA derived from both 0.5 μg and 2 μg of total RNA.

The EtdBr-staining pattern obtained from one of the four independent experiments is shown in Figure 4A. The sizes of the IL-1β target and IL-1β MIMIC PCR products were 0.80 and 0.65 kb, respectively.

The amount of change in IL-1β mRNA can be estimated by visually noting how much more of the MIMIC must be added to achieve an equimolar amount of products. In the experiment shown, this corresponded to a 2-fold to 4-fold change in the amount of IL-1β MIMIC. The theoretical value is 4.0-fold.

In order to more accurately determine the amount of IL-1β MIMIC necessary to achieve equimolar amounts of products, we spiked the PCRs with ^{32}P-dCTP. Following gel electrophoresis, the bands corresponding to the target and MIMIC products were excised from the gel and the amount of

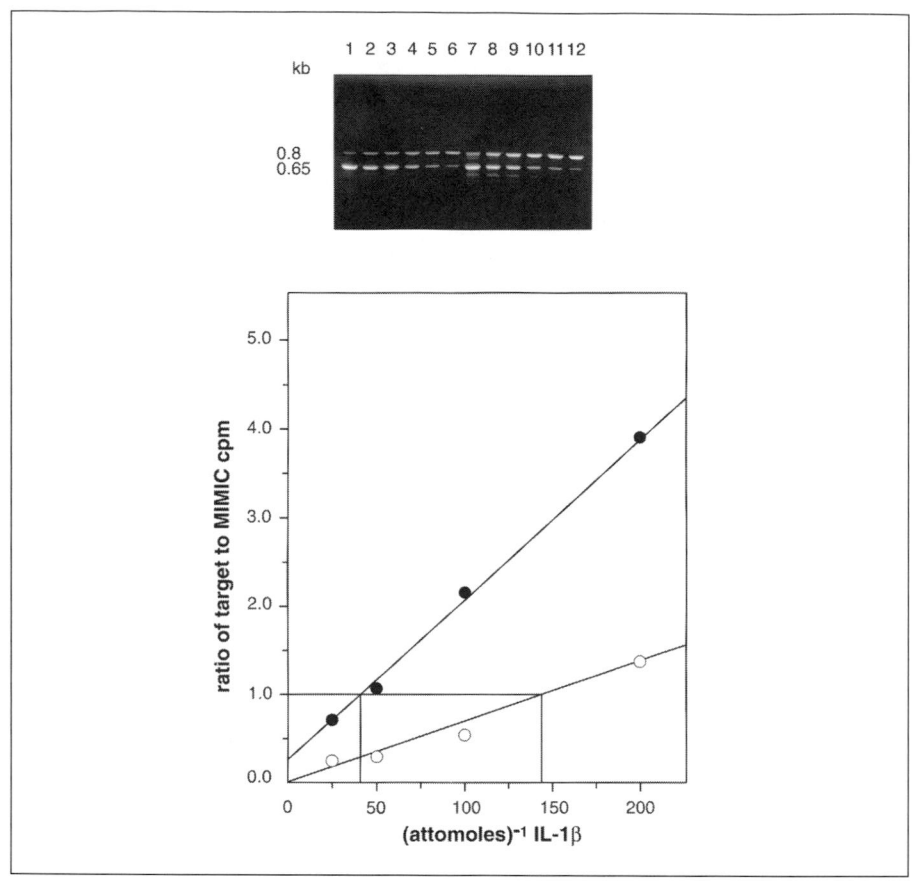

Figure 4. Analysis of relative changes in IL-1β mRNA levels by competitive PCR. 0.5 μg and 2 μg of human lung total RNA were used as a template for cDNA synthesis in four independent experiments. 10% portions of the cDNAs were then amplified in the presence of 2-fold dilutions of the IL-1β MIMIC. 10% portions of the PCR products were then resolved on a 1.6% EtdBr-agarose gel. **A: EtdBr-staining pattern.** Lanes 1–6 and lanes 7–12 contain PCR products derived from 0.5 μg and 2 μg of RNA, respectively. Lanes 1 and 7 contain 80×10^{-3} and IL-1β MIMIC; lanes 2 and 8, 40×10^{-3}; lanes 3 and 9, 20×10^{-3}; lanes 4 and 10, 10×10^{-3}; lanes 5 and 11, 5×10^{-3}; lanes 6 and 12, 2.5×10^{-3}. Three additional experiments were performed in exactly the same manner starting from the reverse transcription step. **B: Quantitative analysis of the competitive PCR experiment shown in A.** The bands corresponding to the IL-1β target (0.8 kb) and MIMIC (0.65 kb) were excised from the agarose gel and the amount of radioactivity determined by scintillation counting. To correct for "trailing" of unincorporated ^{32}P-dCTP in the gel, two blocks of agarose located slightly below two of the bands were also excised and counted. The average amount of radioactivity in these control agarose blocks was subtracted from the values obtained from the PCR product bands. The ratio of the corrected radioactivity was then plotted against the reciprocal of the molar amount of IL-1β MIMIC added to the PCR. The lines were drawn from a linear regression analysis of the four data points. At the highest and lowest amounts of IL-1β MIMIC, the amount of radioactivity of either the target or MIMIC band was very close to the amount of radioactivity of unincorporated ^{32}P-dCTP in the control agarose gel blocks. Because of this, these experimental values were of low confidence and were therefore omitted from the graph. The open and closed circles denote data derived from 0.5 μg and 2 μg of RNA, respectively. The sequences of the synthetic DNAs were as follows: upstream IL-1β primer, 5′ ATGGCAGAAGTACC-TAAGCTCGC 3′; downstream IL-1β primer, 5′ ACACAAATTGCATGGTGAAGTCAGTT 3′ (1); upstream IL-1β complementary DNA-5′ phosphate GATCTACCGTCTTCATGGATTCGAGCG 3′; downstream IL-1β complementary DNA-5′ phosphate AATTTGTGTTTAACGTACCACTTCAGTCAA 3′.

radioactivity quantitated by counting in a scintillation counter. The ratio of IL-1β target radioactivity to IL-1β MIMIC radioactivity was then plotted against the reciprocal of the molar amount of IL-1β MIMIC. The results from the experiment shown in Figure 4A are plotted in Figure 4B.

The amount of IL-1β target cDNA was calculated by extrapolating from the intersection of the curve with a ratio of 1.0 down to the x-axis. For the experiment shown in Figure 4B, values obtained from the 0.5-μg and 2-μg RNA plots were 7.1×10^{-3} and 25×10^{-3} amol, respectively, giving a change of 3.5-fold. The change in IL-1β mRNA determined from all four experiments ranged from 2.4-fold to 3.5-fold with an average value of 3.1-fold. The predicted value is 4.0-fold.

Because the molar quantity of the competitive PCR MIMICS is known, the actual number of target DNA molecules added to the PCR can be calculated. In turn, the number of mRNA molecules can be calculated in the RNA sample used for reverse transcription if it is assumed that the efficiency of cDNA synthesis is 100%. Of course, the actual efficiency must be less than this value. Thus, such a calculation would give the minimum number of mRNA molecules. While the number of mRNA molecules can only be estimated, it is possible to determine relative changes in mRNA levels.

DISCUSSION

The method for generating competitive PCR MIMICS described here is much easier than placing or deleting a restriction site in the target gene by site-directed mutagenesis (4) and more predictable than isolating a genomic PCR product from the same species (4) or cross-species (9). Another advantage of the method described is that heteroduplex formation (i.e., duplex formation between target and competitor PCR product strands, which can complicate analysis) is not possible, since the PCR MIMIC has no homology to the target (except for the short primer templates). In addition, the generic DNA fragment used to generate the PCR MIMIC can be used as a hybridization probe capable of distinguishing between the target and the MIMIC.

Perhaps the greatest advantage of using competitive PCR is that it is not necessary to assay PCR products exclusively during the exponential phase of the amplification (7). It is only during the exponential phase of the amplification that the amounts of products are proportional to the amount of starting target DNA. Knowing when the amplification is proceeding exponentially is often difficult to predict, necessitating a number of pilot experiments. In some cases, amplification can start to plateau shortly after bands are first detected on the electrophoretic gel (Reference 7 and P. Siebert, unpublished observations). Even under optimal conditions, products from each amplification tube must be assayed after several different numbers of cycles—a laborious process.

Another method of determining relative levels of mRNAs involves co-amplification of a target gene with a reference gene that is invariant in the experimental system (3,7). In our hands "multiplex" PCR has proven problematic, as multiple sets of primers often interfere with amplification of either or both of the target genes. Moreover, unless the reference and target genes are present at comparable levels in the RNA sample, PCR products must be assayed after numerous cycles to ensure that analysis of both of the genes is performed only during the exponential phase of the amplification. Also, in at least one case, we have observed that expression of a putative "stable" housekeeping gene (β-actin) actually varied as much as the target gene, tumor necrosis factor (TNF) (J. Larrick, unpublished results).

In agreement with the results of other competitive PCR studies (3,9) we were able to measure a small change in the relative amounts of a specific mRNA with reasonable accuracy. Clearly, a 2-fold to 4-fold change is discernible, making the method comparable to Northern and slot/dot hybridizations, while having the sensitivity, speed and convenience possible with PCR.

In our hands the method described here for generating PCR MIMICS is very reliable once a PCR primer set has been successfully designed. The simplicity of this method for obtaining competitive PCR fragments should broaden the use and applications of quantitative PCR.

ACKNOWLEDGMENTS

We thank Dr. Ted Hung for his helpful advice, Dr. Megan Brown and Alexi Miller for critical reading of the manuscript, Marcie Jordan for the artwork and Reta Vasquez for secretarial assistance.

REFERENCES

1. **Auron, P.E., A.C. Webb, L.J. Rosenwasser, S.F. Mucci, A. Rich, S.M. Wolff and C.A. Dinarello.** 1984. Nucleotide sequence of human monocyte interleukin precursor DNA. Proc. Natl. Acad. Sci. USA *81*:7907-7911.
2. **Chomczynski, P. and N. Sacchi.** 1987. Single-step method of RNA isolation by acid guanidinium thiocyanate-phenol-chloroform extraction. Anal. Biochem. *162*:156-159.
3. **Gaudette, M.F. and W.R. Crain.** 1991. A simple method for quantifying specific mRNAs in small numbers of early mouse embryos. Nucleic Acids Res. *19*:1879-1884.
4. **Gilliland, G., S. Perrin, K. Blanchard and F.H. Bunn.** 1990. Analysis of cytokine mRNA and DNA: Detection and quantitation by competitive polymerase chain reaction. Proc. Natl. Acad. Sci. USA 87:2725-2729.
5. **Larrick, J.W.** 1992. Message amplification phenotyping (MAPPing): principles, practice and potential. Trends Biotechnol. *10*:146-152.
6. **Li, B., P.K. Sehajpal, A. Khanna, H. Vlassara, A. Cerami, K.H. Stenzel and M. Suthanthiran.** 1991. Differential regulation of transforming growth factor β and interleukin-2 genes in human T cells: demonstration by usage of novel competitor DNA constructs in quantitative polymerase chain reaction. J. Exp. Med. *174*:1259-1262.
7. **Murphy, L.D., C.E. Herzog, J.B. Rudick, A.T. Fojo and S.E. Bates.** 1990. Use of the polymerase chain reaction in the quantitation of mdr-1 gene expression. Biochemistry 29:10351-10356.
8. **Tso, J.Y., X.-H. Sun, T.-H. Kao, K.S. Reece and R. Wu.** 1985. Isolation and characterization of rat and human glyceraldehyde-3-phosphate dehydrogenase cDNAs: genomic complexity and molecular evolution of the gene. Nucleic Acids Res. *13*:2485-2502.

9. **Uberla, K., C. Platzer, T. Diamantstein and T. Blankenstein.** 1991. Generation of competitor DNA fragments for quantitative PCR. PCR Methods Appl. *1*:136-139.

10. **Wang, A.M., M.V. Doyle and D.F. Mark.** 1989. Quantitation of mRNA by the polymerase chain reaction. Proc. Natl. Acad. Sci. USA *86*:9717-9721.

11. **Yamamoto, T., T. Nishida, N. Miyajima, S. Kawai, T. Ooi and K. Toyoshima.** 1983. The v-*erb* B gene of Avian erythroblastosis virus is a member of the Src gene family. Cell *35*:71-78.

Update to:

PCR MIMICS: Competitive DNA Fragments for Use as Internal Standards in Quantitative PCR

Paul D. Siebert and James W. Larrick[1]

CLONTECH Laboratories and [1]Palo Alto Institute for Molecular Medicine, Palo Alto, CA, USA

Since the publication of this paper, we have developed a simpler and more versatile method for generating PCR MIMICS. The obligatory validation experiments still apply.

The generation of PCR MIMICS is now achieved by two successive PCR amplifications of heterologous DNA, now as shown in Figure 1. For the first PCR, two composite primers were used. One composite primer contains the upstream target primer sequence linked to 20 nucleotides that anneal to one strand of the heterologous DNA fragment. The other composite primer contains the downstream target primer sequence linked to 20 nucleotides that anneal to the opposite strand of the heterologous DNA fragment. The two composite primers and a small quantity of the heterologous DNA are added to a PCR. During amplification of the DNA fragment, the target primer sequences are incorporated into the PCR product. A dilution of this PCR product is used to perform a second PCR using the shorter target primers. This ensures that the complete target primer sequences have been incorporated into the PCR product. At this point, the PCR product is purified by passage through a spin column, which removes PCR components, primers, and other artifacts such as primer-dimers. The protocol is provided below.

1. Primary PCR using composite primers

    A. To a PCR tube, add:

37.6 µL	distilled water
5 µL	10× PCR buffer
1 µL	50× dNTP mixture
4 µL	neutral DNA fragment (2 ng)
1 µL + 1 µL	each composite primer (20 µM each)
0.4 µL	AmpliTaq® DNA Polymerase (5 U/µL)

        Total volume is 50 µL.

    B. Perform 16 cycles of PCR in an automated thermocycler:

        Denature (94°C for 45 s)

Anneal (60°C for 45 s)
Polymerize (72°C for 90 s)
Perform the final polymerization step for an additional 7 min.

C.   Run 5 µL of the reaction of a 1.8% EtBr-agarose gel. If a strong band of the expected size is obtained, proceed; if not, perform another 4 cycles of PCR.

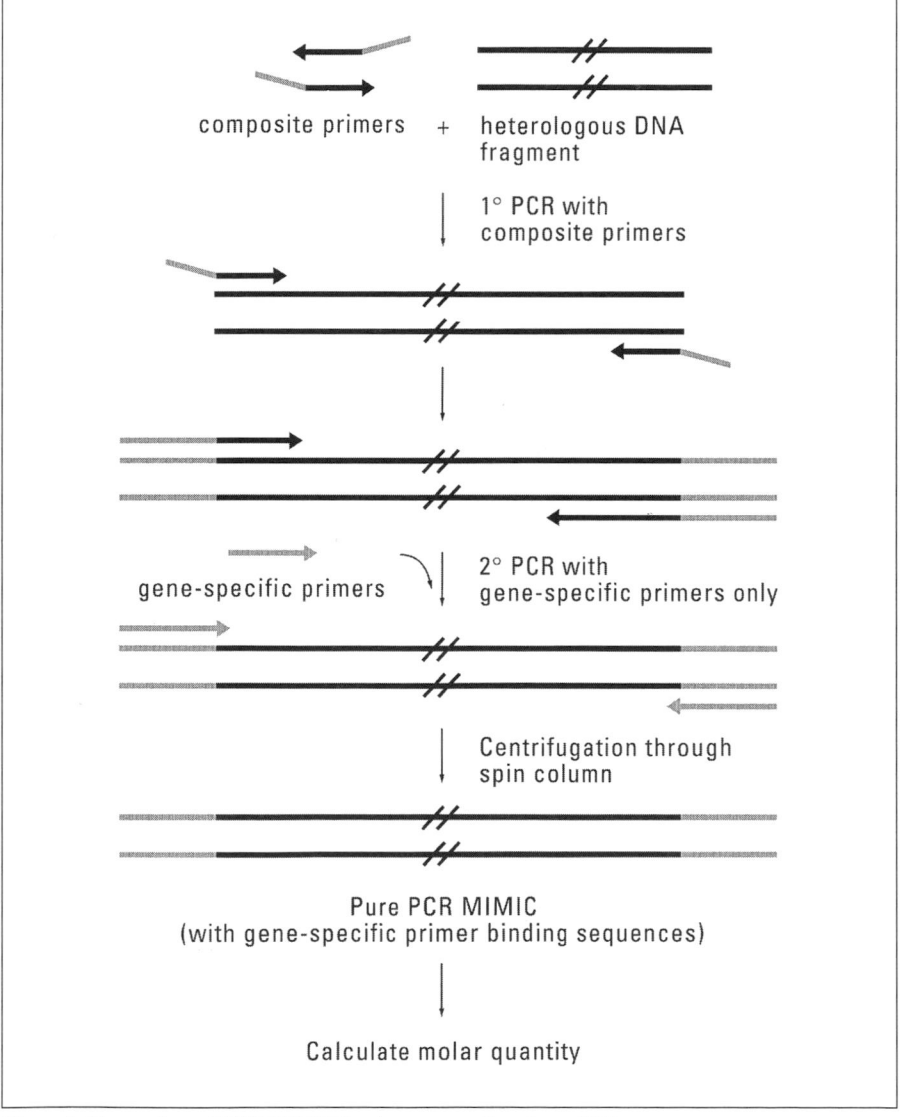

Figure 1. Generation of PCR MIMICS using composite primer PCR.

2. Secondary PCR using gene-specific primers

   A. Remove 2 µL of the primary PCR (from step B) above) and dilute to 200 µL with water.

   B. Add 2 µL of this dilution to a PCR tube containing:

83.4 µL	distilled water
10 µL	10× PCR buffer
2 µL	50× dNTP mixture
2 µL + 2 µL	each gene specific primer (20 µM)
0.6 µL	Amplitaq® DNA Polymerase

   Total volume is 100 µL.

   C. Perform 18 cycles of PCR using the same cycle parameters as for the primary PCR (Part 1, step B).

   D. Run 5 µL of the reaction on a 1.8% EtBr-agarose gel. If a strong band of the expected size is obtained, proceed; if not, perform another 4 cycles of PCR.

3. Purification of the MIMIC

   A. To remove primers and reaction components, pass the reaction through two Chroma Spin™ - 100+ TE columns (CLONTECH). Load 45 µL into each of two pre-spun columns and centrifuge following the manufacturer's protocol.

   B. Check the quality of the MIMIC by running 5 µL on a 1.8% EtBr-agarose gel. Make sure that the salt concentration of the sample is adjusted to that of the agarose gel. If primers are still detected, repeat the chromatography step.

CHAPTER 6

Quantitation of Targets for PCR by Use of Limiting Dilution

P.J. Sykes, S.H. Neoh, M.J. Brisco, E. Hughes, J. Condon and A.A. Morley

Flinders Medical Center, South Australia

ABSTRACT

We describe a general method to quantitate the total number of initial targets present in a sample using limiting dilution, PCR and Poisson statistics. The DNA target for the PCR was the rearranged immunoglobulin heavy chain (IgH) gene derived from a leukemic clone that was quantitated against a background of excess rearranged IgH genes from normal lymphocytes. The PCR was optimized to provide an all-or-none end point at very low DNA target numbers. PCR amplification of the N-ras gene was used as an internal control to quantitate the number of potentially amplifiable genomes present in a sample and hence to measure the extent of DNA degradation. A two-stage PCR was necessary owing to competition between leukemic and non-leukemic templates. Study of eight leukemic samples showed that approximately two potentially amplifiable leukemic IgH targets could be detected in the presence of 160 000 competing non-leukemic genomes.

The method presented quantitates the total number of initial DNA targets present in a sample, unlike most other quantitation methods that quantitate PCR products. It has wide application, because it is technically simple, does not require radioactivity, addresses the problem of excess competing targets and estimates the extent of DNA degradation in a sample.

INTRODUCTION

Quantitation of DNA or RNA by the polymerase chain reaction (PCR) is a problem that is presently under active investigation by many workers. Nearly all methods reported to date have used co-amplification of reporter DNA in the same tube and some form of quantitation of the amplified material (3,4,11). It is assumed that the efficiency of amplification of the reporter DNA is the same as that of the target DNA, and calculation of the amount of target DNA initially present is based on the amount of reporter DNA added or

(Reprinted from BioTechniques 13:444-449, 1992)

originally present and on the ratio of the quantities of target and reporter DNA as determined in the amplified material by various methods.

We have been using the rearranged immunoglobulin heavy chain (IgH) gene as target DNA in the PCR to study patients with acute lymphoblastic leukemia (ALL) in order to detect and quantitate a minor population of leukemic cells within a larger population of normal lymphoid and non-lymphoid cells (1,2). In a particular patient, all leukemic cells will have the same rearranged IgH gene that can act as a genetic marker to distinguish leukemic cells from normal non-lymphoid cells and T lymphocytes, which have not rearranged their IgH genes, and from normal B lymphocytes, which have undergone various and different rearrangements of their IgH genes. Quantitation of the unique rearrangement of the leukemic clone poses two problems. First, only a few copies may be present in a tissue sample taken from a patient in remission following treatment. Second, germ-line IgH genes from cells, other than B lymphocytes, and rearranged IgH genes from normal B lymphocytes will be present and, owing to homology with the leukemic sequence, may compete with it in the PCR. Our approach to quantitation has differed from that of other workers in avoiding use of reporter DNA and quantitation of amplified material. Rather, we have used the principle of limiting dilution, which is based on the use of a qualitative all-or-none end point and on the premise that one or more targets in the reaction mixture give rise to a positive end point. Accurate quantitation is achieved by performing multiple replicates at serial dilutions of the material to be assayed. At the limit of dilution, where some end points are positive and some are negative, the number of targets present can be calculated from the proportion of negative end points by using Poisson statistics.

In this paper we illustrate the use of limiting dilution analysis to quantitate a target for the PCR, using the rearranged IgH gene from a leukemic clone as an example. We also discuss four related issues: optimization of the PCR to detect one or a few DNA targets; the effect of excess competing targets in the PCR; interference in the PCR by primers for an unrelated DNA segment; and the problem of DNA degradation in the sample. Although data from a small number of patients are included in this paper, the clinical and biological information obtained by limiting dilution analysis from a large number of patients with ALL will be described elsewhere (unpublished data).

MATERIALS AND METHODS

DNA Samples

PBL1 DNA was obtained from normal blood cells separated by Lymphoprep™ (Nycomed Pharma AS, Oslo, Norway) to contain predominantly normal lymphocytes, and Ho DNA was from a bone marrow sample of a patient with ALL obtained at diagnosis. Ho DNA provided a source of specific leukemic IgH targets and normal N-*ras* targets, and PBL1 DNA provided a source of normal IgH and N-*ras* targets. Ho DNA contained virtually 100% leukemic cells, whereas PBL1 was estimated to contain approximately 15% normal B lymphocytes.

The optimization experiments relied on DNA concentration of the Ho DNA sample (90 ng/μL) to estimate the number of PCR targets present in the dilutions. Due to the small amount of material available, an OD_{260} was not possible. The concentration of the DNA was obtained by ethidium bromide spotting against known DNA standards (Reference 8, Appendix E, p.6). Because one human diploid cell contains 6 pg of DNA (5), 1 μg of PBL1 DNA would contain approximately 3.3×10^5 N-*ras* genes, approximately 2.4×10^4 rearranged IgH genes and 3×10^5 unrearranged IgH genes.

DNA from seven other patients with ALL was extracted from fresh bone marrow aspirate samples for patients 1, 2, 3 and 7, from frozen Ficoll®-Paque separated lymphocytes for patient 4 and from stained, fixed bone marrow slides for patients 5 and 6. The DNA concentration of samples 1, 2, 3, 4 and 7 was determined by OD_{260} and for patients 5 and 6 by ethidium bromide spotting.

PCR

PCRs (7) contained 16.6 mM $(NH_4)_2SO_4$, 67 mM Tris-HCl, pH 8.8, 10 mM β-mercaptoethanol, 200 μg/mL gelatin, 2 mM $MgCl_2$, 0.1 mM each of deoxyadenosine triphosphate, deoxyguanosine triphosphate, deoxycytidine triphosphate and deoxythymidine triphosphate, 100 ng of each primer, varying amounts of template DNA (0.45 pg–1 μg), and 0.4 U of *Taq* DNA Polymerase (AmpliTaq®; Perkin-Elmer, Norwalk, CT, USA) in a volume of 25 μL, overlaid with 25 μL light mineral oil. The samples were subjected to an initial 5-min denaturation at 94°C followed by varying numbers of cycles of 1 min annealing at 55°C, 1 min extension at 72°C and 1 min denaturation at 94°C. A final 20-min extension at 72°C was performed at the end of each round of PCR.

In all experiments, negative controls containing no template DNA were subjected to the same procedures to detect any possible contamination.

PCR Primers

Primers were synthesized on an Applied Biosystems Model 371 automated synthesizer (Foster City, CA, USA). Consensus primers (2) used to amplify all IgH genes in the first round of PCR were FR3A - 5′ ACACGGC(C/T)(G/C)TGTATTACTGT 3′ for the 3′ end of the V region; LJH - 5′ TGAGGAGACGGTGACC 3′ for the 3′ end of the J region.

Patient-specific primers (1) sited between LJH and FR3A were used to amplify Ho IgH genes only in the second round for the Ho DNA: Ho1 - 5′ TGTGCGAAAGAATCTCTGCC 3′ for the 3′ end of V; Ho2 -

5′ CCAGTAGTCAAGGGTGGGTA 3′ for the 5′ end of J.

Patient-specific IgH primers for the other seven patients will be published elsewhere.

Specific primers were used for N-*ras* in both first and second rounds. First round: 231 (intron 1) - 5′ AAGCTTTAAAGTACTGTAGAT 3′; NB12 (exon 1) - 5′ CTCTATGGTGGGATCATATTCA 3′. Second round: NA12 (exon 1) - 5′ ATGACTGAGTACAAACTGGTGGTG 3′ that lies between 231 and NB-12 and in the second round is used with NB-12.

For single-round PCR, N-*ras* was amplified by 231 and NB12 and Ho IgH by Ho1 and Ho2. For two-round PCR, N-*ras* was amplified by 231 and NA12 followed by NB12 and NA12 and leukemic IgH genes by FR3A and LJH, followed by patient-specific primers in the second round for patients 1–7. PCR products were electrophoresed on 6% polyacrylamide gels in 0.5× TBE buffer (44.5 mM, Tris-HCl, 44.5 mM boric acid, 1 mM EDTA) at 260 V for 1.5 h, stained in (0.5 μg/mL) ethidium bromide, and visualized using short wavelength UV light.

Quantitation

Threefold dilutions of samples were prepared in water or PBL1 DNA and 10 replicates of each dilution were analyzed using the optimal PCR protocol presented in this paper. The mean number of targets required to give a PCR product was determined using the method of Taswell (10), which finds the Poisson distribution that best fits the data and provides an estimate (χ^2 test) of how well the data conform to that Poisson distribution.

RESULTS

Preliminary Experiments

Although one or a few N-*ras* or IgH targets could be detected by an optimized single-stage PCR, a mixing experiment showed that the addition of 1 μg of PBL1 DNA to provide excess competing non-leukemic templates decreased detection of Ho templates to 18% of that otherwise obtainable. A two-round PCR strategy using nested primers was therefore developed to improve sensitivity and specificity of amplification.

Second-Round PCR—Number of Cycles Required

Serial dilutions were made to produce aliquots that contained varying numbers of Ho IgH targets in 1 μg of PBL1 DNA. Each aliquot was amplified for 45 cycles in the first round, and 10^{-4} to 10^{-9} dilutions of the product were made in water to produce a dilution containing one or a few copies of amplified DNA as starting templates for a second round of 20–50 cycles. For IgH targets, amplified product could be detected from the 10^{-4} dilution after 20 cycles, from the 10^{-5} dilution after 30 cycles and from the 10^{-6} and 10^{-7}

dilutions after 40 cycles. Amplified product could not be detected for the 10^{-8} and 10^{-9} dilutions. These results indicated that approximately 10^7 IgH targets for the second round were being produced by the first-round PCR and suggested that 45 cycles should be adequate for second-round amplification.

First-Round PCR—Number of Cycles Required and Effect of Competing Templates

An initial experiment was performed using varying numbers of Ho IgH targets in 1 μg PBL1 DNA, 20–50 first-round PCR cycles, a 1:1000 dilution between the first and second round PCR and 45 second-round PCR cycles. In this experiment, 30 first-round cycles were sufficient to enable an average of 1.5 IgH targets to be detected.

The quantitative aspects of the first round were then studied in more detail in order to determine both the optimal number of cycles and also the appropriate dilution to use between the first and second rounds of the PCR. Serial 10-fold dilutions of Ho DNA were made in water or in 1 μg of PBL1 DNA. Between 20 and 50 cycles of first-round PCR were performed, each aliquot of amplified material was diluted 10^{-2} to 10^{-12} in water and second-round PCR of 45 cycles was performed on each dilution.

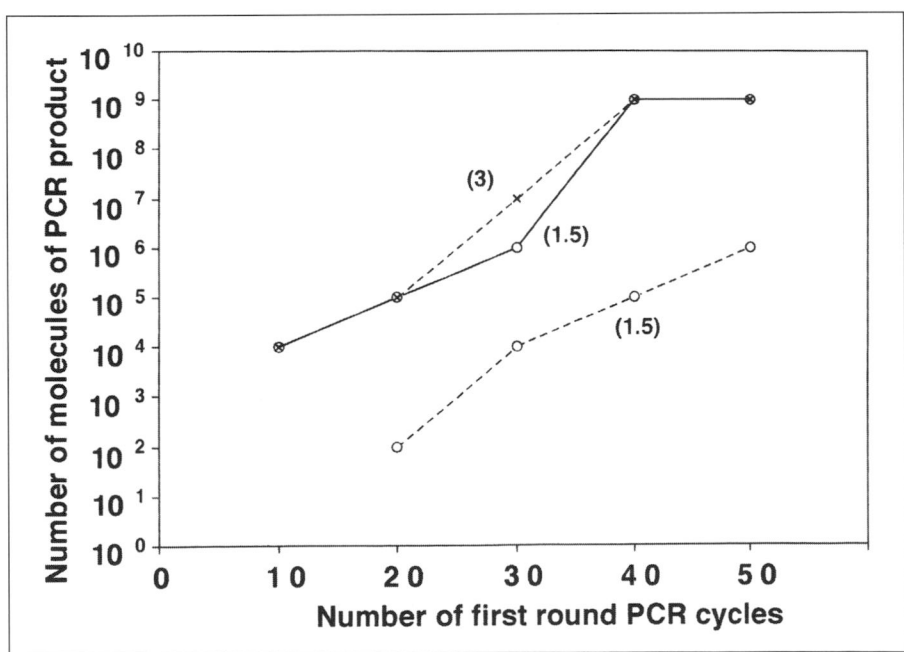

Figure 1. Number of N-*ras* and Ho IgH PCR products produced by a first-round PCR after varying PCR cycle number, in the presence/absence of competing PBL1 DNA. The numbers in parentheses indicate the average number of original PCR targets expected in a single sample of Ho-diluted DNA. N-*ras* PCR product (-PBL1), ---x---; Ho IgH PCR product (+PBL1), ---o---; Ho IgH PCR product (-PBL1), —o—.

Table 1. Sensitivity of Detection (Copies) of N-*ras* or IgH in the Presence or Absence of Competing Template

Number of Copies Added*		Number of Positives in 10 Tubes		
IgH	N-*ras*	N-*ras*	IgH (-PBL1)	IgH (+PBL1)
37.50	75.00	10	10	10
11.25	22.5	10	10	10
3.75	7.5	9	6	3
1.125	2.25	6	3	1
0.375	0.75	4	0	1
0.113	0.225	-	1	0
Mean No. copies detected		2.6	3.3	6.1
*Estimated from DNA concentrations				

The results are shown in Figure 1. Because the 45-cycle, second-round PCR gives detectable amplified DNA from one or a few targets, the maximum dilution of the first-round material that still leads to amplification in the second round indicates the approximate number of copies produced in the first-round amplification. A dilution of Ho DNA containing an average starting target number of 1.5 (IgH) and 3 (N-*ras*) genomes, in the first-round PCR, resulted in a plateau of 10^9 copies after 40 cycles of amplification. However, when other IgH targets were present, having been provided by dilution of starting material in PBL1 DNA, amplification of Ho DNA was less efficient and only 10^6 Ho IgH were produced after 50 cycles of amplification.

Based on these results, we decided to use 45 cycles for first-round amplification and a 10^{-3} dilution of amplified material between the first and second round. These conditions would be expected to produce approximately 1000 targets as starting material for the second round.

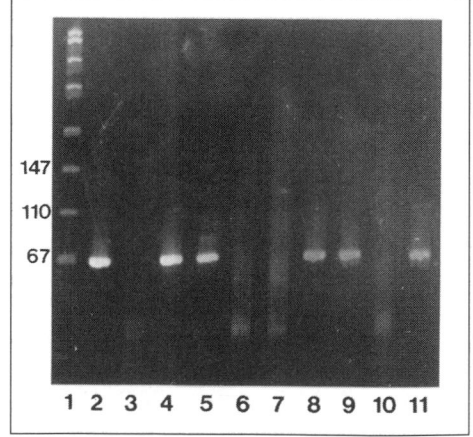

Quantitation by Limiting Dilution Analysis and Poisson Statistics

Serial dilutions of Ho DNA each involving 10 replicates were analyzed by the optimized, two-round PCR (45 cycles, 10^{-3} dilution, 45 cycles). Each tube was

Figure 2. An ethidium bromide stained polyacrylamide gel of PCR products. Lane 1, molecular weight DNA markers; lanes 2–11, amplified products from 10 replicate dilutions of Ho DNA, each containing 3.75 IgH targets.

Table 2. Sensitivity of Detection (Copies) in Diagnostic ALL Patient Samples Using Two-Round PCR

Patient	Minimum Mean No. of Copies Detected			Ratio N-*ras*/IgH	
	N-*ras*	IgH (-PBL1)	IgH (+PBL1)	-PBL1	+PBL1
Ho[a,c]	2.1	3.3	6.0	0.64	0.35
1[b]	1.2	9.7	3.8	0.12	0.32
2	1.2	1.3	2.1	0.92	0.57
3	1.6	0.7	0.9	2.28	1.78
4	1.3	1.5	3.5	0.87	0.37
5	86	211	614	0.41	0.14
6	541	550	621	0.98	0.87
7	23	3	22	7.67	1.05
Geometric mean of ratios				0.90	0.52

[a]10 replicates of dilutions studied.
[b]5 replicates of dilutions studied for patients 1–7.
[c]Geometric mean of all experiments in Ho are shown.

scored as positive or negative for amplification (Figure 2). The detailed results of one such experiment are shown in Table 1. In two experiments, 1.7 and 2.6 copies of N-*ras* could be detected, and in three experiments, 3.5, 6.1 and 10 copies of Ho IgH could be detected in the presence of 1 µg of PBL1 DNA. The data in each experiment were consistent with a Poisson distribution (χ^2 test, $P > 0.05$ for each).

We studied sensitivity of detection in 7 additional diagnostic ALL patient samples. The results of all 8 samples are summarized in Table 2, and the integrity of these DNA samples was investigated by electrophoresis on a 1.3% agarose gel (Figure 3). In this table, the results in the columns referring to minimum mean number of copies detected are calculated from the total number of copies present as based on the estimated DNA concentration. High sensitivity for detection of N-*ras* and IgH genes was observed in 5 patients and lesser sensitivity in 3, patients 5, 6 and 7. Because the number of N-*ras* copies detected depends on both the total number of copies which are present and the proportion which are amplifiable, the ratios of the number of N-*ras* targets detected to the number of IgH targets detected give the proportion of potentially amplifiable IgH targets that were actually amplified. In the absence of competing non-leukemic IgH targets, it proved possible to amplify a mean of 90% of the leukemic targets,

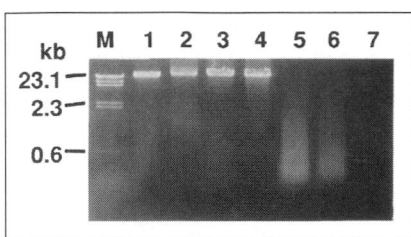

Figure 3. Extracted DNA from diagnostic ALL patient samples. Approximately 250 ng of DNA were loaded onto the gel except for patient 7 where only 24 ng were available. M - molecular weight markers with the sizes given in kb. Lanes 1–7, patients 1–7.

87

and in the presence of competing targets, it proved possible to amplify a mean of 52%, i.e., to detect approximately two ($1/_{0.52}$) potentially amplifiable leukemic targets.

Electrophoresis confirmed extensive DNA degradation in samples from patients 5 and 6 but not in patient 7, in whom the results, although somewhat variable, suggested a lesser degree of degradation. The DNA seen in Figure 3 for patient 7, although faint, appears to be largely intact.

Factors Affecting Amplification Efficiency

Competing IgH targets. The effect of these in reducing amplification of the specific target has already been illustrated (Figure 1 and Table 1).

Competition between primer pairs. Several methods for quantitation of PCR targets rely on the use of two primer pairs in the same tube (3,4,11). To see whether this interferes with the efficiency of amplification and whether N-*ras* and IgH targets could be amplified in the same tube, we quantitated low numbers of IgH targets in a two-round PCR by amplifying using IgH primers either alone or together with N-*ras* primers and in the absence or presence of PBL1 DNA. In the absence of PBL1 DNA, co-amplification with N-*ras* primers had no effect on the minimum mean number of IgH targets detectable (3.2 targets). However, in the presence of PBL1 DNA, IgH primers alone were able to detect mean target numbers of 3.5 in one experiment and 10 in a second, whereas with co-amplification with N-*ras* primers, the IgH primers could only detect mean target numbers of 18 or 79. This effect of N-*ras* primers presumably resulted from the concurrent amplification of the 10^5-fold excess of N-*ras* targets provided by the PBL1 DNA.

DISCUSSION

We have presented a general method for quantitation of targets by PCR using the principle of limiting dilution and use of Poisson statistics. For this approach, the PCR needs to be optimized so that amplification will take place in an "all-or-none" fashion, and one or a few starting targets will give a positive result. When the optimal conditions are known, target concentration can be estimated by Poisson statistics applied to the results from replicate tubes taken at the limit of dilution.

Limiting dilution analysis is widely used for quantitation in cell biology but is not commonly used for molecular quantitation. In 1990 Simmonds et al. (9) reported its use to detect and quantitate single HIV molecules. We were unaware of their report when we developed the method presented here, and review of the PCR literature suggests that the general approach is either not known or is misunderstood. Several other workers have performed serial dilutions in their PCR studies and have used the term "limiting dilution." However, replicates have not been performed, the results have not been analyzed by Poisson statistics and, importantly, the PCR has not been verifiably all-or-none because the reaction has been stopped after an arbitrary number of cycles, and there has been no knowledge of the number of targets that leads to either a positive or a negative reaction. Single serial dilution with use of an arbitrary end point does not constitute a proper limiting dilution analysis and gives only semiquantitative results.

Some areas of research require a method to quantitate a small number of targets against a background of highly homologous targets, e.g., detection of specific mutations in a population of normal cells. The biological problem in our study was the detection of rare leukemic cells in a large population of normal cells, which in molecular terms became the problem of detection of a rare unique IgH sequence against a background of numerous other IgH sequences. The two-stage PCR system that was developed proved capable of detecting approximately two ($1/_{0.52}$) potentially amplifiable leukemic IgH sequences against a background of approximately 160 000 total genomes. These genomes would provide a vast excess of sequences that would compete with the leukemic IgH sequences for the PCR primers because they would contain approximately 2.4×10^4 rearranged IgH genes from normal B lymphocytes and 3×10^5 germ-line IgH genes, each containing multiple V and J segments.

Quantitation of PCR targets by limiting dilution can be compared with other methods for quantitation which use an added internal or external standard, which is carried through the amplification, and which involve some form of quantitation of the amplified product(s) (3,4,11). Quantitation by limiting dilution does not require the use of an added standard, with the inherent assumptions involved, and the end point is simple, nonquantitative and nonradioactive. Furthermore, the end point is based on an all-or-none signal derived from the terminal plateau phase of the PCR, and the technique is therefore relatively robust, being able to cope with wide variations in amplification efficiency without affecting the estimation of DNA target number. One potential disadvantage is the possibility of contamination, particularly if a two-stage PCR is performed. We use the precautions recommended by Kwok et al. (6) to minimize the risk of contamination. Replicate negative controls are used but also, in effect, dilutions below the critical limit of dilution act as additional negative controls.

When molecules are quantitated by PCR, it is necessary to express the results in terms of a denominator such as mass of DNA studied or number of cells from which DNA was extracted. The assumption is that the extraction process does not modify the DNA. However, chemical modifications to DNA, such as strand breakage, depurination or formation of adducts, may render the DNA incapable of acting as a template for the PCR. Quantitation by limiting dilution may have a unique advantage in overcoming the potential problem of DNA modification because it is possible to also quantitate an endogenous gene yielding a PCR product of similar size which is present in known number in all cells and which also undergoes the extraction procedure. In the present study the N-*ras* gene was selected as the endogenous gene to correct for degradation. The 8 patients studied fell into 2 groups (Table 2). In 5 patients, all or nearly all of the N-*ras* targets could be amplified, whereas in 3 patients a lesser proportion of targets could be amplified. This suggested the presence of a variable degree of DNA degradation in these 3 patients, and this was confirmed in the 2 most obvious cases by electrophoresis (Figure 3). Nevertheless, as seen in Table 2, in all 8 patients there was an approximately constant ratio between the number of amplifiable IgH targets and the number of amplifiable N-*ras* targets. These data suggest that the number of amplifiable N-*ras* genes, rather than the DNA concentration, is the best indicator of the number of amplifiable genomes present, that virtually all potentially amplifiable leukemic IgH genes are amplified in the absence of competing non-leukemic IgH genes and that approximately half of the leukemic IgH genes are amplified in the presence of competing genes. Quantitation in the presence of DNA degradation can thus be performed based on these data. Degradation can be a significant problem especially where fresh samples are unavailable and to our knowledge this is the first reported PCR approach that enables correction to be made for it.

Limiting dilution quantitation is simple, requires few manipulations of samples and has widespread application. Our usual approach is to perform a preliminary serial dilution experiment to determine the approximate point at which some amplifications are likely to be negative and some positive and then perform a detailed experiment, perhaps 40 tubes in all, involving multiple replicates of dilutions around this point. If quantitation is to be performed in terms of the number of amplifiable targets of another gene such as N-*ras*, it should be noted that separate amplification reactions must be performed for each gene and that the variance of the final value will be contributed to additively by the variance of the estimation for each gene. More replicates will therefore be required for a given level of precision.

ACKNOWLEDGMENT

We thank Dr. R. Seshadri and Dr. I. Toogood for providing us with the patient samples used in this study. This work was supported by the National Health and Medical Research Council and the Anti-Cancer Foundation of the Universities of South Australia. M.J.B. is in receipt of a Rotary Peter Nelson Leukemia Research Fund Fellowship.

REFERENCES

1.**Brisco, M.J., J. Condon, P.J. Sykes, S. Neoh and A.A. Morley.** 1991. Sensitive quantitation of neoplastic cells in remission and relapse marrows in B-lymphoproliferative disease by use of the polymerase chain reaction. Br. J. Haematol. *79*:211-217.

2. **Brisco, M.J., L.W. Tan, A.M. Orsborn and A.A. Morley.** 1990. Development of a highly sensitive assay, based on the polymerase chain reaction, for rare B-lymphocyte clones in a polyclonal population. Br. J. Haematol. *75*:163-167.

3. **Chelly, J., D. Montarras, C. Pinset, Y. Berwald-Netter, J-C. Kaplan and A. Kahn.** 1990. Quantitative estimation of minor mRNAs by cDNA-polymerase chain reaction. Application to dystrophin mRNA in cultured myogenic and brain cells. Eur. J. Biochem. *187*:691-698.

4. **Gilliland, G., S. Perrin, K. Blanchard and H.F. Bunn.** 1990. Analysis of cytokine mRNA and DNA. Detection and quantitation by competititive polymerase chain reaction. Proc. Natl. Acad. Sci, USA *87*:2725-2729.

5. **Jeffreys, A.J., V. Wilson, R. Neumann and J. Keyte.** 1988. Amplification of human minisatellites by the polymerase chain reaction: toward DNA fingerprinting single cells. Nucleic Acids Res. *16*:10953-10971.

6. **Kwok, S. and R. Higuchi.** 1989. Avoiding false positives with PCR. Nature *339*:237-238.

7. **Saiki, R.K., D.H. Gelfand, S. Stoffel, S.J. Scharf, R. Higuchi, G.T. Horn, K.B. Mullis and H.A. Erlich.** 1988. Primer-directed enzymatic amplification of DNA with a thermostable DNA polymerase. Science *239*:487-491.

8. **Sambrook, J., E.F. Fritsch and T. Maniatis.** 1989. Molecular Cloning: A Laboratory Manual, 2nd Edition, Cold Spring Harbor Laboratory, Cold Spring Harbor, NY.

9. **Simmonds, P., P. Balfe, J.F. Peutherer, C.A. Ludlam, J.O. Bishop and A.J.L. Brown.** 1990. Human immunodeficiency virus-infected individuals contain provirus in small numbers of peripheral mononuclear cells and at low copy numbers. J. Virol. *64*:864-872.

10. **Taswell, C.** 1981. Limiting dilution assays for the determination of immunocompetent cell frequencies. J. Immuno. *126*:1614-1619.

11. **Wang, A.M., M.V.L. Doyle and D.F. Mark.** 1989. Quantitation of mRNA by the polymerase chain reaction. Proc. Natl. Acad. Sci. USA *86*:9717-9721.

91

Update to:
Quantitation of Targets for PCR by Use of Limiting Dilution

Pamela J. Sykes and Alec Morley

Flinders Medical Center, South Australia

The method of quantitative limiting dilution PCR analysis was developed in our laboratory for the purpose of detecting minimal residual disease in acute lymphoblastic leukemia patients; however, the method is a general method that can be applied to the quantification of any DNA or RNA target. A major advantage of our method is that it incorporates amplification of an endogenous control gene (N-*ras*) which enables the quantification of amplifiable targets present in a DNA sample, thus providing an estimate of the integrity of the DNA being amplified. We have made very few technical changes to the method since the original publication. A single-round PCR for the N-*ras* gene internal control was developed that has reduced the time required to perform the assay (described in detail in Reference 7).

Since writing the original article we have used the method to quantify bone marrow and blood samples from acute lymphoblastic leukemia patients during induction therapy and from myeloma patients at the time of autologous transplant for detection of minimal residual disease (1,2,4,8). The levels of disease detected ranged from 1.7×10^{-1} to 1.4×10^{-6} and the level of disease at the end of induction therapy is highly predictive of outcome in childhood and adult leukemia. Duplicate quantifications of the same sample for 22 leukemic samples gave 95% confidence limits for a single estimate as equivalent to multiplication or division of the observed value by a factor of 4.9 (8). The quantitative PCR limiting dilution method has been used by other workers (5,6) to study residual disease and has been applied to the quantification of microsatellite markers (our laboratory, unpublished). The potential application of the quantitative limiting dilution PCR technique to the study of RNA is detailed in Sykes and Morley (7) and has been used by Gniadecki and Serup (3) to quantify transforming growth factor-β1 mRNA.

REFERENCES

1. Brisco, M.J., J. Condon, E. Hughes, S.H. Neoh, P.J. Sykes, R. Seshadri, I. Toogood, K. Waters, G. Tauro, H. Ekert and A.A. Morley. 1994. Outcome prediction in childhood acute lymphoblastic leukaemia by molecular quantification of residual disease at the end of induction. Lancet *343*:196-200.

2. Brisco, M.J., E. Hughes, S.H. Neoh, P.J. Sykes, K. Bradstock, A. Enno, J. Szer, K. McCaul and A.A. Morley. 1996. Relationship between minimal residual disease and outcome in adult acute lymphoblastic leukemia. Blood 87:251-256.

3. Gniadecki, R. and J. Serup. 1994. Enhancement of the granulation tissue formation in hairless mice by a potent vitamin D receptor agonist - KH 1060. J. Endocrinol. 141: 411-415.

4. Henry, J.M., P.J. Sykes, M.J. Brisco, L.B. To, C.A. Juttner and A.A. Morley. 1996. Comparison of myeloma cell contamination of bone marrow and peripheral blood stem cell harvests. Br. J. Haematol. 92:614-619.

5. Morley, A.A. 1995. Measurement of minimal residual disease in acute leukemia. Leukemia 10:920.

6. Ouspenskaia M.V., D.A. Johnston, W.M. Roberts, Z. Estrov and T.F. Zipf. 1995. Accurate quantitation of residual B-precursor acute lymphoblastic leukemia by limiting dilution and a PCR-based detection system: a description of the method and the principles involved. Leukemia 9:321-328.

7. Sykes, P.J. and A.A. Morley. 1995. Limiting dilution polymerase chain reaction, p. 150-165. In J.W. Larrick and P.D. Siebert (Eds.), Reverse Transcriptase PCR. Ellis Horwood, Hemel Hempstead, Herts, UK.

8. Sykes, P.J., L.E. Snell, M.J. Brisco, S-H. Neoh, E. Hughes, G. Dolman, L M. Peng, A. Bennett, I. Toogood and A.A. Morley. 1997. The use of monoclonal gene rearrangement for detection of minimal residual disease in acute lymphoblastic leukemia n childhood. Leukemia 11:153-158.

CHAPTER 7

New Approach to mRNA Quantification: Additive RT-PCR

A. Reyes-Engel, J. García-Villanova, J.L. Dieguez-Lucena, N. Fernández-Arcás and M. Ruiz-Galdón

University of Málaga, Málaga, Spain

The quantification of a determined mRNA by reverse transcription polymerase chain reaction (RT-PCR) is based on the linear fitting of the "exponential" growth obtained after different concentrations of the template are amplified. Usually a modified internal standard, which allows the quantification by comparing bands or extrapolating in a standard curve, is used. We propose a method that uses an unmodified internal standard identical to the target molecule, which simplifies and improves the methodology. We also propose a logistic fitting of the standard curve as an alternative to the linear-exponential adjustment.

For this purpose, we have studied Scavenger receptor genes and quantified their expression. These receptors mediate the binding and uptake of chemically modified lipoproteins and are therefore involved in atherosclerosis. To obtain specific Scavenger receptor mRNAs to be used as an internal standard, total cytoplasmic RNA from 1.5×10^8 THP-1 cells derived from macrophages was isolated by the guanidinium isothiocyanate method (2). Fibroblast RNA was used as a negative control. The RNA was hybridized with 150 pmol of a specific biotinylated oligonucleotide complementary to both scavenger receptor mRNAs 5'-CTAATTTACTGATTTCCTCTTGTTGTTTGAAGGT-ATTCTCTTGGATTTT-3' (4) at 70°C for 5 min. Streptavidin MagneSphere® paramagnetic particles (2 mg; Promega, Madison, WI, USA), washed three times with 0.5× standard saline citrate (SSC), were then added. After 2 min, the beads were captured by using a magnetic rack, and then three stringent washes with 0.5× SSC were done. To elute the specific mRNA, 500 µL of RNase-free water were added. The sample was concentrated by using 3 M ammonium acetate (0.1 vol), 20 mg/mL glycogen (Boehringer Mannheim, Mannheim, Germany) (0.02 vol) and 2-propanol (2 vol) and redissolved in 10 µL of RNase-free water, and the concentration was determined by capillary electrophoresis (CE) (5). Then, 1.2×10^4 copies of the specific mRNA were reverse-transcribed in the presence of 0.8 µg fibroblast total RNA, with 40 U RNasin® (Promega), 1 mM deoxyribonucleoside triphosphates (dNTPs), 60

(Reprinted from BioTechniques 21:202-204, 1996)

U of Moloney murine leukemia virus reverse transcriptase (Promega) and 2.5 mM of random hexamers in 20 μL RT buffer (Promega) at 42°C for 45 min and then incubated at 95°C for 10 min. cDNA was amplified with the addition of 10 pmol of the upper primer 5'-CCACATTGCTTGATTTGCAGCTC-3', 10 pmol of the lower primer 5'-CTCCCACCGACCAGTCGAAC-TTTCGTAA-3' and 2.5 U *Taq* DNA Polymerase (Promega) in 80 μL of 1× PCR buffer. PCR primers were selected by using the "Oligo" program (Med-Probe, Oslo, Norway). The amplification protocol was: 20 s at 95°C, 15 s at 60°C and 40 s at 72°C for 30 cycles, then 7 min at 72°C in a DNA Thermal Cycler 480 (Perkin-Elmer, Norwalk, CT, USA). A 513-bp band was obtained as expected and was confirmed by sequencing. To check for possible contamination, we also amplified by RT-PCR the specific mRNA by using primers for IL-1α andβ-actin genes, and no bands were observed. A standard curve of 5, 10, 50, 100 and 500 fg of the captured specific mRNA was obtained after 30 amplification cycles. Ten femtograms were selected as the internal standard for the additions (see below).

Total RNA from 10 mL peripheral blood mononuclear cells (PBMC), of different individuals, was isolated as previously described. Duplicated tubes containing 0.8 μg of total RNA each were set up. Ten femtograms of the internal standard were added to one of them (Figure 1) and then were amplified by RT-PCR. PCR products were electrophoresed on an agarose gel and stained with ethidium bromide. The optical density of the 513-bp band, in a digitalized picture, was measured by using a Sun Spark Station (Visage software; Millipore, Bedford, MA, USA). In addition, 15 μL of each RT-PCR product was also quantified by measuring the intensity of fluorescence using Hoechst reagent 33258 (Polysciences, Warrington, PA, USA) as the dye (λexc: 365 nm and lem: 460 nm) with a Hitachi F-2000 Fluorometer (Hitachi, Tokyo, Japan).

mRNA concentration was calculated by applying two fittings to the standard curve obtained, a linear-exponential and a logistic mathematical adjustment defined by the equation $y = [(a-d)/(1 + (x/c)^b)] + d$, in which a = maximal value of the signal, b =

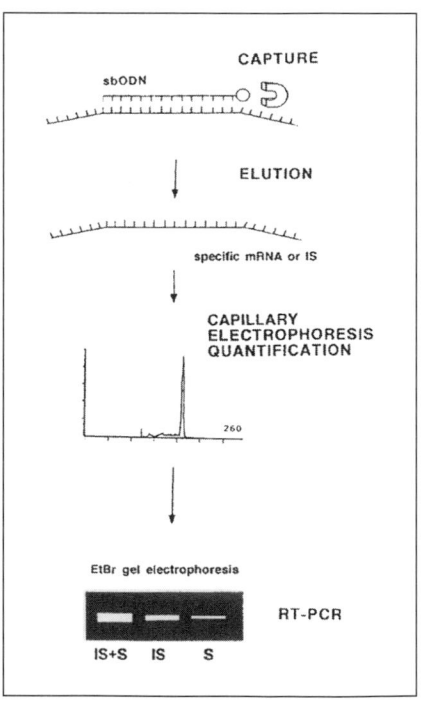

Figure 1. Schematic view of the procedure used to obtain the internal standard and its application to additive RT-PCR. sbODN: selected biotinylated oligonucleotide; IS: internal standard; S: sample of total RNA.

slope parameter, c = value at inflection point, d = minimal value signal (SigmaPlot™; Jandel Scientific, San Francisco, CA, USA). It was observed that the PCR process with *Taq* DNA polymerase is better fitted with a logistic than a linear-exponential adjustment when a wide range of concentrations is studied. Results were extrapolated to a standard curve based on a logistic equation and compared with the linear-exponential curves. These fittings showed similar results, but the logistic one was more accurate when applied to a wider range of concentrations (Figure 2). This behavior of *Taq* DNA polymerase has also been shown by other authors who have studied activity vs. increasing cycles and template concentrations (1).

By definition, the best internal standard to quantitate PCR products is a template identical to the target sequence to be studied, since this ensures an identical amplification efficiency. The efficiency and detection signal of PCR products change when an internal standard of a different size or sequence from the wild-type is used (3). Methods used so far employ modified internal standards to differentiate both molecules. This procedure has led to the necessity of producing, in vitro, large amounts of the internal standard that has

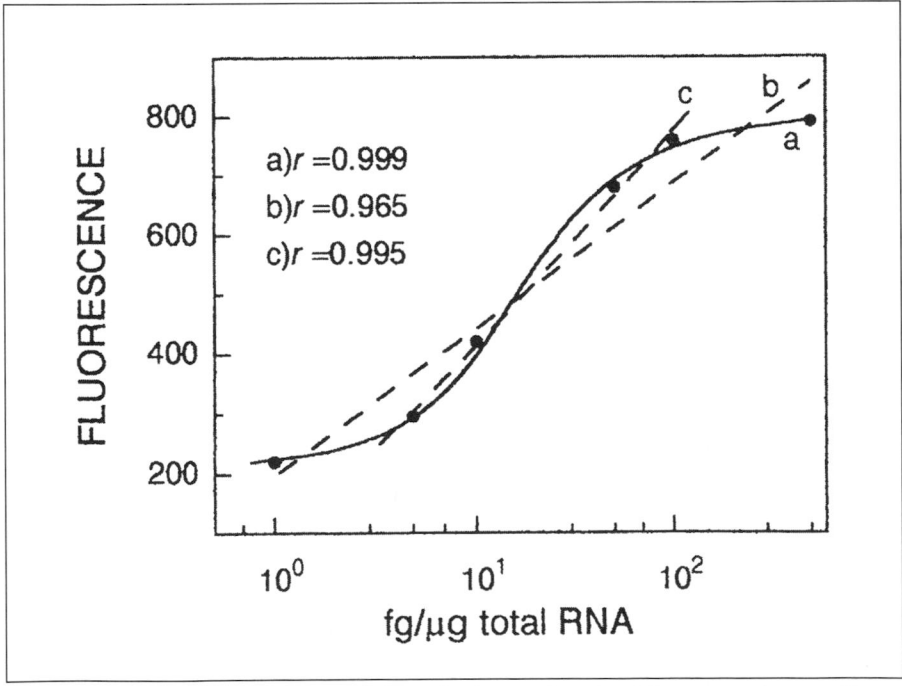

Figure 2. Scavenger receptor type I mRNA standard curve. Intensity of fluorescence (measured by using Hoechst reagent 33258 and a Hitachi F-2000 fluorometer) vs. different amounts of the specific mRNA in cycle 30. Different adjustments were applied: (a) logistic, defined by equation $y = [(a-d)/(1 + (x/c)^b)] + d$, in which a = 800.5, b = -1.32, c = 16.2 and d = 201.3, (b) linear-exponential, to a wide range (1–500) and (c) linear zone adjustment to a narrow range (4–100) where each point was assayed in triplicate, with a CV <9.2% (10-fg point [*n* = 10] CV of 6.3%).

to be modified; therefore, values have to be corrected by at least two factors in these competitive assays. First, the correction here relates to the efficiency of both RT and PCR itself. Second, the correction of the signal, when an internal standard of a different size is applied, is irrespective of the detection system (fluorescence, radioactivity or optical density). In the method described in this work, the addition of an identical internal standard has two practical features. It increases the measurable signal to a range of the curve with higher sensitivity and controls the accuracy in each single sample. Duplicated aliquots with and without an internal standard can be used to calculate the recovery of the standard in each sample (Table 1). Moreover, this protocol is simpler than the classical competitive PCR because no standards with different concentrations are used for each assay, as well as the unnecessary correction of the efficiency and signal detection. Additive quantitative RT-PCR has also been applied to other mRNAs such as the angiotensin II receptor and human immunodeficiency virus (HIV)-1 RNA (data not shown). Some variations in the method could be related to the internal standard since three different ones can be used, (*i*) specific mRNA from a gene (real concentration of the sample), (*ii*) specific cDNA (subjected to variations in retrotranscription efficiency) and (*iii*) DNA such as a purified PCR product, which depends on RT and PCR efficiency mainly in the first cycles. Although each one could be used to quantify gene expression, we have concluded that a specific mRNA is the optimal choice since it is subjected to less variations during the process and therefore leads to more accurate results.

ACKNOWLEDGMENTS

This work was supported by two personal grants, Grant Nos. 94/5377 and 94/5334, from FIS (Ministerio de Sanidad y Consumo, Spain).

REFERENCES

1. **Berry, N., K. Ariyoshi, O. Jobe, P.T. Ngum, T. Corrah, A. Wilkins, H. Whittle and R. Tedder.** 1994. HIV type 2 proviral load measured by quantitative polymerase chain reaction correlates with CD4+ lymphopenia in HIV type 2-infected individuals. AIDS Res. Hum. Retroviruses *10*:1031-1037.
2. **Chomcznsky, P. and N. Sacchi.** 1987. Single-step method of RNA isolation by acid guanidinium thiocyanate-phenol-chloroform extraction. Anal. Biochem. *162*:156-159.
3. **Gemen, B.V., T. Kievits, R. Shukkink, D.V. Strijp, L.T. Malek, R. Sooknanan, H.G. Huisman and P. Lens.** 1993. Quantification of HIV-1 RNA in plasma using NASBA™ during HIV-1 primary infection. J. Virol. Methods *43*:177-188.
4. **Matsumoto, A., M. Naito, H. Itakura, S. Ikemoto, H. Asaoka, I. Hayakawa, H. Kanamori, H. Aburatani, et al.** 1990. Human macrophage scavenger receptors: primary structure, expression and localization in atherosclerotic lesions. Proc. Natl. Acad. Sci. USA *87*:9133-9137.
5. **Reyes-Engel, A. and J.L. Dieguez-Lucena.** 1993. Direct quantification of specific mRNA using a selected biotinylated oligonucleotide by free solution capillary electrophoresis. Nucleic Acids Res. *21*:759-760.

Update to:

New Approach to mRNA Quantification: Additive RT-PCR

Armando Reyes-Engel

University of Málaga, Málaga, Spain

The additive RT-PCR has been used to quantify the scavenger receptor involved in the process of atherogenesis and the angiotensin II AT1 receptor related to blood pressure (1–3). Observations on the quantification of HIV-1 RNA are still unpublished.

REFERENCES

1. **Dieguez-Lucena, J.L., P. Aranda-Lara, M. Ruiz-Galdon, J. Garcia-Villanova, M. Morell-Ocana and A. Reyes-Engel.** 1996. Angiotensin I-converting enzyme genotypes and angiotensin II receptors. Response to therapy. Hypertension *28*:98-103.
2. **Garcia-Villanova, J., J.L. Dieguez-Lucena, P. Aranda-Lara, N. Fernandez-Arcas and A. Reyes-Engel.** 1996. Cuantificacion por RTPCR de la expresion genica de los receptores scavenger tipos I y II en celulas mononucleares de sangre periferica de pacientes con hipertension esencial. Relacion entre genotipos de la ECA y respuesta a tratamiento. Iberoamerican J. Hypertens. *1*:191-197.
3. **Garcia-Villanova, J., J.L. Dieguez-Lucena, N. Ferdandez-Arcas and A. Reyes-Engel.** 1996. Increased expression of Scavenger receptor type I gene in human peripheral blood from hypercholesterolemic patients. Biochim. Biophys. Acta *1300*:135-141.

CHAPTER 8

A Versatile ELISA-PCR Assay for mRNA Quantitation from a Few Cells

P. Alard[1], O. Lantz[1,2], M. Sebagh[1,3], C.F. Calvo[1], D. Weill[1], G. Chavanel[1], A. Senik[1] and B. Charpentier[1,2]

[1]Institut de Recherches Scientifiques sur le Cancer, [2]Hôpital de Bicêtre and Université Paris-Sud, [3]Hôpital Paul-Brousse and Université Paris-Sud

ABSTRACT

Gene expression studies require a sensitive and quantitative assay of mRNA amounts present in small samples. We describe a general method of quantifying specific mRNA quickly and easily from purified RNA or directly from a few cells by PCR and enzyme-linked immunosorbent assay (ELISA) revelation of the resulting products (sensitivity of the last step: <0.1 fmol). Cells are digested and the total cellular RNA is reverse-transcribed and then amplified with 5' and 3' primers; the former being 5' biotinylated. The amplification product is captured on avidin-coated microplates and quantified by hybridization with a digoxigenin-labeled internal oligonucleotide probe. After revelation with an anti-digoxigenin alkaline phosphatase coupled antibody (anti-DIG-AP[1]), the amount of hybridized probe is determined by optical reading. The results can be easily converted to absolute values by comparison with an external DNA standard curve. An internal DNA or RNA standard can also be used. The method we describe can be adapted to any cellular or viral gene of known sequence in a matter of days. Since it uses nonradioactive probes, commercially available reagents and standard microplate readers, it is inexpensive and could be automated easily. In this study, interleuken-2 mRNA expression could be studied in as few as 40 Jurkat cells. It was also possible to quantify human immunodeficiency virus (HIV) DNA from 1500 to 1.5 copies out of 1.5×10^5 human genomic DNA copies.

INTRODUCTION

Studies on the regulation of gene expression in various tissues or after in vitro modulation need quantitative assays to measure the amounts of several mRNAs or proteins present in numerous samples. Polymerase chain reaction (PCR) has been used by several authors to qualitatively study the lymphokine gene expression in pathological situations (20) or during T cell ontogeny (5).

(Reprinted from BioTechniques 15:730-737, 1993)

In these situations, a 10-fold to 100-fold difference in gene expression was observed (5,20), and direct PCR could be used meaningfully. However, to study gene regulation where differences in expression are much smaller, simple PCR cannot be used because of a supposed variability in the PCR amplification rate (6,17). To overcome this problem, the use of an internal standard has been suggested (1,2,7,21), and quantitative PCR can thus be performed by competition (1,7) during the exponential phase of the amplification (21) or in end-point (2). However, all semiquantitative or quantitative PCR assays proposed (1–4,7,9, 13,21) have one or more of the following disadvantages: the large number of test tubes in competitive PCR (1,7); time-consuming dot blots or Southern blots (1–4,7,21); absence of easy digitalization of the results and use of radioactive isotopes (1,2,4,7,21); and use of noncommercially available (13), expensive custom-made reagents (9) or expensive material such as the phosphor imager machine (1–3,21).

Ideally, a quantitative PCR method should be simple, fast and inexpensive. In order to provide numeric data essential for any quantitation, the results should be digitalized and not represented solely as a spot on a radiogram. Moreover, the technique should allow the simultaneous processing of many samples. Finally, it should be applicable to any gene of known sequence and should not need special reagents or equipment. Automation should be sought from the outset of the development of the assay.

In this report, we describe a quantitative PCR method that uses only commercially available reagents (5′ biotinylated oligonucleotide, digoxigenin labeling and detection kits) and widely available equipment such as the enzyme-linked immunosorbent assay (ELISA) reader. The ELISA-based detection method of amplified products is sensitive enough (below 15 pg of amplified product) to quantify several aliquots of the PCR mixture sampled at regular intervals during amplification. Since the method is based on a liquid-phase hybridization step, an internal standard can be used. After an optical density (OD) reading, the results can be handled easily by computers. In a matter of days the method can be applied to any newly discovered gene.

MATERIALS AND METHODS

Primers, Probes and PCR Product Size

All primers were purchased from Genset (Paris, France) and are listed in Table 1. The oligonucleotides were digoxigenin-labeled using the digoxigenin oligonucleotide 3′-end labeling kit as indicated by the manufacturer (Boehringer Mannheim, Meylan, France).

Internal Standard Construction

A synthetic interleukin (IL)-4 DNA standard (308 base long) was obtained by recombinant PCR by amplifying an IL-2 R β plasmid (kindly provided by

Table 1. Primers Used in This Study

Denomination	Sequence[a]	Position on cDNA	Expected Size
5'-GAPDH-biotinylated	GGTGAAGGTCGGAGTCAACGGA	69-90	240
3'-GAPDH	GAGGGATCTCGCTCCTGGAAGA	287-308	
5'-IL2-biotinylated	GAATGGAATTAATAATTACAAGAATCCC	182-209	229
3'-IL-2	TGTTTCAGATCCCTTTAGTTCCAG	387-410	
5'-IL-4-biotinylated	CAGAGCAGAAGACTCTGTGCAC	188-209	247
3'-IL-4	GGACAGGAATTCAAGCCCGCCA	413-434	
5'-IL-10-biotinylated	GCAACCTGCCTAACATGCTTCG	134-155	389
3'-IL-10	GAAGATGTCAAACTCACTCATGGC	499-522	
5'-GAG-biotinylated	TAATCCACCTATCCCAGTAGGAGA	1095-1118[b]	111
3'-GAG	GGTCCTTGTCTTATGTCCAGAATGCTG	1178-1205[b]	
5'-POL-biotinylated	TTCTGGGAAGTTCAATTAGGAATACC	2394-2418[b]	316
3'-POL	TCAGATCCTACATACAAATCATCCATGTA	2681-2709[b]	
5'-TAT-biotinylated	GGGTGTCGACATAGCAGAATAGG	5367-5388[b]	203
3'-TAT	CGCTTCTTCCTGCCATAGGAGAT	5547-5569[b]	
5'-IL-4 standard construct	CAGAGCAGAAGACTCTGTGCACAGCAT GGAGGAGACGTCCAGAA		308
3'-IL-4 standard construct	GGACAGGAATTCAAGCCCGCCAAGTAA AGTACACCTGGCAGGCC		
GRAPDH-probe	AAAGCAGCCCTGGTGACC	111-128	
IL-2-probe	GTGGCCTTCTTGGGCATG	242-259	
IL-4-probe	GCAGCCCTGCAGAAGGTT	264-281	
IL-10-probe	GCTGAAGGCATCTCGGAG	160-177	
GAG-prob	TTTAATCCCAGGATTATCCATCTTTTA	1124-1150[b]	
POL-probe	CACATCCAGTACTGTTACTGATTT	2444-2467[b]	
TAT-probe	CTGGATGCTTCCAGGGCTCTA	5437-5457[b]	
Standard probe	TGTCCTGCTGCAGGAGCA		

[a]5' to 3' sequences as synthetized are shown
[b]Position relative to the sequence of HIJV-BRU in GenBank®

Dr. M. Hatakeyama, Osaka, Japan) with 2 hybrid primers (the 5' end was the 22 bases of the IL-4 primers, and the 3' end was the IL-2 R sequence). The double-stranded DNA standard was gel-purified, quantified by ethidium bromide staining and comparison to molecular weight markers, and aliquoted and frozen until use.

RNA Extraction and cDNA Synthesis

Total cellular RNA was extracted using the method of Chomczynski et al. (4) as modified in the RNAzol® kit (Tel-Test, Friendswood, TX, USA). When RNA was prepared without extraction (10), the indicated number of cells were washed once in phosphate-buffered saline (PBS) and resuspended in 7.5 μL PBS containing RNase inhibitor (RNasin®; Promega, Madison, WI, USA), 5 U per reaction mixture and 1 mM dithiothreitol (DTT). Tween® 20 (0.5%) and proteinase K (100 μg/mL) were then added. Incubation was at 55°C for 30 min and then for 10 min at 90°C to inactivate protease. The purified total RNA (1 μg) or cell lysates prepared as described above were incubated in 20 μL reverse transcriptase (RT) buffer (Promega), 100 μM hexanucleotides and 4 U per reaction of avian myeloblastosis virus reverse transcriptase (AMV-RT). Incubation was for 10 min at room temperature and 1 h at 42°C. The reaction was stopped by incubation at 95°C for 5 min and at 99°C for 1 min.

PCR Amplification

PCR amplification was carried out in 50 μL of amplification buffer (Promega) containing 0.25 μM 5' and 3' primers with 1.25 U per reaction *Taq* DNA polymerase and overlayed with mineral oil. The thermal profile was initially 93°C for 3 min, followed by the indicated number of cycles at 94°C for 30 s, at 60°C for 1 min and at 72°C for 1 min in a 480 DNA Thermal Cycler (Perkin-Elmer, Norwalk, CT, USA). At the indicated number of cycles, 5 μL of the reaction mixture were sampled through oil. All the pre-amplification steps were carried out under laminar flow hoods. In every experiment, numerous controls were set up to detect any possible contamination. To use the "hot start" technique without opening the reaction tubes, we followed a protocol for preparing PCR that kept all the samples on ice before placing all the tubes almost simultaneously into the thermal cycler, which had been set beforehand at 93°C.

Quantitation of Amplification Products

The procedure is summarized in Figure 1. The microplates (Maxisorp; Nunc, Roskilde, Denmark) were coated with 100 μg/mL avidin (Sigma Chemical, St. Louis, MO, USA), and the free sites were saturated with bovine serum albumin (BSA). The amplification products were transferred onto avidin-coated microplates and incubated at room temperature for 1 h. The DNA was denatured by adding 0.25M NaOH at room temperature for 10 min. After 3 washes, a digoxigenin-labeled oligonucleotide probe (0.2 pmol/well) was added in 100 µL of 0.5× sodium chloride sodium phosphate EDTA (SSPE) and incubated for 2 h at 42°C. After 3 washes, anti-digoxigenin alkaline phosphatase coupled antibody (anti-DIG-AP) (1/7500) was added and incubated for 1 h at room temperature. After 4 additional washes, paranitrophenyl (PNP) substrate (1 mg/mL) was added in 1 M diethenolamine buffer. The OD at 405 nm was measured after incubation at 37°C.

HIV Reconstruction Experiment

We used the ANRS (Agence Nationale de la Recherche sur le SIDA) panel, kindly provided by C. Brechot, Hybridotest, Paris, France. DNA was extracted from the 8E5-LAV cell line, which contains a defective integrated human immunodeficiency virus (HIV) (1 copy per cell). Serial dilutions of purified DNA were mixed with a constant amount (1 µg) of purified genomic human DNA. Thus, a variable number of HIV copies (1500 to 0.15) were mixed in 1.5×10^5 human genomes. Amplification was carried out as above with the indicated primers, chosen according to References 11, 15 and 18 with slight modifications.

RESULTS

The general scheme of the assay is depicted in Figure 1: RNA, either purified or not, is reverse transcribed, and the PCR amplification is carried out using a biotinylated 5′ primer and a non-biotinylated 3′ primer. The PCR product is captured by avidin-coated microplates, is alkaline-denatured and

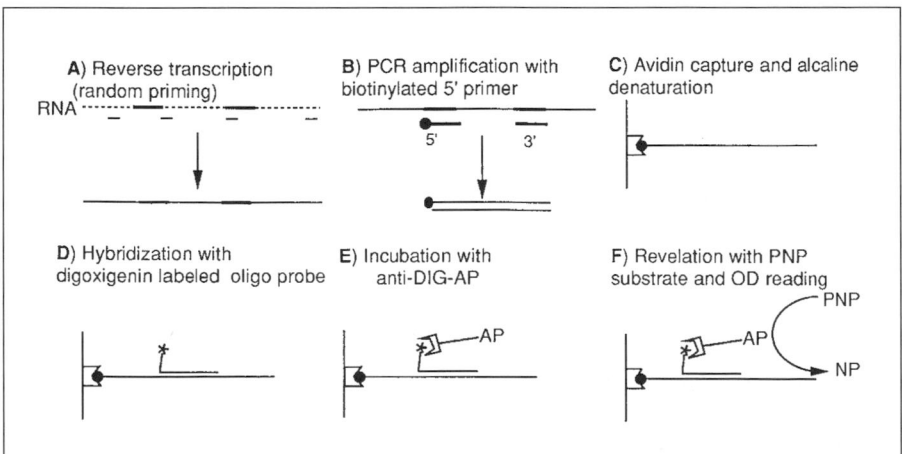

Figure 1. General outline of assay. The RNA (purified or not) is reverse-transcribed and PCR-amplified with a biotinylated 5′ primer. Biotinylated amplified products are captured on avidin-coated microplates, alkaline-denatured and hybribized with a digoxigenin-labeled oligonucleotide probe, which is revealed by anti-DIG-AP.

quantified by hybridization with a digoxigenin-labeled probe. The amount of hybridized probe is revealed by anti-DIG-AP-coupled antibody, PNP substrate and OD reading at 405 nm.

We will critically examine each step of this process, starting from the last

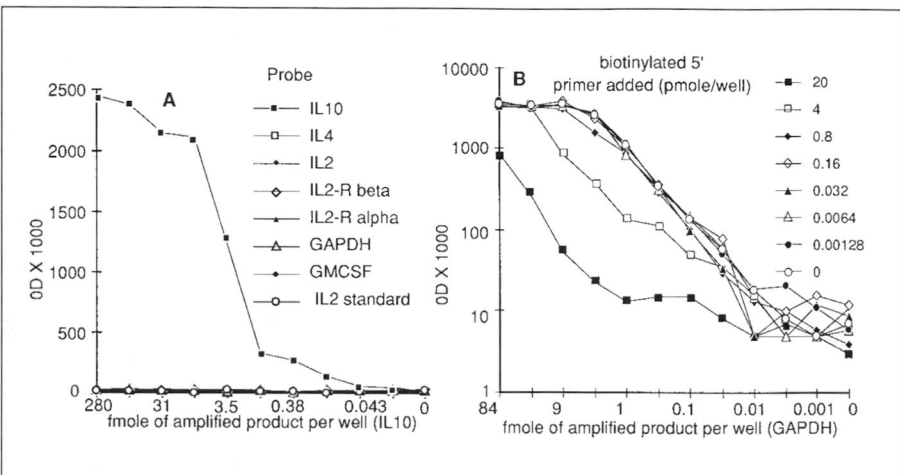

Figure 2. Sensitivity and specificity of the revelation step. A) Specificity: serial dilutions of the amplified products (IL-10) were avidin-captured, alkaline-denatured and hybridized with homologous or irrelevant probes (0.2 pmol/well). B) Sensitivity: purified amplified products (GAPDH) were avidin-captured with or without several concentrations of biotinylated 5′ primers and then hybridized with homologous probe.

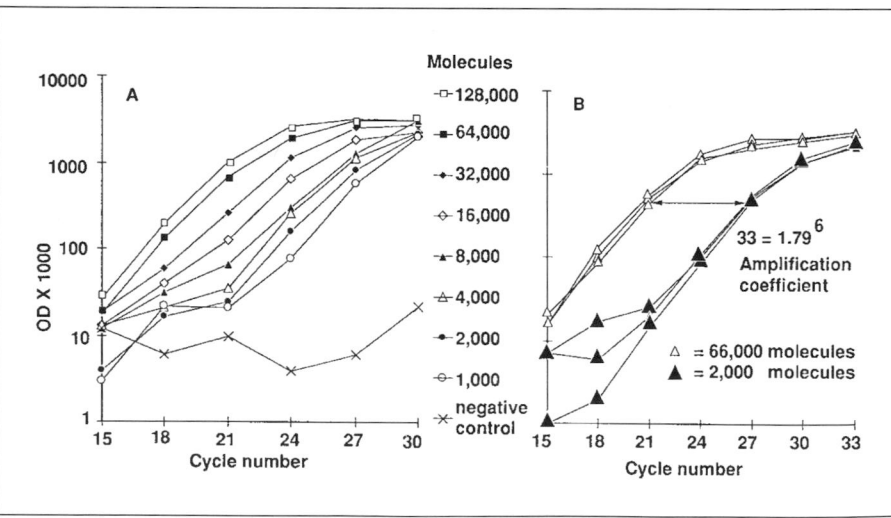

Figure 3. Quantitation of DNA with ELISA-PCR. A) Serial dilutions of purified products (IL-2) were processed as described in Materials and Methods. OD values of samples at the indicated cycle number are shown. B) Two aliquots of 66 000 and 2000 molecules of previously amplified products (IL-2) were amplified separately in triplicate. The two groups of OD curves were parallel, and an amplification coefficient could be computed.

106

one, working backward, and validating each step in turn.

1) Detection of amplified products by ELISA. Using 5′ biotinylated and 3′ digoxigenin-labeled oligonucleotides, we estimated the sensitivity of the detection step to be below 0.1 fmol, which is equivalent to 16 pg of a 250-bp PCR product (not shown). Indeed, as shown in Figure 2, when known amounts of PCR products (IL-10 in Figure 2A and GAPDH in Figure 2B) were hybridized with the DIG probe, the sensitivity was below 0.1 fmol. The hybridization was specific as shown by the absence of signals when using irrelevant probes (Figure 2A). Figure 2B shows that no competition due to an excess of biotinylated 5′ primer was observed below 1 pmol of 5′ primer per well. Thus, when assaying PCR products, unincorporated 5′ biotinylated primers will not interfere significantly in the detection step, since the amplification is carried out with 1.25 pmol/5 μL of the 5′ primer, which is the volume of the PCR used in each well of the ELISA.

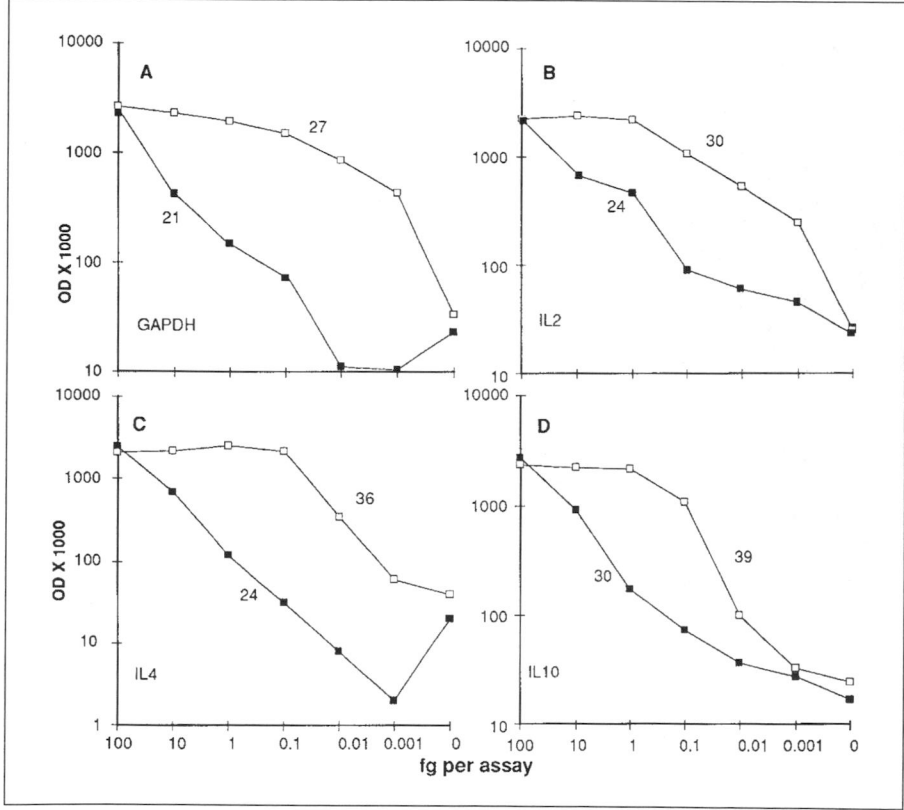

Figure 4. Quantitation of DNA with ELISA-PCR. Serial dilutions of previously amplified products as indicated were subjected to PCR amplification and sample at regular intervals. Only the two most informative cycles are shown for each experiment: the first cycle allows the quantitation of the 100–0.1-fg range and the second cycle allows the quantitation of the 0.1–0.001-fg range. The cycle number is indicated on each curve.

2) Quantitative considerations concerning the dynamic range of PCR-ELISA. To examine the reproducibility and the dynamic range of the ELISA-PCR method, we amplified serial dilutions of known amounts of a previously amplified DNA product (IL-2). Aliquots of the reaction mixture were sampled from the cycles 15 to 30. The amplified product was then quantified as above. As shown in Figure 3A, the OD increased regularly with the number of cycles. Parallel curves were obtained according to the initial amount of DNA used until a plateau was reached. Figure 3B shows one example of a calculation of the amplification coefficient and the good reproducibility of the triplicates. As shown in Figure 4, the dynamic range of the ELISA-PCR assay was large because sampling at regular intervals during the amplification allows the analysis of different concentration ranges. Indeed, the 100-fg–0.1-fg range could usually be quantified at 24 cycles. The 0.1-fg–1-ag range needed 30 to 39 cycles according to the amplification coefficient.

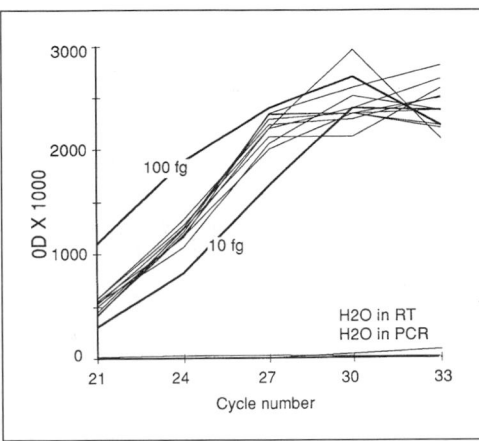

Figure 5. Reproducibility of the RT and amplification steps. Ten aliquots (100 ng) of purified total RNA from PHA + PMA-activated Jurkat cells were reverse-transcribed and amplified with IL-2 primers as above. The external DNA scale (the number shown refers to the amount of double-stranded DNA input) is also shown.

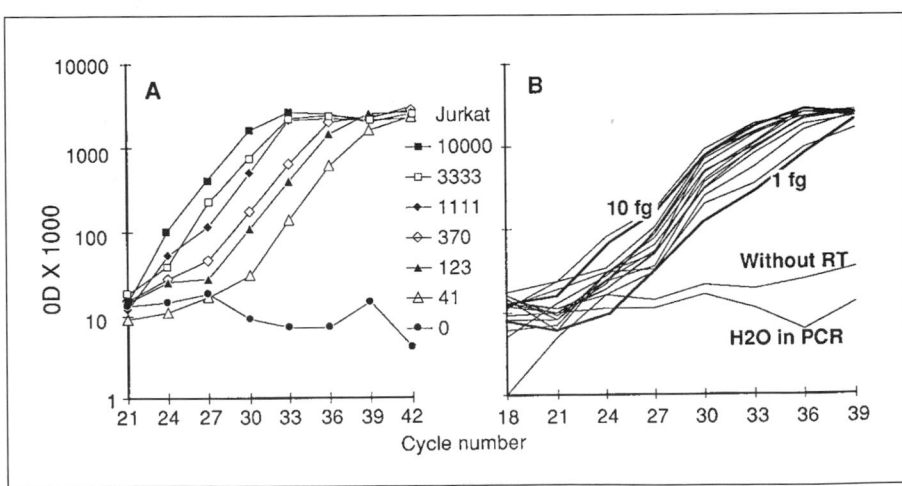

Figure 6. IL-2mRNA quantitation in activated Jurkat cells. A) Serial numbers of Jurkat cells (from 10000 to 41) were extracted, reverse-transcribed and amplified as described in the Materials and Methods section. B) Fourteen aliquots of 1000 Jurkat cells (simulated by PHA + PMA for 4 h) were extracted, reverse-transcribed and amplified as above.

It should be noted that 1 ag corresponds approximately to 4 molecules of the 4 genes studied (GAPDH, IL-2, IL-4 and IL-10) (Figure 4, A–D, respectively).

3) Quantitation of mRNA in purified RNA or after rapid extraction from cells. To quantify mRNA, we first tested the reproducibility of the whole process (RT and amplification) by reverse transcribing 10 aliquots (100 ng) of purified RNA from Jurkat cells that were activated for 4 h by phytohemaglutinin (PHA) and phorbol myristate acetate (PMA). One-twentieth of the 10 cDNA aliquots were amplified with IL-2 primers. As shown in Figure 5, the reproducibility of the RT and amplification steps was good (coefficient of variation [CV] = 6% at cycle 24). Comparison to the external DNA standard indicates that there were 31 fg of IL-2 mRNA out of 5 ng of total Jurkat RNA, which represents 83 IL-2 molecules per cell (1 µg of total RNA corresponds to 3×10^5 cells).

To quantify mRNA in a few cells (10 to 10 000) without RNA purification, we used the following protocol: the cells were treated with proteinase K and Tween 20, and the whole extract was reverse-transcribed. We investigated the expression of IL-2 mRNA in serial 1/3 dilutions of PMA + PHA-induced Jurkat cells. As shown in Figure 6A, according to the number of cells seeded, parallel and quite equidistant curves were obtained. However, it is apparent that the reproducibility of the whole process on cells was not as good as it was with purified RNA. As shown in Figure 6B, 14 aliquots of 1000 PMA + PHA-induced Jurkat cells were digested and reverse-transcribed separately and then 1/20 of each RT product was amplified with IL-2 primers. The CV of OD at 30 cycles was 39%. There were about 3 fg of IL-2 mRNA (221 bases long) per 50 Jurkat cells, which is equivalent to 220 copies per cell.

This variability could be related partly to the cellular clumps in PMA + PHA-stimulated cells, to uneven RNA degradation or to inhibition of RT. To increase reproducibility, it is necessary to normalize the lymphokine expression according to the amount of total RNA recovered. The recovery of RNA and the RT yield can be normalized by comparison with the expression of housekeeping genes. In our method, the genomic DNA is still present, and although primers are chosen to

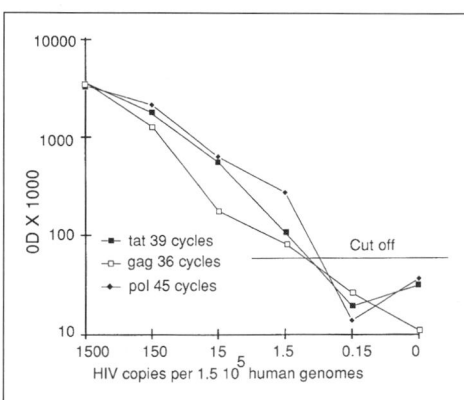

Figure 7. HIV sequence quantitation. Variable amounts of DNA extracted from the 8E5 cell line were mixed with a constant amount (1 µg corresponding to 1.5×10^5 copies) of human genomic DNA. Aliquots containing 1 µg of DNA were amplified separately with gag (16), pol (17) or tat (18) specific primers. The results corresponding to the optimum cycle number are shown (the first cycle with a positive signal for the lowest concentration without saturation for the highest).

encompass several introns, interferences can occur through pseudogenes. Indeed, most of the housekeeping genes [HPRT (19), actine (16) and GAPDH (8)] display many pseudogenes, which are very homologous to mRNA and thus prevent normalization. DNase treatment after cell digestion and before the RT step maintained the possibility of normalization (data not shown).

4) Detection of viral DNA. Figure 7 shows that this method is also suitable for quantitation of HIV DNAs. Serial dilutions of DNA obtained from the 8E5-LAV cell line were mixed with 1 µg of genomic human DNA. Thus, 1500 to 0.15 HIV copies were mixed with a constant amount of genomic DNA (150 000 copies) and separately amplified with gag, pol and tat primers. It is apparent that a quantitation could be done quite accurately and that the sensitivity was about one copy per 1.5×10^5 human genomes.

5) Use of internal standard (Figure 8). The method can also be used with internal standard in the exponential phase according to Wang et al. (21) as shown in the following reconstruction experiment. A standard was constructed by recombinant PCR (cf methods). Serial dilutions of previously amplified and purified IL-4 cDNA were co-amplified with 25 fg of this IL-4 standard. Aliquots were sampled at regular intervals as above. The microplates were duplicated and hybridized with either a cDNA (IL-4) or a standard (IL-2R) specific probe. The OD for cDNA or standard probe are shown with their ratio by a dashed line. If the standard helps to normalize any variation in the amplification rate, this latter curve should be straighter. At cycle 21 the use of internal standard did not increase the accuracy, but at cycle 24 it corrected for the slightly lower amplification seen in the 6.2-fg tube. It should also be noted that if one looks at the concentration where the signal is equivalent for standard and cDNA, it is analogous to a competition experiment. Thus, quantification can be done either in the exponential phase or by competition.

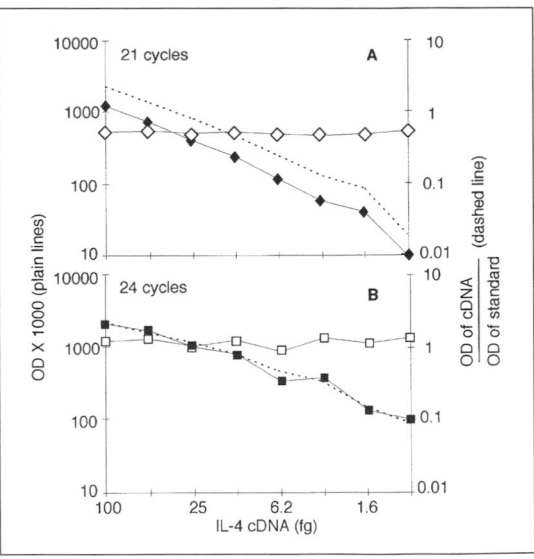

Figure 8. Internal standard reconstruction experiment. Serial dilutions of previously amplified and quantified IL-4 cDNA were co-amplified with 25 fg of IL-4 DNA standard. Beginning at cycle 18 and every 3 cycles thereafter, each reaction tube was sampled. Only cycle 21 (A) and 24 (B) are shown. The microplates were duplicated and each replicate was alkaline-denatured and hybridized with either a cDNA or a standard probe. On the left side the crude OD obtained with both probes are shown. On the right side the ratio of the 2 OD is shown., which normalizes for any variations in amplification efficiency.

DISCUSSION

The method described allows direct and quick digital results that reflect the amount of PCR products at different moments of amplification. Thus, the amount of products obtained could be quantified during the exponential phase and quantitative PCR performed easily and accurately. Our assay could be applied to the study of the expression of lymphokine mRNA after mitogenic stimulation of a few Jurkat cells and to the detection and the quantitation of as few as one HIV DNA copy per 1.5×10^5 DNA cell equivalents. Since there is a hybridization step in the detection method, one can also use any internal standard (DNA or RNA) based quantitation method in either exponential (21), competition (1,7) or end point, as recently described (2).

There are five sources of variability when measuring mRNA expression by quantitative PCR on a few cells: the cell numbers, the RNA extraction yield, the RNA degradation during the cell extraction, the RT efficiency and amplification rate. Our results demonstrate that reproducibility is excellent in the RT and amplification steps (CV = 6% in Figure 4). In the first steps, the reproducibility was rather good as seen by the parallel curves obtained with serial cell dilutions (Figure 6A) and the CV of 39% (Figure 6B). It should be stressed that controlling for cell numbers, RNA integrity and yield by measuring the expression of housekeeping gene mRNAs requires the elimination of genomic DNA before RT when RNA is not extracted and purified because of the presence of numerous homologous pseudogenes for the commonly used loci [HPRT (19), actine (16) and GAPDH (8)]. DNase treatment of the cell extract before RT allows the normalization of the results according to the amount of GAPDH mRNA present in each sample. However, the cell digestion protocol became quite cumbersome with the addition of this DNase step and was almost as complicated as the classical RNAzol technique, which is more reliable if one uses a carrier during the precipitation step and normalizes results to GAPDH (SD <25% of the mean for duplicate cell preparations; data not shown).

To increase the accuracy of quantitation during the amplification step, the use of internal standards has been proposed (i.e., DNA that could be amplified with the same primers but differs in size or in internal sequence) (1,2,7,21). Provided that the internal DNA standard has a nonhomologous sequence with the cDNA, our assay can be used with an internal standard either in the exponential phase or in competition, by hybridizing PCR products with either a standard or a cDNA-specific probe, after duplicating the microplates as shown in Figure 8. Since the reproducibility of the assay was already good, the use of an internal standard marginally increased the accuracy of the quantitation. The high sensitivity and the direct digitalization of the results also allow competitive (1,7) or end-point (2) PCR quantitation to be done in an easier way than with the densitometry of ethidium bromide colored gels or of Southern blots, or with cutting and counting gel slice radioactivity.

In contrast, if our methodology was reproducible enough to compare mRNA amounts precisely in various samples, absolute values were not obtained. The use of external (14) or internal (2,21) RNA standards is necessary if one wants to express the results in molecules per sample. According to the same principles described above for DNA standards, our assay can easily be modified to this end: T7 promoter can be adapted to the DNA standards seen above, and after in vitro transcription, known amounts of standard RNA could be distributed with the cells before digestion. This would take into account the variability in RNA degradation and RT yield. Thus, Wang's method (21) could be applied to a few cells without using radioisotopes or carrying out numerous and tedious hot electrophoreses. But, in our opinion, internal RNA standards are required only for viral RNA targets since RNA extraction and RT yield can be controlled in eukaryote cells by studying the expression of housekeeping genes.

Our technique compares favorably to others that study small amounts of mRNA. For instance, the Northern blot technique requires at least 10 µg of RNA per lane, which is sometimes very difficult to obtain. Moreover, it is at most a semiquantitative technique. By contrast, our assay was both sensitive and quantitative enough to study the lymphokine gene expression in a few hundred cells and to scale down to one copy of HIV DNA in 1.5×10^5 human genomes. Moreover, these results could be obtained quickly and easily without carrying out the numerous reactions of competitive PCR (1,7).

In conclusion, this method uses only commercially available reagents and readily available equipment. The method is also fast, not labor-intensive, inexpensive and does not use radio-isotopes. It can be adapted to any genes very quickly and could be automated without difficulty using robotic stations that are already available. By using an adapted thermal cycler, the handling could be minimized and throughput increased significantly. Clinical applications (viral load quantitation, expression of lymphokines or other genes in biopsy specimens, etc.) are currently being developed.

ACKNOWLEDGMENTS

We are grateful to Pr. C. Brechot (Institut Pasteur, Paris) who gave us the reconstituted HIV panel through the auspices of the ANRS (Agence Nationale de la Recherche sur le SIDA). M. Kress and C. Boucheix are thanked for helpful discussions. We are grateful to O. Acuto and M. Tosi for helpful criticisms and for reviewing an earlier version of this manuscript. This work was supported by grants from the CNAMTS, the Assistance Publique de Paris, the ARC and the Université de Paris-Sud. P.A. is supported by Sandoz France SA.

REFERENCES

1.**Becker-André, M. and K. Hahlbrock.** 1989. Absolute mRNA quantification using the polymerase chain reaction (PCR). A novel approach by a PCR aided transcript titration assay (PATTY). Nucleic Acids Res. *17*:9437-9446.

2. **Bouaboula, M., P. Legoud, B. Pessegué, B. Delpech, X. Dumont, M. Piechaczyk, P. Cassellas and D. Shire.** 1992. Standardization of mRNA titration using a polymerase chain reaction method involving co-amplification with multispecific internal control. J. Biol. Chem. *267*:21830-21838.

3. **Choi, Y., B. Kotzin, L. Herron, J. Callahan, P. Marrack and J. Kappler.** 1989. Interaction of Staphylococcus Aureus toxin "superantigen" with human T cells. Proc. Natl. Acad. Sci. USA *86*:8941-8945.

4. **Chomczynski, P. and N. Sacchi.** 1987. Single-step method of RNA isolation by acid guanidinium thiocyanate-phenol-chloroform extraction. Anal. Biochem. *162*:156-159.

5. **Elhers, S. and K.A. Smith.** 1991. Differentiation of T cell lymphokine gene expression: the in vitro acquisition of T cell memory. J. Exp. Med. *173*:25-36.

6. **Erlhich, H.A., D. Gelfand and J.J. Sninsky.** 1991. Recent advances in the polymerase chain reaction. Science *252*:1643-1651.

7. **Gilliland, G., S. Perrin, K. Blanchard and H.F. Bunn.** 1990. Analysis of cytokine mRNA and DNA: detection and quantitation by competitive polymerase chain reaction. Proc. Natl. Acad. Sci. USA *87*:2725-2729.

8. **Hanauer, A. and J.L. Mandel.** 1984. The glyceraldehyde 3 phosphate dehydrogenase gene family: structure of a human cDNA and of an X chromosome linked pseudogene; amazing complexity of the gene family in mouse. EMBO J. *11*:2627-2633.

9. **Holodny, M., D.A. Katzenstein, S. Sengupta, A.M. Wang, C. Casipit, D.H. Schwartz, M. Konrad, E. Groves and T. Merigan.** 1991. Detection and quantification of human immunodeficiency RNA in patient serum by use of the polymerase chain reaction. J. Infect. Dis. *163*:862-866.

10. **Kawasaki, E.S.** 1990. Sample preparation from blood, cells, and other fluids, p. 146-152. In M.A. Innis, D.H. Gelfand, J.J. Sninsky and T.J. White (Eds.), PCR Protocols. Academic Press, London.

11. **Laure, S., V. Courgnaud, C. Rouziou, S. Blanche, S. Veber, M. Burgard, C. Jacomet, S.C. Griscelli and C. Brechot.** 1988. Detection of HIV-1 DNA in infants and children by means of polymerase chain reaction. Lancet *iii*:538-541.

12. **Lewis, D.B., K.S. Prickett, A. Larsen, K. Grabstein, M. Weaver and C.B. Wilson.** 1988. Restricted production of interleukin 4 by activated human T cells. Proc. Natl. Acad. Sci. USA *85*:9743-9747.

13. **Mantero, G., A. Zonaro, A. Albertini, P. Bertolo and D. Primi.** 1991. DNA enzyme immunoassay: general method for detecting products of polymerase chain reaction. Clin. Chem. *37*:422-429.

14. **Michael, N.L., M. Vahey, D.S. Burke and R.R. Redfield.** 1992. Viral DNA and mRNA expression correlate with the stage of human immunodeficiency virus (HIV) Type 1 infection in humans: Evidence for viral replication in all stages of HIV disease. J. Virol. *66*:310-316.

15. **Meyerhans, A., R. Cheynier, J. Albert, M. Seth, S. Kwok, J. Sninsky, L. Morfeld-Manson, B. Asjo and S. Wain-Hobson.** 1989. Temporal fluctuations in HIV quasispecies in vivo are not reflected by sequential HIV isolations. Cell *58*:901-910.

16. **Moos, M. and D. Gallwitz.** 1982. Structure of human β-actin-related pseudogene which lacks intervening sequences. Nucleic Acids Res. *10*:7843-7849.

17. **Mullis, K.B.** 1991. The polymerase chain reaction in an anemic mode: how to avoid cold oligodeoxyribonuclear fusion. PCR Methods Appl. *1*:1-4.

18. **Ou, C.Y., S. Kwok, S.W. Mitchell, D.H. Mack, J.J. Sninsky, J.W. Krebs, P. Feorino, D. Warfield and G. Schoetman.** 1988. DNA amplification for direct detection of HIV-1 in DNA of peripheral blood mononuclear cells. Science *238*:295-297.

19. **Patel, P.I., R.L. Nussbaum, P.E. Framson, D.H. Ledbetter, C.T. Caskey and A.C. Chinault.** 1984. Organization of the HPRT gene and related sequences in the human genome. Somatic Cell Mol. Genet. *10*:483-493.

20. **Yamamura, M., K. Uyemura, R.J. Deans, K. Weinberg, T.H. Rea, B.R. Bloom and R.L. Modlin.** 1991. Defining protective responses to pathogens: cytokine profiles in leprosy lesions. Science *254*:277-279.

21. **Wang, A.M., M.V. Doyle and D.F. Mark.** 1989. Quantification of mRNA by the polymerase chain reaction. Proc. Natl. Acad. Sci. USA *86*:9717-9721.

Update to:

A Versatile ELISA-PCR Assay for mRNA Quantitation from a Few Cells

Oliver Lantz

Unité INSERM 267, Villejuif and Hopital de Bicêtre, Faculté de Médecine Paris-Sud, Le Kremlin-Bicêtre, France

Examples of Applications

The ELISA-PCR assay described in our original paper was based on the ELISA measurement of the amount of amplicons produced by the PCR at successive cycles. It has been used in several reports (2–5,7–9,10) dealing either with T cell repertoire analysis, lymphokine gene expression or microchimerism. Analysis at the single cell level has been also carried out (5). Analysis of low copy numbers in high complexity mixtures (eukaryote genomes) such as that carried out in microchimerism analysis required special precautions to achieve high sensitivity (3).

Modifications to the ELISA System

A few minor modifications have been brought to the ELISA system: herring sperm and sodium dodecyl sulfate (SDS) are added during the hybridization step to decrease the background encountered with certain probes; FITC-labeled probe made during the oligo-synthesis are used to avoid the proprietary, and thus more expensive, digoxigenin system. The main improvement has been the use of luminometry to reveal the assay; we have been using an alkaline phosphatase substrate, the dioxiethane derivative CSPD that has given us much better results than the peroxidase substrate luminol. The sensitivity is of the same order as that of colorimetry but the saturation threshold is much higher, giving a much wider dynamic range. We tried fluorometry but got a poor sensitivity with methylumbelliferyl phosphate substrate.

Based upon our method, a commercial kit has also been developed by Tropix (Bedford, MA, USA) to allow quantitation of amplicons using luminometry. Another company uses a slightly different ELISA format in which the amplicons are labeled during the PCR with digoxigenin-dUTP and captured afterward by hybridizing to a biotinylated oligo-probe, which in turn binds to avidin-coated microplates. High binding capacity avidin-coated microplates are required because an excess of capturing probe is needed. In this format, there is a 3 molecule reaction where the non-hybridized amplicon strand can compete with the capturing oligo-probe, which may decrease either the sensitivity or the dynamic range of the revelation step. Actually, the sensitivity seems to be of the same magnitude as that of our format but the cost is higher because of the high price of digoxigenin-dUTP and the dynamic range of the revelation step has not been reported.

Source of Variability in Analysis of Gene Expression by Quantitative PCR

In our experience, if a true hot start is used (most easily done with the anti-Taq antibody from CLONTECH and now with the thermally activated *Taq* DNA polymerase from Perkin-Elmer) the main source of variability for quantitative PCR of RNA sequences is not the reverse transcriptase (RT) or the PCR steps but the amount of starting material and the RNA extraction. Therefore, all the samples are tested in duplicates beginning at the RNA extraction step. For cellular genes, the amount of starting material and the efficiency of RNA extraction can be checked by quantifying a housekeeping gene such as GAPDH, actin or HPRT. However, for extracellular targets, such as viruses, one has to use an RNA or DNA (in case of DNA viruses) standard spiked in the samples at the time of the RNA (DNA) extraction to verify the extraction efficiency and RT yield.

Kinetic Quantitative PCR vs. End-Point Quantitative PCR with Internal Standard

There are several formats for performing quantitative PCR: kinetically without internal standard or at end-point using an internal standard. In this latter case, the standard can be homologous or heterologous and the measure of the amplicon amounts can be carried out during the exponential phase or at the plateau. ELISA with luminometry readings can be used to quantify the amount of amplicons in all of these methods and its wide dynamic range has allowed us to use the ELISA in the setting of a quantitative PCR method run to the plateau phase using homologous internal standards specially devised for ELISA revelation. Samples are divided into two parts, hybridized either with the target or the internal standard probe. The ratio of the signal is compared to a scale amplified in the same experiment.

The use of luminometry has allowed us to measure the amount of initial target over 4 to 5 log10 with only one dose of standard (Figure 1C). Indeed, as an application of our quantitative PCR assay, we quantitated IL-2 mRNA expression by stimulated peripheral blood lymphocyte cells (PBLs) of HIV-infected patients, before and during anti-retroviral bi-therapy treatment. Figure 1A displays the results of the signals obtained at cycle 42 with either the IL2 cDNA or the IS probe. In Figure 1B, the ratios of these two signals are shown. These ratios are converted into absolute values by linear regression using the standard curve generated in the same experiment by amplifying serial dilutions of purified IL2 amplicons together with IS (Figure 1C). To correct for variations in RNA extraction and RT yield the same procedure was carried out with GAPDH primers (1). Final results are expressed as a ratio IL2/GAPDH (1). Figure 1, D–F display the normalized IL2 results obtained in 3 patients.

The main drawback of using an internal standard is the decrease in dynamic range when compared with kinetic quantitative PCR (1,6). It is clear

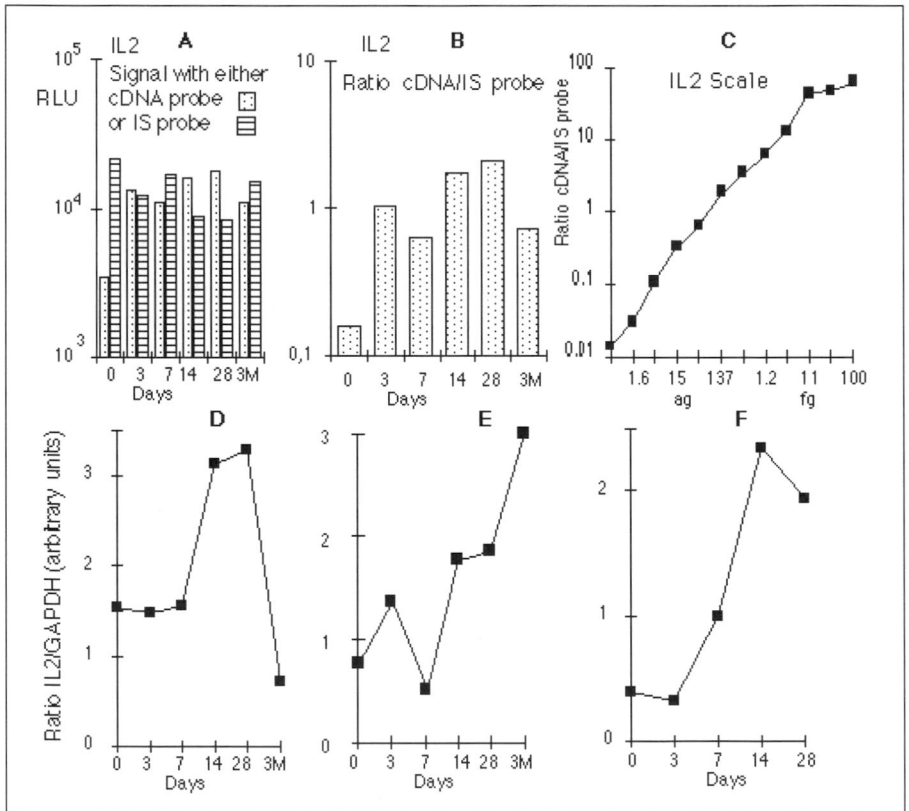

Figure 1. End-point quantitative PCR using internal standard and ELISA with luminometry readings to study IL-2 production in 3 patients undergoing anti-reverse transcriptase bi-therapy. PBLS harvested at the indicated time after beginning the treatment were stimulated with ionomycine and PMA. After 4 h, the RNA was extracted, reverse transcribed and the amount of cDNA for several lymphokines quantified by PCR with IS. Panel A displays the crude RLU data for IL-2 amplification with the cDNA and the IS probe for 1 patient. The ratio is shown in panel B and compared to the ratios of the scale shown in panel C. The same procedure is carried out for GAPDH and the normalized results are displayed in panels D–F for the 3 patients.

116

that kinetic quantitative PCR will always have the best dynamic range (1) and allows one to get meaningful results without previous knowledge of the target amount and without having to construct and validate any internal standard.

Choice of a Quantitative PCR Method

For the time being, kinetic quantitative PCR with ELISA and luminometry readings is probably the most versatile and inexpensive method available: there is only one PCR per sample and the cost of the ELISA assay is small compared to the cost of the PCR itself. This method is well adapted to research settings where one wants to quantitate the level of several mRNA in a limited number of samples and to be able to rapidly change the targets.

However, kinetitc quantitative ELISA-PCR may not be well adapted to routine hospital labs because of higher cost and labor. Here, the use of ELISA-PCR with internal standard is very attractive because it allows high throughput for a reasonable cost and amount of work. However, it requires devising and validating internal standards and fine tune-up of the system that may require many preliminary experiments. Luminometry readings are required if one studies targets varying over a wide quantitative range.

Finally, the question of which quantitative PCR method to use has been recently renewed because of the availability of the ABI PRISM 7700 apparatus that monitors directly the amount of amplicons produced, on the cycler itself every 5 s. This homogenous phase assay TaqMan™ is based on the 5′ to 3′ nuclease activity of *Taq* polymerase, which allows the release of a fluorescent reporter during the PCR. A probe labeled with both a reporter and a quencher dye is spiked into the PCR mix at the beginning of the reaction. The reporter fluorescence is quenched because of the spatial proximity between the two dyes. The probe anneals specifically to the amplicons during the annealing/extension phase and is cleaved when the *Taq* polymerase synthesizes the new DNA strand. The reporter dye fluorescence intensity increases with the amount of cleaved probe, which is proportional to the amount of amplicons. The 7700 apparatus monitors the amount of fluorescence generated in every well during the PCR on the thermal cycler.

In conclusion, the kinetic TaqMan as implemented in the ABI PRISM 7700 is certainly the best quantitative PCR method for the moment because it has the wide dynamic range and robustness of kinetic PCR and the advantages of homogenous phase assays. However, because of the high price of the ABI prism 7700, kinetic ELISA PCR remains still very attractive.

REFERENCES

1. **Alard, P., O. Lantz, M. Sebagh, C.F. Calvo, D. Weill, G. Chavanel, A. Senik and B. Charpentier.** 1993. A versatile ELISA-PCR assay for mRNA quantitation from a few cells. BioTechniques. *15*:730-737.

2. **Alard, P., J.A. Matriano, S. Socarras, M.A. Ortega and J.W. Streilein.** 1995. Detection of donor-derived cells by polymerase chain reaction in neonatally tolerant mice. Microchimerism fails to predict tolerance. Transplantation *60*:1125-1130.

3. **Bonney, E.A. and P. Matzinger.** 1997. The maternal immune system's interaction with circulating fetal cells. J. Immunol. *158*:40-47.

4. **Lantz, O., L.I. Sharara, F. Tilloy, Å. Andersson and J.P. DiSanto.** 1997. Intrathymic selection of IL-4 producing NK-T cells in the absence of NK1.1. J. Exp. Med. (In press).

5. **Lantz, O. and A. Bendelac.** 1994. An invariant T cell receptorα chain is used by a unique subset of MHC class I-specific CD4+ and CD4-8- T cells in mice and humans. J. Exp. Med. *180*:1097-1106.

6. **Lantz, O, E.A. Bonney, S. Umlauf and Y. Taoufik.** 1997. Kinetic ELISA-PCR: a versatile quantitative PCR method. *In* F. Ferré (Ed.), Gene Quantitation. Birkauser Boston, Cambridge (In press).

7. **Lenaour, F., C. Francastel, M. Prenant, O. Lantz, C. Boucheix and E. Rubinstein.** 1997. Early transcriptional activation of the CD9 gene during megakaryocytic differentiation of K562 cells. Leuk. Res.

8. **Suberville, S., A. Bellocq, B. Fouqueray, C. Philippe, O. Lantz, J. Perez and L. Baud.** 1996. Regulation of interleukin-10 production by β-adrenergic agonists. Eur. J. Immunol. *26*:2601-2605.

9. **Taoufik, Y., O. Lantz, C. Vallon, A. Charles, E. Dussaix and J.F. Dlefraissy.** HIV gp120 inhibits IL12 production by human monocytes: an indirect IL10 mediated efffect. Blood (In press).

10. **Umlauf, S., Beverly, B. Lantz, O. and R. Schwartz.** 1994. Regulation of IL-2 gene expression by CD28 costimulation in mouse T cell clones. Both nuclear and cytoplasmic RNA are regulated with complex kinetics. Mol. Cell. Biol. *15*:3197-3205.

CHAPTER 9

Video Image Analysis of Quantitative Competitive PCR Products: Comparison of Different Evaluation Methods

Heidi E. Chehadeh, Gerold Zerlauth and Josef W. Mannhalter

Immuno AG, Vienna, Austria

The use of quantitative competitive polymerase chain reaction (QC-PCR) to examine levels of gene transcripts has become very popular due to its speed and sensitivity. It is now possible to calculate the ratio between amplified target and competitor DNA directly on ethidium bromide-stained agarose gels using computer image analysis. Several different PCR conditions with internal standards have been published (2,4,5), but little attention has been paid to the different methods for determining copy numbers of the amplified PCR products. Most of the available software systems offer various procedures for scanning gel bands. In this study, we compared one-dimensional (1-D) vs. two-dimensional (2-D) scanning of PCR products using both direct gel scans and scans of the corresponding photographic images (Polaroid). We conclude that QC-PCR is sufficiently accurate to discern a twofold difference in mRNA levels, irrespective of the conditions used.

We used the THP-1 cell line and the constitutively expressed gene for glyceraldehyde 3-phosphate dehydrogenase (G3PDH) as a test system. G3PDH is often used as a control because its expression remains unchanged under most conditions (3,6). Total cellular RNA was prepared from 1×10^6 THP-1 cells as described (1) and reverse-transcribed into double-stranded cDNA using the cDNA synthesis kit (Boehringer Mannheim, Mannheim, Germany) and oligo(dT)$_{15}$ primer. The resulting cDNA was dissolved in 10 µL of double-distilled water per µg of RNA. Three microliters (equivalent of 0.3 µg RNA) of this reverse transcript, as well as 3 µL of a 1:2, a 1:5 and a 1:10 dilution, were co-amplified with competitor DNA (Catalog No. 5880-1, G3PDH Mimic; CLONTECH Laboratories, Palo Alto, CA, USA) using serial 1:2 dilutions ranging from 1.2×10^7 to 7.5×10^5 copies. These PCRs were performed on three different occasions (runs 1–3).

Reaction conditions were based on the recommendations of the reagent supplier (Perkin-Elmer, Norwalk, CT, USA), except that each reaction was performed in a total volume of 50 µL (5). A typical reaction contained 2 µL

(Reprinted from BioTechniques 18:26-28, 1995)

119

Table 1. Comparison of the Copy Numbers of the Target cDNA Generated by a 1-D or 2-D Analysis of the PCR Products

	1-D	2-D	*P*
undiluted	9.40 ± 2.12	8.63 ± 1.82	≥0.2
1:2	5.34 ± 1.31	3.89 ± 0.24	≥0.2
1:5	2.14 ± 0.03	1.95 ± 0.20	≥0.2
1:10	1.09 ± 0.12	0.95 ± 0.083	≥0.2

Values given are mean copy numbers $\times 10^6$ ± standard deviation ($n = 3$).
P is the significance of differences.

Table 2. Comparison of Various Dilution Factors Determined by 1-D and 2-D Analysis

Dilution Factor of cDNA		
	Found	
Calculated	1-D Analysis	2-D Analysis
2	1.8 ± 0.04	2.2 ± 0.44
5	4.8 ± 1.0	4.9 ± 1.3
10	9.4 ± 1.1	9.9 ± 0.85

Values given are mean dilution factors of cDNA ± standard deviation ($n = 3$).
The significance of differences (*P*) in all cases is ≥0.2 as calculated by the
Wilcoxon-Mann-Whitney test.

of appropriate competitor dilution, 3 µL of test sample and 45 µL of reaction cocktail consisting of 30.75 µL of distilled water, 5 µL of 10× reaction buffer, 4 µL of deoxyribonucleoside triphosphates (dNTP) (each 200 µM), 2.5 µL of each primer and 0.25 µL of *Taq* DNA polymerase (1.25 U). The primers used were taken from the Human G3PDH Control Amplimer Set (Catalog No. 5406-1; CLONTECH) and have the following sequences:

5′ primer: 5′TGAAGGTCGGAGTCAACGGATTTGGT 3′
3′ primer: 5′CATGTGGGCCATGAGGTCCACCAC 3′

The amplified G3PDH fragment is 983-bp long, and the size of the competitor DNA product is 630 bp.

Reactions were incubated in a thermal cycler (Model No. 60/2; Biomed, Theres, Germany) for 35 cycles (denaturation: 1 min at 94°C, annealing: 2 min at 60°C; extension: 3 min at 72°C). Twenty percent of each reaction mixture was run on a 2% NuSieve® 3:1 agarose gel (FMC BioProducts, Rockland, ME, USA) in TBE buffer (0.5 M Tris, 0.05 M boric acid, 1 mM EDTA). Ethidium bromide-stained DNA bands were visualized under UV light, and a digital image was produced by a video camera interfaced to a computer video image analyzer system (ATS 800; Hirschmann, Taufkirchen,

Table 3. Comparison of the Direct Agarose Scans and of Photographic Image Scans Using 2-D Scanning

Dilution Factors	Polaroid Photo	Agarose Gel
Undiluted	9.40 ± 2.12	9.81 ± 2.17
2	5.34 ± 1.31	5.13 ± 0.59
5	2.14 ± 0.03	2.17 ± 0.11
10	1.09 ± 0.12	0.97 ± 0.014

Values given are mean copy numbers $\times 10^6$ of three PCR runs ($n = 3$) ± standard deviation.
The signification of differences (P) is ≥0.2 in all cases.

Germany). Since the comparison and equivalence point determination in the QC-PCR technique are based on molar ratios, the fluorescence associated with the 630-bp competitor product was corrected by multiplication with a factor corresponding to the base pair ratio of both fragments.

For determination of target copy numbers, the $\log_{10}$ of the ratio of the fluorescence intensities of the competitor band and the target band was plotted as a function of the $\log_{10}$ of the number of competitor molecules added. Graphically, this transformation yields a linear plot. Interpolation on the regression for a Y value of 0 ($\log_{10} 1 = 0$) gives the number of copies corresponding to the number of target template in the sample (4). The significance of differences (P) between 1-D and 2-D systems was calculated by the Wilcoxon-Mann-Whitney test.

The Hirschmann software for analysis of ethidium bromide-stained gels allows the choice between 1-D scanning (scan follows the flow direction of the gel through the bands) and 2-D scanning (measurement of the whole band area).

The copy numbers of the target cDNA and several dilutions thereof, calculated by scanning with the 1-D as well as the 2-D procedure, are shown in Table 1. There were no significant differences in the copy numbers found. Moreover, there was no significant difference in the copy numbers (standard deviation below 3%), regardless of whether we led only one line through the bands or scanned the whole band width for the 1-D system. Band "smiling" or band-edge "tailing" did not yield different results, because the ratio between target and competitor remained unchanged in these cases (data not shown).

As a THP-1 cell lysate contains an unknown number of G3PDH mRNA copies, we used twofold dilutions of the cDNA in a semiquantitative system to compare 1-D and 2-D analysis. The relative amounts of the 1:2, 1:5 and 1:10 dilutions of the cDNA sample are given in Table 2. No significant difference between the dilution factors calculated by 1-D and 2-D analysis was detected. When three QC-PCR runs of the same cDNA sample were compared, the standard deviations ranged from 1%–25% for 1-D and 6%–21% for the 2-D system.

The three PCR runs were evaluated directly from the gel as well as from photographic images using the 2-D system. The film used was Polaroid Type 667 (Cambridge, MA, USA), coaterless black and white; contrast, medium; resolution, 12–14 line pairs/mm; image size, 7.3 × 9.5 cm, under the following conditions: exposure time, 1/4 s and processing time, 45 s at 21°C. Results are listed in Table 3. There was no significant difference between the results obtained with the Polaroid photo and with the agarose gel.

The results of our study show that QC-PCR can be analyzed by computer video image analysis using 1-D as well as 2-D scanning methods without significant differences in the results. Analysis of direct scans does not differ significantly from analysis of photographic images. Our results show that template copy numbers can be determined with an accuracy of 7%–25%. For semiquantitative determinations, e.g., the comparison of induced and untreated cells, the variability of the relationships also remained at 2%–26%. Twofold differences were measurable, and the copy numbers obtained were not dependent on the method of evaluation.

REFERENCES

1. **Chomczynski, P. and N. Sacchi.** 1987. Single-step method of RNA isolation by acid guanidinium thiocyanate-phenol-chloroform extraction. Anal. Biochem. *162*:156-159.
2. **Diviacco, S., P. Norio, L. Sentilin, S. Menzo, M. Clementi, G. Biamonti, S. Riva, A. Falaschi and M. Giacca.** 1992. A novel procedure for quantitative PCR by coamplification of competitive templates. Gene *122*:313-320.
3. **Edwards, D.R. and D.T. Denhardt.** 1985. A study of mitochondrial and nuclear transcription with cloned cDNA probes. Changes in relative abundance of mitochondrial transcripts after stimulation of quiescent mouse fibroblasts. Exp. Cell. Res. *157*:127-143.
4. **Piatak, M., K.-C. Luk, B. Williams and J. Lifson.** 1993. Quantitative competitive polymerase chain reaction for accurate quantitation of HIV DNA and RNA species. BioTechniques *14*:70-81.
5. **Siebert, P.D. and J.W. Larrick.** 1993. PCR MIMICS: Competitive DNA fragments for use as internal standards in quantitative PCR. BioTechniques *14*:244-249.
6. **Zentella, A., F.M. Weis, D.A. Ralph, M. Laiho and J. Massague.** 1991. Early gene responses to transforming growth factor-beta in cells lacking growth-suppressive RB function. Mol. Cell. Biol. *11*:4952-4958.

Update to:

Video Image Analysis of Quantitative Competitive PCR Products: Comparison of Different Evaluation Methods

Heidi E. Chehadeh, Gerold Zerlauth and Josef W. Mannhalter
Immuno AG, Vienna, Austria

Since the publication of our paper, gel electrophoresis and video image analysis have continued to be widely used for separation and quantitation of amplified PCR products. The advantage of this method is that it takes less time and does not require additional manipulations, which could be a source of inaccuracy in quantitation. However, other methods that are more automatizable have also been reported on (reviewed in Reference 1, "Detection and Analysis of PCR Products").

REFERENCE

1. **Zimmermann, K. and J.W. Mannhalter.** 1996. Technical aspects of quantitative competitive PCR. BioTechniques *21*:268-279.

CHAPTER 10

Rapid Quantification of Gene Expression by Competitive RT-PCR and Ion-Pair Reversed-Phase HPLC

Amanda Hayward-Lester, Peter J. Oefner[1] and Peter A. Doris

Texas Tech University Health Sciences Center, Lubbock, TX, and [1]Stanford University, Stanford, CA, USA

ABSTRACT

Competitive reverse transcription PCR (RT-PCR) techniques for quantification of gene expression employ titrations in which the products of multiple PCRs must be separated, analyzed and quantified to compute gene expression in a single sample. We have employed a novel, ion-pair reversed-phase HPLC (IP-RP-HPLC) system to analyze and quantify RT-PCRs performed with mutant RNA internal standards. PCR products could be separated and quantified in 6 minutes per reaction using the absorbance signal from an on-line UV detector. Crude PCR products can be analyzed without further processing and without the addition of radioactive or fluorescent markers to reactions. Analysis of titration regression and slope values approached mathematical ideals, indicating that amplification of native and competitor RNA occurred with equal efficiency. Further, serial dilution of input RNA over three orders of magnitude did not affect the calculated level of gene expression or the slope of the titration. IP-RP-HPLC appears to offer important advantages to quantitative measurements of gene expression. These include rapid sample analysis and column re-equilibration, reduced sample handling and opportunity for introduction of quantification error, avoidance of fluorescent or radioactive tracers, high detector sensitivity and linearity and excellent quantitative reliability.

INTRODUCTION

The polymerase chain reaction (PCR), in conjunction with reverse transcriptase, has provided a convenient and highly sensitive method for examining gene expression. In many settings, accurate quantitative information concerning the level of gene expression is desirable (7,15,22). Modifications of

(Reprinted from BioTechniques 20:250-257, 1996)

the reverse-transcription PCR (RT-PCR) method to accomplish this goal have been introduced. Becker-André and Hahlbrock were the first to propose the use of competition assays employing a competitor mutant RNA of the gene of interest (1). In this system, the RNA homolog shares the same sequence that is recognized by the reaction primers but is modified either to alter the existence of a restriction site or to alter the length of the sequence intervening between the primer binding sites. Mutant internal standard RNA is added in a range of known quantities to multiple reactions each containing uniform amounts of the RNA preparation to be quantified. Gene expression is estimated by observing the relative amount of native and homolog products resulting from RT-PCR. Numerous applications of this approach in varying modifications have now been published (6,23).

As this technique gains wider application, there is a growing need to develop methods which provide accurate quantitative analysis of competitive RT-PCRs but which limit the labor-intensive analysis of the products of multiple titration assays. Current methods employed to quantify the products of competitive RT-PCRs center on gel electrophoresis, though some exceptions exist (12). After separation of the reaction products, quantification is based on the use of radiolabeled or fluorescent PCR components. Reaction products have been measured by excising the bands and counting the radioactivity contained in them (15) or performing phosphor imager analysis of the resolved products (5). Some groups have hybridized labeled DNA probes to blots made from gels resolving PCR products (1), while others have employed fluorescent primers in the PCR and measured fluorescence of the resulting bands (14). Each of these approaches suffers the disadvantage that two steps are necessary to measure the reaction products. Each can be time-consuming and each introduces the possibility of errors that may impair accuracy. The ideal approach would reduce post-reaction analysis to a single step, eliminate the time-consuming activities of gel preparation, loading, running and analysis, eliminate the need to use radioactive components, lend itself to convenient automation and produce accurate and reproducible measurements of gene expression.

Several reports have appeared describing the application of HPLC to the analysis of PCR products. These reports have employed nonporous resin-based, anion-exchange column packings that have been adapted for rapid analysis but require re-equilibration intervals between runs (3,4,13,24). Recently, IP-RP-HPLC on alkylated nonporous poly(styrene-divinylbenzene) particles has been shown to be well suited to rapid, automated quantitative analysis of double-stranded DNA (dsDNA) fragments in the pico- to femto-mole range by on-line UV absorbance (9) or fluorescence (16) detection, respectively, with calibration curves exhibiting linearity over three orders of magnitude. Building on these features, as well as on the size-dependence of elution (10), the applicability of IP-RP-HPLC to the quantitation of competitive RT-PCR products is demonstrated.

MATERIALS AND METHODS

Production of Competitive Mutant RNA

Rat sodium, potassium-ATPase alpha 1 isoform (NKAα1) cDNA was generously provided by Dr. Jerry Lingrel (University of Cincinnati). The partial cDNA was provided in a pUC18 construct. It was removed from this construct by digestion with *Eco*RI and *Hin*dIII. The NKAα1 partial cDNA was isolated by agarose gel electrophoresis of the restriction digest products, cut from the agarose gel and ligated into the multiple cloning site of pGEM®-4Z (Promega, Madison, WI, USA).

Rat angiotensinogen cDNA (a gift of Dr. Kevin Lynch, University of Virginia) was digested with *Ppu*M1, the resulting fragments were separated by agarose slab gel electrophoresis and the 145-bp fragment was excised. The pGEM-4Z construct was digested with *Ppu*M1, and the 145-bp angiotensinogen cDNA fragment was ligated to produce a construct in which the NKAα1 cDNA was interrupted by the presence of 145 bp of angiotensinogen cDNA. Synthetic RNA (541 bases) was synthesized from the T7 polymerase promoter (T7 kit; Novagen, Madison, WI, USA) contained in this construct after digestion with *Eco*RI. Synthetic sense RNA was recovered by isopropanol precipitation at room temperature, washed and then quantified by spectrophotometry. The recovered RNA was checked for purity and quality on a denaturing agarose (MetaPhor®; FMC BioProducts, Rockland, ME, USA) gel and stored at -70°C in diethylpyrocarbonate (DEPC)-treated water.

Primer Design

We employed the Primer 0.5 program generously provided by the Lander laboratory, Whitehead Institute for Biomedical Research, MIT, to design primers for PCR amplification. The primers selected were designed to avoid dimerization and to span an intron so as to distinguish amplification of cDNA from any genomic DNA that may have contaminated samples. The forward primer sequence was an 18-mer:
5'- CCCTAGTTCCCGCCTCTC, the reverse primer was a 21-mer: 5'- TG-GTCGTCCATAGACACTTCC. These primers yield PCR products of 245 bp (native) and 390 bp (insertion mutant). Primers were synthesized by the Texas Tech University Biotechnology Core Facility and were checked for homogeneity by HPLC.

RNA Preparation

Total RNA was prepared from whole rat blood by the addition of 100–200 μL of blood to 800 μL RNAzol™ B (Biotecx Laboratories, Houston, TX, USA; currently available from Tel-Test, Friendswood, TX, USA). After extraction by the manufacturer's recommended procedure, the integrity of the

RNA was checked by denaturing gel electrophoresis, dissolved in DEPC-treated water and quantitated by spectrophotometry; the remaining RNA was stored frozen at -70°C.

PCR Conditions

Competitor mutant RNA was diluted in a solution of 20 µg/mL yeast total RNA from a working stock solution of competitor RNA (100 ng/µL). Yeast total RNA was added as carrier to reduce nonspecific binding of mutant RNA to the dilution tubes. The range of dilutions of competitor that were added to the reverse transcription reactions was 1, 5, 10, 20, 50 and 100 fg/µL. One microliter of these dilutions was added to approximately 0.1 µg of blood total RNA. The mixture was reverse-transcribed in a 10-µL reaction volume using Moloney murine leukemia virus (MMLV) reverse transcriptase and random hexamer primers (Perkin-Elmer, Norwalk, CT, USA) The RT reaction was performed at 42°C for 25 min, followed by 5 min at 99°C and then cooled rapidly to 5°C. The RT reaction was overlaid with Chill-out 14 wax™ (MJ Research, Watertown, MA, USA), and incubations were performed in 200 µL tubes in a Peltier-effect Thermal Cycler (MiniCycler™; MJ Research).

PCRs were initiated by the addition of 40 µL of PCR master mixture containing AmpliTaq® DNA polymerase (Perkin-Elmer) at a final concentration of 1 U/50 µL and 0.3 µM NKAα1 forward and reverse primers. The reaction cycle sequence comprised 2 min at 95°C, followed by 36 cycles comprising 45 s at 94°C, 45 s at 56°C and 1 min at 72°C, with a final extension for 5 min at 72°C. Samples were then rapidly cooled and held at 0°C before removal for storage at -20°C.

IP-RP-HPLC of PCR Products

The HPLC system was comprised of the binary Rainin Rabbit HPLC pumps, a MacRabbit™ system controller software, a 1.2-mL volume dynamic solvent mixer, column oven (50°C), a LDC-Milton Roy Spectromonitor 3000 variable wavelength UV detector set at 256 nm (all obtained from Rainin, Woburn, MA, USA), a pre-column filter (0.5-µm PEEK; Upchurch Scientific, Oak Harbor, WA, USA) and a Hewlett-Packard 3390A calculating integrator (Hewlett-Packard, Kennett Square, PA, USA). Solvent A was 0.1 M triethylammonium acetate (TEAA), pH 7.0, and solvent B was 25% acetonitrile in 0.1 M TEAA, pH 7.0. A gradient profile was used for elution which started with 39% solvent B, which was increased linearly to 60% solvent B over 3 min and was then followed by a linear increase to 67% solvent B over 4 min. The profile was completed by a further linear gradient to 85% solvent B over 1 min, 1 min further at 85% solvent B and then a return to the starting conditions over an additional 1 min. This gradient can be modified for products of specific known sizes to reduce elution time but retain separation of products. Flow rate was 1 mL/min.

The column used for analysis (50× 4.6 mm i.d.; Sarasep, Santa Clara, CA,

USA) was packed with nonporous alkylated poly(styrene-divinylbenzene) particles prepared according to previously published protocols (2). Sample injection volume was 10 µL. All samples were injected without further processing.

RESULTS

Nonporous alkylated poly(styrene-divinylbenzene) packing materials are very well suited for analysis of dsDNA molecules. Rapid separation and high resolution of analytes were obtained. Further information on system optimization and size dependence of elution times is provided elsewhere (10, 11,16).

The competitive PCR products have been resolved by agarose gel electrophoresis and show the separation of increasing amounts of competitor product as the amount of competitor input is increased (Figure 1). Corresponding to the increase in competitor product, the native product is reduced. Accurate quantification of the amounts of these products is necessary when using this titration to calculate the amount of native NKAα1 RNA in the sample.

Figure 2 shows chromatograms obtained from PCRs whose products were previously resolved (Figure 1) by agarose electrophoresis. The chromatograms indicate a clear separation of reaction products. Increases in the ratio of competitor to native products are reflected in the chromatograms and can be readily quantified by on-line integration of the area of peaks detected in the absorbance signal (256 nm) from a UV detector attached to the column outlet.

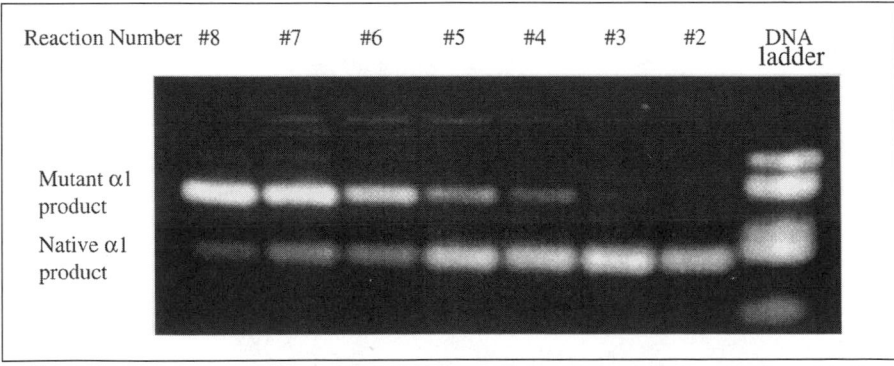

Figure 1. Products from competitive RT-PCR to quantitate sodium, potassium-ATPase α1 isoform gene expression in dissected kidney tissue, analyzed by agarose gel electrophoresis. From left to right, the first 7 lanes represent 15 µL of products generated in sequential 50-µL reactions containing equal starting amounts of native RNA and decreasing amounts of competitor mutant RNA. Amounts of mutant competitor RNA added to each reaction (#2–#8) were 500 ag, 1 fg, 5 fg, 10 fg, 50 fg, 100 fg and 200 fg, respectively. The native product is 245 bp and the mutant product is 390 bp in length. The right lane contains size markers, consisting of pUC18 DNA digested with *Hae*III (visible band sizes are 587, 458/433, 298, 267/257 and 174 bp).

Competitive RT-PCR using homologous RNA standards may lead to the production of heteroduplexes formed between native and mutant products which share sequences that are partially complementary (8). Methods to separate and analyze products of these reactions must take into consideration the possibility of heteroduplex formation and account for such products in the subsequent calculations. We noted that some reactions produced an additional product peak in our chromatograms. To determine the likelihood that this product reflected heteroduplex formation, we digested reactions with S1 nuclease. When the additional reaction product was present, S1 nuclease treatment resulted in its selective removal. This finding is compatible with a heteroduplex in which S1 nuclease digests the single-stranded loop formed by the unmatched competitor insert and subsequently digests the remaining nicked double-stranded product.

To confirm the identity of the third peak, native and competitor RNA were reverse-transcribed and amplified in separate reactions and then mixed together. IP-RP-HPLC analysis of this mixture revealed only two peaks (Figure 3). Heating the mixture to 97°C for 3 min and cooling to 4°C led to the

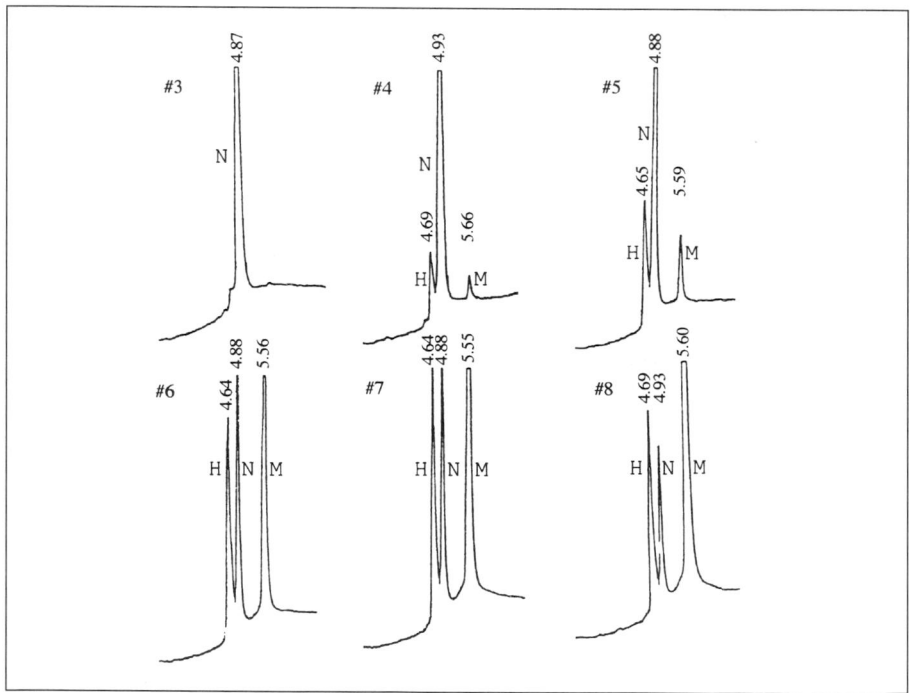

Figure 2. Products from competitive RT-PCR to quantitate sodium, potassium-ATPase α1 isoform gene expression in dissected kidney tissue, analyzed by IP-RP-HPLC. Chromatograms #3–#8 represent 10 μL of products generated in sequential 50-μL reactions containing equal starting amounts of native RNA and increasing amounts of competitor mutant RNA. Reactions are the same reactions analyzed by gel electrophoresis in Figure 1. Peaks represent UV absorbance at 256 nm of the following products: N denotes native product, M mutant product and H heteroduplex products, elution times are indicated.

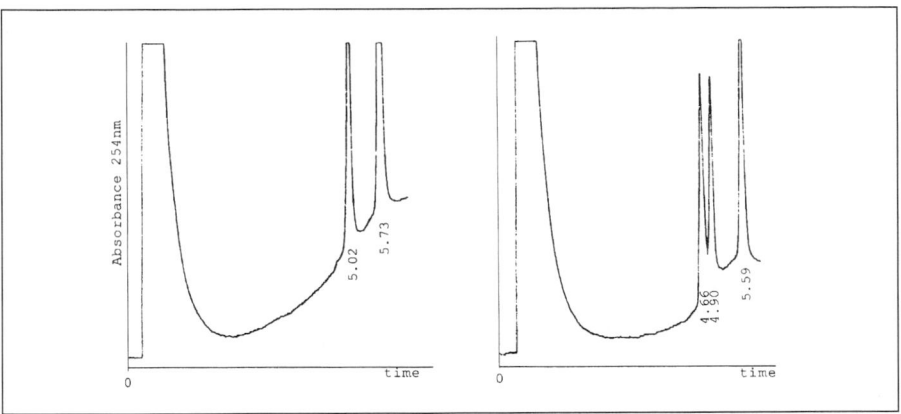

Figure 3. Chromatograms generated by IP-RP-HPLC analysis of RT-PCR products illustrating the formation of the heteroduplex products. The left-hand chromatogram was produced on IP-RP-HPLC analysis of a 10-μL aliquot of a mixture of products from two separate RT-PCRs, one generating only native product, the other only mutant product. The right-hand chromatogram was produced by analysis of a 10-μL aliquot of the same mixture, heated to 97°C then cooled to 4°C, resulting in the appearance of a third peak representing the formation of heteroduplex products. Elution times of the products are marked on the chromatograms.

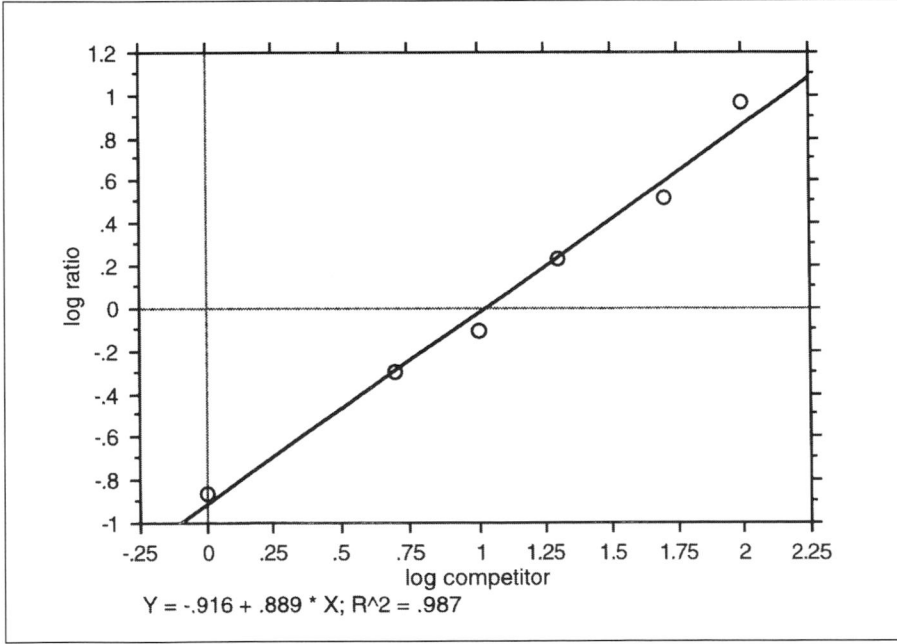

$$Y = -.916 + .889 * X; R^2 = .987$$

Figure 4. Regression plot generated by a representative competitive RT-PCR titration reaction to quantitate sodium, potassium-ATPase α1 isoform gene expression in rat blood. For each reaction tube, log ratio of native to mutant product amounts generated was plotted against the log of the corresponding initial amount of competitor RNA added. The initial amount of total blood RNA added to each reaction tube was 96 ng; the competitor input levels were 1, 5, 10, 20, 50 and 100 fg. The initial amount of native RNA in the blood sample was determined from the regression at the point where the log ratio = 0. This corresponds to an expression level of 112 fg of mutant RNA input/μg blood total RNA.

131

appearance of a new product. The retention time of the third product corresponded to that of the unaccounted peak in the chromatograms of the initial set of RT-PCRs (Figure 2). There was a concomitant decrease in the peak area of the native and competitor products. Further studies indicated that the amount of the heteroduplex formed was influenced by the rate of cooling from 96°C. Similarly, reduction in salt concentration reduced the amount of heteroduplex formed but did not prevent its formation.

The presence of heteroduplex formation is a consequence of similarity between the competitor and native products. However, such similarity is essential for the efficiency of amplification of native and competitor products to be equal. Amplification by PCR of both products using the same primers with identical efficiency is a requirement to meet the mathematical constraints of absolute quantification of gene expression (18). Quantification was performed by calculating the log of the ratio of native to competitor products in reactions that varied in the initial amount of competitor. A plot of the log ratio of products against the log of the initial amount of competitor should generate a straight line with a slope of 1. We calculated the slope and regression coefficients resulting from quantitative titrations of NKAα1 isoform message from RNA extracted from whole blood. Blood was chosen to examine the robustness of the system because of the low levels of expression that is limited to nucleated cells. We obtained a mean (± standard error of the mean) coeffi-

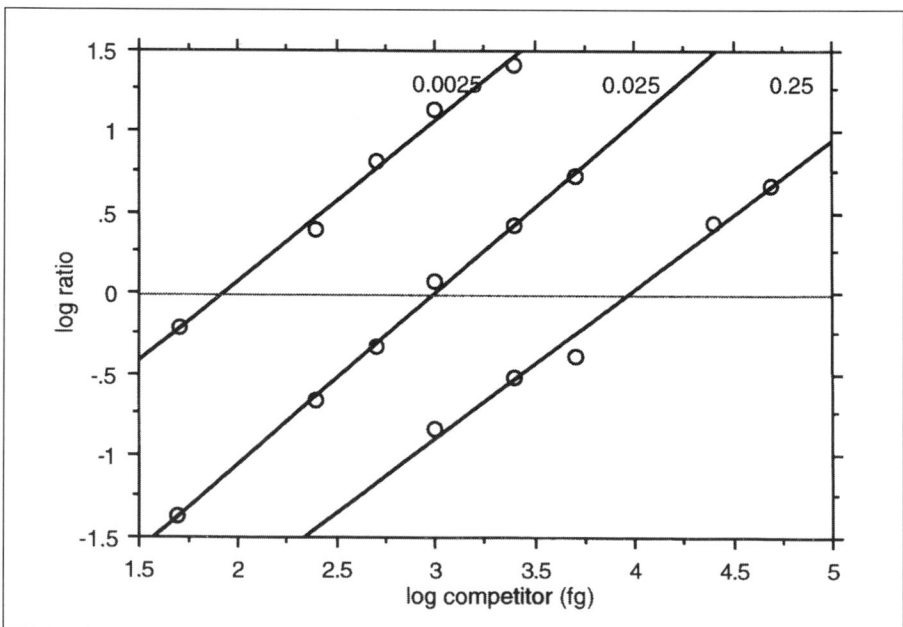

Figure 5. Regression plots from 3 RT-PCR titration reactions with different starting levels of rat brain total RNA. From left to right, plots generated by reactions initially containing 0.0025, 0.025 and 0.25 μg brain total RNA are shown superimposed to illustrate their relationship.

cient of determination (R^2) of 0.986 ± 0.007 and a slope value of 0.88 ± 0.023 ($n = 5$). Figure 4 shows a plot of the relationship between the ratio of reaction products and the amount of competitor added per reaction in one randomly selected titration.

We have also examined the effect of altering the amount of native RNA input on the calculated level of gene expression. This was performed using RNA from tissue with abundant NKAα1 expression (brain). Titrations against competitor were performed at three starting levels of native RNA obtained by diluting 1/10 and 1/100. Figure 5 shows the regression plots obtained at each of these input levels. The calculated amounts of gene expression determined (at each dilution) were 37.5 pg/mg brain RNA (0.25 μg brain RNA/reaction), 39.2 pg/μg brain RNA (0.025 μg RNA) and 33.9 pg/μg brain RNA (0.0025 μg RNA), respectively.

DISCUSSION

Competitive titration assays for the quantification of gene expression by RT-PCR result in the generation of multiple samples because of the necessity to examine the amplification of native mRNA at several levels of competitor RNA input. Traditional methods for the analysis of these samples have emphasized slab gel electrophoresis that requires the preparation, loading, running and subsequent analysis of gels. This task is labor-intensive and time-consuming. Additionally, it introduces new opportunities to enter error into a quantitative system. Furthermore, to detect reaction products, DNA must be stained with potentially toxic dyes, or reaction products must be labeled or hybridized to labels so as to visualize them by fluorescence, electrochemiluminescence, autoradiography or phosphor imaging. HPLC approaches offer the possibility to avoid these time-consuming and labor-intensive activities and also to improve accuracy of quantification by reduction of sample handling. In the system reported in the present study, PCR products can be analyzed directly by IP-RP-HPLC without further sample handling. Quantification of reaction products using on-line UV absorbance detection provides accurate information concerning the amount of reaction products present in each peak. Our analyses typically employ only 10 μL of a 50-μL reaction volume, permitting replication of analysis. To perform these analyses with HPLC equipment is far more cost-effective than fluorescence or isotopic detection and quantitation systems that are now available for gel analysis. Reduction in the overall dead volume of HPLC and optimization of the elution profile for products within a defined size range will permit the analysis of up to 30 samples per hour with little operator involvement beyond loading samples into an autosampler tray (9). This effectively minimizes an inherent limitation of HPLC, which is the necessity of sequential sample analysis.

Sensitivity of the system is illustrated by the ability to quantify expression in small volumes of whole blood in which input abundance is low (Figure 4). We have not attempted to quantitate very low abundance inputs because as input copy number declines, the biological relevance of differences between samples is progressively reduced. The quantification of gene expression in dissected renal tissue used total RNA extracted from 0.5 mm of isolated nephron tissue in each reaction (Figures 2 and 3). This indicates the suitability of the method to determine quantitative expression differences using RNA from very small tissue samples.

Additional advantages emerge from the application of this system to competitive RT-PCRs that use partially homologous internal competitor RNA. Partially homologous competitors are selected for RT-PCR quantification because they can be amplified by the same primers that amplify the native gene product. However, the similarity between competitor and native sequences can result in products that have potential for heteroduplex formation (8). This occurrence has been considered a disadvantage to the accuracy of the quantitation system; however, this exists only when the analytical method fails to accurately detect and quantify heteroduplexes.

The potential for heteroduplex formation leads to a design decision whether to generate competitor RNA which differs in size from the native signal or which differs in the presence of a selected restriction site, but not in size. We have chosen to generate size variation because restriction mutants require an additional step in PCR product analysis. Furthermore, restriction enzymes are unable to digest heteroduplexes that contain recognition sequence for the restriction enzyme on only one strand. Such undigested material would not be distinguishable based on size and would therefore be mistakenly added to the balance of the undigested reaction product used to calculate the final product ratio. This problem may be reduced if heteroduplexes form with almost the same efficiency as homoduplexes in such systems. In such a system, De Kant and colleagues have shown that analysis of only the digested homoduplex product can yield reliable quantification of PCR inputs if appropriate mathematical analysis of the proportion of digestible products is performed (4). In such a system, it is also necessary to verify that restriction digestion is complete for each batch of samples.

Our experience with heteroduplex formation in the present study indicates that heteroduplexes readily formed even though our reaction products differed in size by 145 bp and shared only 63% homology. Formation of heteroduplexes was apparent through the appearance of an additional reaction product in IP-RP-HPLC chromatograms. This product eluted with a retention time less than the 245-bp homoduplex native product. Since retention behavior is determined by the ability of dsDNA to bind the ion-pair reagent, such retention is consistent with a product containing somewhat less dsDNA than the native 245-bp product. Formation of a heteroduplex between a 390-bp competitor strand and the 245-bp native strand permits matching of a maximum of 245 bp with a reduction in total base pairing around the location of the unmatched strand. The migration of such a heteroduplex under non-denaturing gel electrophoresis conditions is unpredictable due to the effect on mobility of the flexible single-stranded portion of the duplex and quantification of heteroduplex using dye staining of the products appears to be unreliable (20). We confirmed the identity of our heteroduplex by experiments involving melting and reannealing of a mixture of pure homoduplexes of both the native and competitor products. Heteroduplexes were more rapidly digested by S1 nuclease than homoduplexes, and extensive alterations in primer concentrations failed to affect heteroduplex formation (data not shown). Comparison of Figures 1 and 2 indicates that heteroduplexes are not readily identifiable when PCR products are analyzed by agarose (MetaPhor) gel electrophoresis.

The importance of analysis of heteroduplex formation to the quantitative accuracy of competitive RT-PCR cannot be overemphasized. Titrations are calculated from the ratio of reaction products. Since heteroduplexes subtract equimolarly from both homoduplex reaction products, the only circumstance when their formation will not change the ratio of homoduplex products is when the ratio is unity. By determining the amount of heteroduplex present in a reaction and reallocating each of its components back to the corresponding homoduplex, the error generated by heteroduplex formation is corrected. Thus, our experiments reveal the ease with which heteroduplexes can form even when native and competitor sequences have relatively low homology. However, they also reveal the utility of IP-RP-HPLC analysis for the detection of and correction for these products.

Raeymaekers has analyzed some of the mathematical considerations concerning competitive mutant RT-PCR (18). He has emphasized that such reactions generate log-log plots of product ratio against an initial amount of competitor which should form a straight line with a slope of unity. As demonstrated by the data obtained in Figures 4 and 5, our reactions meet linearity requirements almost ideally. Furthermore, the slopes generated by these reactions are consistently close to unity. This appears to be an important advantage of IP-RP-HPLC quantification compared with other methods that have generated slope values greatly different from unity (8,17,19,21). Finally, by diluting input RNA preparations over three orders of magnitude and performing titration analysis on each of the dilutions, we have shown that the calculated initial level of gene expression is essentially unaffected by the level of initial input. The slopes of the titration lines for these reactions are indistinguishable (analysis of covariance), and the effect of dilution is to shift the titration line by the order of magnitude difference in the input RNA amount.

Ion-pair reversed-phase HPLC analysis of competitive mutant RT-PCRs provides a convenient and efficient method of reaction product quantification. The IP-RP-HPLC method can also demonstrate whether heteroduplex formation has occurred and permits correct reallocation of heteroduplex

135

components. Quantitative accuracy may benefit from the minimal amount of post-PCR product handling. Competition reactions using a competitor that is 159% greater in size results in titrations that closely approach mathematical ideals, thus indicating similar efficiency of amplification of native and competitor products. Dilution of input RNA over three orders of magnitude further confirms the reliability of absolute quantification by this method.

ACKNOWLEDGMENTS

This work was supported in part by Grants NIH DDK45538 (P.A.D.) and IPO1-HG00205 (P.J.O.). We are grateful to Dr. Jerry Lingrel (University of Cincinnati) and Dr. Kevin Lynch (University of Virginia) for cDNAs used in the present study.

REFERENCES

1. **Becker-André, M. and K. Hahlbrock.** 1989. Absolute mRNA quantification using the polymerase chain reaction (PCR). Nucleic Acids. Res. *17*:9437-9446.
2. **Bonn, G., C. Huber and P. Oefner, inventors.** 1994. Verfahren zur trennung von nucleinsaeuren (Methods for the separation of nucleic acids). Austrian patent 398 973.
3. **Chang, A., J. Zhao and M. Krajden.** 1994. Polymerase chain reaction kinetics when using a positive internal control target to quantitatively detect cytomegalovirus target sequences. J. Virol. Methods *48*:223-236.
4. **deKant, E., C.F. Rochlitz and R. Herrmann.** 1994. Gene expression analysis by a competitive and differential PCR with antisense competitors. BioTechniques *17*:934-942.
5. **Dostal, D.E., K.N. Rothblum and K.M. Baker.** 1994. An improved method for absolute quantification of mRNA using multiplex polymerase chain reaction: determination of renin and angiotensinogen mRNA levels in various tissues. Anal. Biochem. *223*:239-250.
6. **Ferré, F., A. Marchese, P. Pezzoli, S. Griffin, E. Buxton and V. Boyer.** 1994. Quantitative PCR: an overview. *In* K.B. Mullis, F. Ferré and R.A. Gibbs (Eds.), The Polymerase Chain Reaction. Birkhauser, Boston.
7. **Ferré, F., A.L. Marchese, S.L. Griffin, A.E. Daigle, S.P. Richieri, F.C. Jensen and D.J. Carlo.** 1993. Development and validation of a polymerase chain reaction method for the precise quantitation of HIV-1 DNA in blood cells from subjects undergoing a 1-year immunotherapeutic treatment. AIDS 7 (2 Suppl): S21-S27.
8. **Gilliland, G., S. Perrin, K. Blanchard and H.F. Bunn.** 1990. Analysis of cytokine mRNA and DNA: detection and quantitation by competitive polymerase chain reaction. Proc. Natl. Acad. Sci. USA *87*:2725–2729.
9. **Huber, C.G., P.J. Oefner and G.K. Bonn.** 1993. Rapid analysis of biopolymers on modified non-porous polystyrene-divinylbenzene particles. Chromatographia *37*:653-658.
10. **Huber, C.G., P.J. Oefner and G.K. Bonn.** 1995. Rapid and accurate sizing of DNA fragments on alkylated nonporous poly(styrene-divinylbenzene) particles. Anal. Chem. *67*:578-585.
11. **Huber, C.G., P.J. Oefner, E. Preuss and G.K. Bonn.** 1993. High-resolution liquid chromatography of DNA fragments on non-porous poly(styrene-divinylbenzene) particles [published erratum appears in Nucleic Acids Res., May 11, 1993; *21*:2284]. Nucleic Acids Res. *21*:1061-1066.
12. **Ikonen, E., T. Manninen, L. Peltonen and A.-C. Syvanen.** 1994. Quantitative determination of rare mRNA species by PCR and solid-phase minisequencing. PCR Methods Appl. *1*:234-240.
13. **Katz, E.D. and L.A. Haff.** 1990. Rapid separation, quantitation and purification of products of polymerase chain reaction by liquid chromatography. J. Chromatogr. *512*:433-444.
14. **Landgraf, A., B. Reckmann and A. Pingoud.** 1991. Quantitative analysis of polymerase chain reaction (PCR) products using primers labeled with biotin and a fluorescent dye. Anal. Biochem. *193*:231-235.
15. **Noonan, K.E., C. Beck, T.A. Holzmayer, J.E. Chin, J.S. Wunder, I.L. Andrulis, A.F. Gazdar, C.L.**

Willman, B. Griffith, D.D. Von Hoff and I.B. Roninson.** 1990. Quantitative analysis of MDR1 gene expression in human tumors by polymerase chain reaction. Proc. Natl. Acad. Sci. USA *87*:7160-7164.

16.**Oefner, P.J., C.G. Huber, E. Puchhammer-Stöckl, F. Umlauft, K. Grünewald, G.K. Bonn and C. Kunz.** 1994. High-performance liquid chromatography for routine analysis of hepatitis C virus cDNA/PCR products. BioTechniques *16*:898-908.

17.**Piatak Jr., M., K.-C. Luk, B. Williams and J.D. Lifson.** 1993. Quantitative competitive polymerase chain reaction for accurate quantification of HIV DNA and RNA species. BioTechniques *14*:70-80.

18.**Raeymaekers, L.** 1993. Quantitative PCR: theoretical considerations with practical implications. Anal. Biochem. *214*:582-585.

19.**Schmitt, J.F., M. Guthridge, C. Economou, J. Bertolini and M.T.W. Hearn.** 1992. A new quantitative polymerase chain reaction-high performance ion exchange liquid chromatographic method for the detection of fibroblast growth factor-beta gene amplification. J. Biochem. Biophys. Methods *24*:119-133.

20.**Schneeberger, C., P. Speiser, F. Kury and R. Zeillinger.** 1995. Quantitative detection of reverse transcriptase-PCR products by means of a novel and sensitive DNA stain. PCR Methods Appl. *4*:234-238.

21.**Siebert, P.D. and J.W. Larrick.** 1992. Competitive PCR. Nature *359*:557-558.

22.**Simmonds, P., P. Balfe, J.F. Peutherer, C.A. Ludlam, J.O. Bishop and A.J. Leigh Brown.** 1990. Human immunodeficiency virus-infected individuals contain provirus in small numbers of peripheral mononuclear cells and at low copy numbers. J. Virol. *64*:864-872.

23.**Volkenandt, M., A.P. Dicker, D. Banerjee, R. Fanin, B. Schweitzer, T. Horikoshi, K. Danenberg, P. Danenberg and J.R. Bertino.** 1992. Quantitation of gene copy number and mRNA using the polymerase chain reaction. Proc. Soc. Exp. Biol. Med. *200*:1-6.

24.**Wei, J., E.A. Walton, A. Milici and J.J. Buccafusco.** 1994. m1-m5 Muscarinic receptor distribution in rat CNS by RT-PCR and HPLC. J. Neurochem. *63*:815-821.

Update to:

Rapid Quantification of Gene Expression by Competitive RT-PCR and Ion-Pair Reversed-Phase HPLC

Amanda Hayward-Lester, Peter J. Oefner[1] and Peter A. Doris

Texas Tech University Health Sciences Center, Lubbock, TX, and [1]Stanford University, Stanford, CA, USA

Our original paper describes the development of a gene quantification system that combines competitive RT-PCR with reaction product analysis by HPLC. Since this report we have employed the same methods to examine expression of a number of other rat genes. These include the alpha 2 and 3 and gamma isoforms of Na, K-ATPase, GAPDH, calcineurin A alpha and rat cyclophilin-like protein. The development of these additional quantitation systems has revealed the general utility of the method and its relative independence of target sequence.

A subsequent report from this lab has demonstrated that quantification by this method is highly accurate, that is, the copy number estimated by the assay system accurately reflects the copy number present in the sample (1). This was demonstrated by performing PCR and RT-PCR with known quantities of template. Plasmid containing the DNA target sequence was used in the PCR experiments and RNA transcribed in vitro from this plasmid was used in the RT-PCR experiments. Our experiments revealed that the difference in sequence between the native and competititors used in competitive RT-PCR had no effect on PCR amplification efficiency. However, sequence differences clearly were able to affect RT efficiency. This was revealed in experiments which showed that the system was able to accurately estimate known DNA inputs. However, when the quantity of RNA was known, a consistent discrepancy between known input and estimated input was observed.

This discrepancy was shown to be due to differences in RT efficiency in a series of studies (currently under submission for publication) in which we examined the effect of creating various mutations on RT efficiency. The studies have lead us to conclude that mutation to create the competitor can, but does not always, lead to differences in RT efficiency between the native and competititor sequences. Some mutations are without effect, while others can increase or decrease the relative efficiency of transcription of the competitor compared with the native template. Presumably such differences are due to the formation of differences in secondary structure or possibly different intermolecular interactions between native and competitor sequences. In either

138

case, the RT efficiency difference for a given native and competitor sequence is highly consistent. This means that accurate quantification remains possible even in the face of such RT efficiency differences if this difference is first estimated by comparing the relative efficiency of transcription of the two templates.

Finally, regulated alternative splicing is another area that has yielded to the important qualities of HPLC analysis of reaction products. We have shown that the transcript of a novel DNA binding protein involved in the control of uteroglobin expression can exist in two alternative spliced variants (2). The splice variation results in two transcripts that differ by only 57 bases. By designing PCR primers that span the alternatively spliced site, RT-PCRs can be performed which demonstrate whether one or both transcripts are expressed. Such reactions can easily produce heteroduplexes between the amplicons resulting from the two transcripts. Here the ability of HPLC to identify not only the two native amplicons, but also the heteroduplexes formed between them has provided a novel means to quantify the relative amounts of each transcript present. We have shown that the relative levels of one transcript compared with the other can be controlled by signaling pathways involving steroid hormones.

REFERENCES

1. **Hayward-Lester, A., P.J. Oefner, S. Sabatini and P.A. Doris.** 1995. Accurate and absolute quantitative measurement of gene expression by single tube RT-PCR and HPLC. Genome Res. *5*:494-499.
2. **Hayward-Lester, A., A. Hewetson, E.G. Beale, P.J. Oefner, P.A. Doris and B.S. Chilton.** 1996. Cloning, characterization, and steroid-dependent posttranscriptional processing of RUSH-1alpha and beta, two uteroglobin promoter-binding proteins. Mol. Endocrinol. *10*:1335-1349.

CHAPTER 11

Rapid Quantitative Analysis of Differential PCR Products by High-Performance Liquid Chromatography

R. Zeillinger, C. Schneeberger, P. Speiser and F. Kury[1]

Molecular Oncology Division, First Department of Obstetrics & Gynecology, University of Vienna Medical School, [1]Labordiagnostika GesmbH "Vienna Lab", Vienna, Austria

ABSTRACT

We describe the use of high-performance liquid chromatography (HPLC) for the rapid quantitative analysis of short DNA fragments generated in differential PCR, where a target gene and a control gene are in vitro co-amplified in one single reaction. Using an anion-exchange nonporous column, both separation and quantitation of the differential PCR products are achieved in about 5 min per sample. The performance of this technique proved to be superior to that of conventional gel electrophoresis and subsequent analysis by a laser densitometer, a solid-state scanner and a charge-coupled device video camera imaging system. The usefulness for clinical testing is described in the example of the quantitative analysis of the c-erbB-2 oncogene copy number of human breast carcinomas by differential PCR. The combined use of differential PCR and automated HPLC analysis of the PCR products may well substitute for classical Southern blot hybridization in routine clinical analysis of oncogene amplification.

INTRODUCTION

Previously a modification of the polymerase chain reaction (PCR) (16–18), so-called differential PCR has been described (6). The principle of differential PCR is the in vitro co-amplification of a target gene and a reference gene in one single tube. The amplification of a single gene cannot be controlled for varying reaction efficiencies. However, the general sensitivity to cycle numbers and possible errors in quantifying input DNA can be overcome by the co-amplification of the target and a control sequence in the same reaction tube. Differential PCR can detect gene amplification independent of

(Reprinted from BioTechniques 15:89-95, 1993)

the quantity of input genomic DNA, and the results are independent of PCR cycle numbers, since every variation in the reaction conditions will influence the PCR product amount of both the target gene and the control gene to the same extent (6). In the case of gene amplification, an increase in the PCR product from the target sequence and a decrease in the amount of PCR product from the control gene (a single-copy reference gene that is known not to be amplified in the tissue type to be analyzed) is seen. The level of the target gene amplification is reflected in the ratio between the amounts of the two PCR products. Differential PCR has proven to be an alternative to classical hybridization using Southern blots or slot blots to detect gene amplification. Less quantities of input sample DNA are needed and also DNA from formalin-fixed, paraffin-embedded tissue may be used. The first application of this technique was the determination of amplification of the c-*erbB-2* oncogene (6,10,11,14). The gene for interferon gamma was chosen as the single-copy reference gene.

The c-*erbB-2* oncogene (also termed "HER-2" or "neu"; located on chromosome 17) has been found to be amplified in several different types of tumors. In human breast cancer and ovarian cancer, the amplification status of this oncogene has proven to be of strong predictive value for relapse-free and overall survival of the patients (3,4,22,23,25,28). Therefore, c-*erbB-2* amplification should be assessed routinely in breast and ovarian cancer. Besides c-*erbB-2*, other candidate oncogenes have been described to be of possible diagnostic value for human breast cancer (1,2,15,20,24). However, the classical methods of Southern blot or dot blot hybridization are rather time-consuming and complicated. They also require reasonable amounts of high-quality DNA and therefore are not applicable to samples with low DNA recovery (e.g., fine needle aspirates and single frozen sections), integrity or purity.

Differential PCR is the alternative method of choice to detect

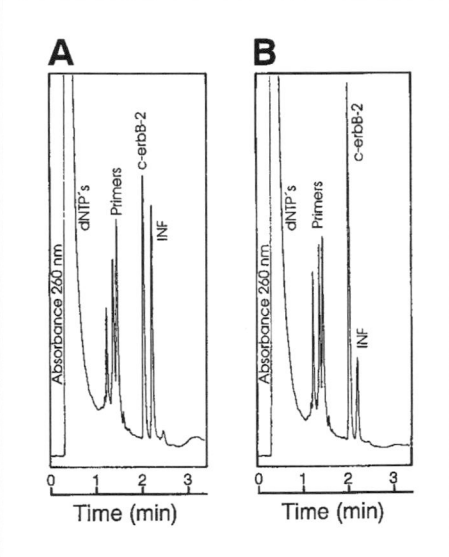

Figure 1. HPLC separation of the products of differential PCR for the detection of c-*erB-2* amplification. Chromatograms showing the results for (A) placenta DNA (single copy control) and (B) for DNA from BT-474 breast carcinoma cell line (high copy control). Eighty microliters of PCR products (diluted 1:5) were injected. dNTPs and primers eluted first, followed by the PCR products for the c-*erbB-2* oncogene and the control gene (interferon gamma, INF). Whereas the peak areas for placenta DNA (A) were almost equal (one c-*erbB-2* oncogene copy per chromosome 17), the area for the c-*erbB-2* PCR product was about 6-fold higher than for the INF PCR product for the high copy control (6 oncogene copies per chromosome 17).

R. Zeillinger et al.

Table 1. Amplification of the c-erbB-2 Oncogene in Selected 9 Primary Breast Carcinomas, in the Single Copy Control and in the High Copy Control

Sample #	HPLC	Densitometer Negative Type 55	CCD System Negative Type 55	Scanner Negative Type 55	Scanner Positive Type 55	Scanner Positive Type 57
3	3	2	2	2	2	2
4	1	1	1	1	1	1
5	1	1	1	1	1	1
6	6	4	5	3	3	3
7	2	2	2	2	2	2
8	1	1	1	1	1	1
9	6	4	5	3	3	3
10	4	3	3	3	3	2
11	1	1	1	1	1	1
12 (SC)	1	1	1	1	1	1
13 (HC)	6	5	5	4	4	4

Sample numbers correspond to lane numbers in Figure 3. The degree of amplification for the amplified tumor samples ranged from 2-fold up to 6-fold ("1" corresponds to normal copy number, as seen for placenta DNA that was used as single copy control, SC). The breast carcinoma cell line BT-474 (high copy control, HC) showed 6-fold amplification. Amplification was calculated (see Materials and Methods) either from the ratio of the peak areas for the c-$erbB$-2 oncogene to the areas for the control gene obtained by HPLC or from the ratio of the band intensities as determined by laser densitometry (Personal Densitometer), CCD imaging (Bio-2D Videodensitometer) and scanning (Bio Image System 50) of negatives (Polaroid type 55) and black and white prints (Polaroid type 55 and type 57).

quantitative alterations of genes in small samples (6,14). Typically, the analysis of differential PCR products is done by agarose gel electrophoresis and densitometry of photographs. However, nonlinear film sensitivity and unstable fluorescence generated by ethidium bromide staining of the gels are the major drawbacks.

As an alternative to gel electrophoresis, high-performance liquid chromatography (HPLC) may be used for the separation of PCR products. The use of HPLC for rapid analysis and purification has previously been described only for single DNA fragments generated in PCR (8,12,26,27). Here we present the separation of the products of differential PCR by HPLC and simultaneous quantitation by UV-detection. We describe the usefulness of this method for the determination of c-$erbB$-2 oncogene amplification in human primary breast carcinomas.

MATERIALS AND METHODS

DNA Isolation

Tumor tissue specimens were frozen in liquid nitrogen immediately after surgery and stored until preparation of nuclei, membranes and cytosols was carried out. They were homogenized in a dismembrator (Braun-Melsungen, Melsungen, Germany) for 2 min. The resulting powders were suspended in homogenization buffer (0.01 M K_2HPO_4/KH_2PO_4, 0.0015 M K_2EDTA, 0.003 M NaN_3, 0.01 monothioglycerol, pH 7.5, 10% vol/vol glycerol). Nuclei were collected by centrifugation at 800× g at 4°C for 15 min. The resulting supernatants were used for preparation of membranes and cytosols. After washing nuclei in homogenization buffer, DNA was isolated by the use of the DNA Extraction System I (ViennaLab, Labordiagnostika GesmbH, Vienna, Austria). Briefly, nuclei were lysed with sodium dodecyl sulfate (SDS) and proteinase K, proteins were precipitated by addition of saturated NaCl solution and DNA was recovered by ethanol precipitation. High molecular weight DNA was spooled on plastic inoculation loops, dissolved in TE buffer [10 mM Tris-HCl (pH 7.4), 1 mM Na_2EDTA] and stored at -80°C.

Differential PCR

For the detection of c-*erbB-2* oncogene amplification, a commercially available differential PCR system (ViennaLab, Labordiagnostika GesmbH) was used. This kit contains a "ready-to-use" mastermix composed of a reagent buffer [10 mM Tris-HCl (pH 9.0), 50 mM KCl, 1.5 mM $MgCl_2$, 0.1% vol/vol Triton® X-100, 0.01% wt/vol gelatin], two primer pairs specific for the c-*erbB-2* oncogene (primer erb-2/1:5'-CCT CTG ACG TCC ATC ATC TC-3' and primer *erbB*-2/2:5'-ATC TTC TGC TGC CGT CGC TT-3') and the interferon gamma gene (primer IFN/1: 5'-TCT TTT CTT TCC CGA TAG GT-3' and primer IFN/2: 5'- CTG GGA TGC TCT TCG ACC TC-3'), deoxynucleotides, Hi-Taq™ DNA Polymerase (ViennaLab, Labordiagnostika GesmbH) and control

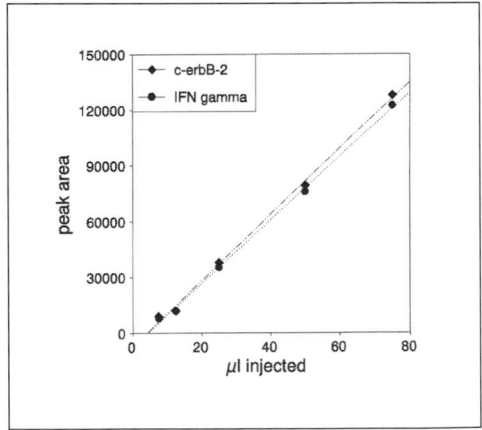

Figure 2. Linear relation between the amount of the products for the c-*erbB-2* oncogene and the control gene (INF gamma) from differential PCR and the peak areas calculated from absorbance at 260 nm. 7.5 µL, 15 µL, 25 µL, 50 µL and 75 µL of PCR product (diluted 1:5, using human placenta DNA (single copy control) as template were injected. The ratio of the peak areas for c-*erbB*-2 to the peak areas for the control gene varied between 1.01 and 1.07 and was independent of the amount of PCR products injected.

DNA. Briefly, 33 µL of the mastermix containing 25 pmol of each of the four primers and 250 µM deoxyribonucleoside triphosphates (dNTPs) were mixed on ice with about 200 ng of chromosomal DNA extracted from tumor samples, and 0.2 U Hi-Taq DNA Polymerase in a final volume of 50 µL. Amplification cycles were as follows: 94°C for 30 s, 50°C for 30 s and 72°C for 45 s. Thirty amplification cycles were preceded by a primary denaturation step (94°C for 1 min) and followed by a final extension step (72°C for 5 min) after the last cycle. A GeneAmp™ PCR System 9600 (Perkin-Elmer, Norwalk, CT, USA) was used.

DNA from human placenta (single copy control) and DNA from the BT-474 (HTB20; ATCC, Rockville, MD, USA) breast carcinoma cell line (high copy control) were used for standardization and as control. The lengths of the PCR products for the c-erbB-2 oncogene and the interferon gamma gene were 98 bp and 150 bp, respectively. Water blanks were included routinely to control for possible DNA carryover and contamination.

HPLC System and Operating Conditions

Products from differential PCR were analyzed by HPLC. The HPLC system used in that study consisted of two 2150 Pumps, a 2152 Controller, a 5156 Solvent Conditioner, a 2151 Variable Wavelength Monitor, a High Pressure Mixer (Pharmacia Biotech, Bromma, Sweden) and a Merck/Hitachi AS2000A Autosampler (Hitachi, Tokyo, Japan). The 3000 Series Chromatography Data System (Perkin-Elmer Nelson Systems, Cupertino, CA, USA) served for data handling. The analytical column was a TSK DEAE-NPR column (hydrophilic resin bonded with DEAE groups, 2.5-µm particle size, 4.6-mm ID, 35 mm long; Perkin-Elmer).

The mobile phase was as follows: solvent A was 25 mM Tris-HCl (pH 9.0) and solvent B was 25 mM Tris-HCl (pH 9.0), 1 M NaCl. A gradient program

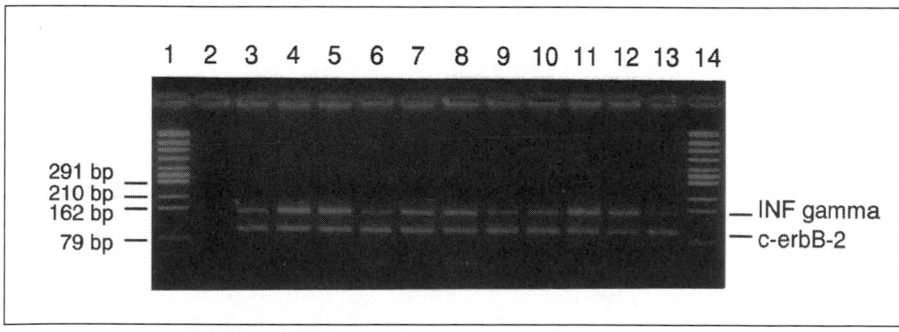

Figure 3. Electropherogram of the PCR products from the c-erbB-2 oncogene and the interferon gamma gene obtained by differential PCR. Lanes 1 and 14, HincII digest of ΦX174 DNA; lane 12, single copy control (placenta DNA); lane 13, high copy control (DNA from breast carcinoma cell line BT-474); lanes 3–11, DNA from primary breast carcinomas; lane 2, blank (water). In tumor DNA on lanes 3, 6, 7, 9 and 10, amplification of the c-erbB-2 oncogene was noted. Lane numbers correspond to sample numbers in Table 1.

consisting of linear segments was employed: 30%–55% solvent B in 0.5 min, 55%–65% solvent B in 2.5 min, 65%–100% solvent B in 0.5 min, 100% solvent B for 0.5 min and then 100%–30% solvent B in 0.5 min. The total injection-to-injection cycle times were 5 min. The column was operated at 1 mL/min at room temperature, and UV-extinction was measured at 260-nm.

Since the time for equilibration of the column between the injections was only 0.5 min, the retention time for the first injection was about 10% longer than for the following consecutive injections. To allow the analysis of samples as fast as possible, 80 µL of solvent A were used for the first injection always, followed by the PCR product samples (80 µL) that had been diluted 1:5 into water.

Agarose Gel Electrophoresis and Analysis

Ten microliters of the PCR mixture were loaded on 4% 3/1 NuSieve®/SeaKem® GTG-agarose (FMC BioProducts, Rockland, ME, USA) gels and the PCR products were separated at 10 V/cm for 20 min. Gels were stained with ethidium bromide and photographs were made on Polaroid type 55 positive/negative (ISO 3000/36°) or Polaroid type 57 (ISO 50/18°; Polaroid, Cambridge, MA, USA) positive instant films using transmitted UV light at 302 nm. Negative films were analyzed by a 2-D laser scanning densitometer (Personal Densitometer; Molecular Dynamics, Sunnyvale, CA, USA), with a solid-state scanner (Bio Image™ System 50; Millipore, Bedford, MA, USA) and a charge-coupled device (CCD) video camera imaging system (Bio-2D Videodensitometer; Vilber Lourmat, Marne, France). In addition, black and white prints were analyzed with the Bio Image System.

Calculation of Oncogene Amplification Degree

The degree of c-erbB-2 amplification (A) in tumor DNA (t) was calculated as follows: $A = Y_t/Y_c$, where Y_t is the ratio of the peak area (obtained by HPLC or analysis of Polaroid films) for the c-erbB-2 PCR product to the peak area for the interferon gamma PCR product, and Y_c is the ratio of the c-erbB-2 signal to the interferon signal obtained with the single copy control DNA.

RESULTS

Excellent separation of the products for the c-erbB-2 oncogene (98 bp) and the interferon gamma gene (150 bp), which served as a single-copy control gene, was achieved by the HPLC method used (Figure 1). The nature of the two PCR product peaks was verified by collecting the eluates and analyzing by agarose gel electrophoresis (data not shown). As expected, peak elution was in the order of increasing chain length of the PCR products. Usually

146

a scale setting of 0.04 absorbance units full scale (AUFS) was chosen for the analysis of 80 µL of 1:5 diluted PCR mixture, i.e., about one third of the total reaction volume. A total injection-to-injection time of 5 min was sufficient for reproducible separation of the PCR products for the c-*erbB-2* oncogene (retention time = 2.021 ± 0.009 min, CV = 0.43%) and the control gene (retention time = 2.263 ± 0.054 min, CV = 2.4%) as derived from consecutive injections of 50 different samples. Maximal carryover between injections of the high copy control and subsequent injections of water blanks was 2.1% for the c-*erbB-2* PCR product. No carryover for the control gene was detectable. However, carryover that may occur between injections did not influence the results of the described analysis, which is given as a ratio of the two peaks for the c-*erbB-2* oncogene and the control gene. Moreover, the total amount of the amplified products that was generated in differential PCR was very similar, even in a case of high oncogene amplification (a 27-fold amplification of the c-*erbB-2* oncogene was the highest seen in more than 1000 primary breast carcinomas so far). The column has been used for about 1000 separations without any noticeable change of performance.

Figure 2 shows a linear correlation between the amount of PCR mixture injected and the calculated peak areas. It is evident from the serial dilutions that sufficiently accurate results were obtained when only 7.5 µL of the diluted PCR mixture were injected, which corresponds to 3% of the total reaction volume. The ratio of the peak area for the c-*erbB-2* PCR product to the peak area for the interferon gamma PCR product for the single copy control was about 1 for all different volumes injected. Injecting variable amounts of PCR products (between 3% and 30% of the total PCR mixture) resulted in the same "calculated degree of oncogene amplification," assuming that possible variations in the total amount of PCR products will not affect the determined oncogene amplification.

In Figure 3 the separation of the two products from the differential PCR by agarose gel electrophoresis is shown. As for the HPLC analysis, about one third of the total reaction volume was loaded onto the gel. However, the intensity of the fluorescent signal decreased rapidly with lower DNA amounts. When analyzing the degree of oncogene amplification with a laser densitometer, a solid-state scanner and CCD video camera imaging system of negatives and black and white prints, CCD imaging proved to be the most sensitive method. In two cases, a higher amplification rate for the c-*erbB-2* oncogene was determined by the CCD image sensor compared to densitometry. Scanning was the less sensitive method. Polaroid type 55 black and white film seemed to be more sensitive for this application than type 57 film (see sample #10; Table 1) that is generally preferred because of its high speed. However, compared to HPLC, the analysis of photographs of agarose gels proved to be less sensitive. The degree of oncogene amplification was underestimated by the described analysis of agarose gels in almost all amplified cases.

DISCUSSION

The data presented demonstrate that the analysis by HPLC provides accurate detection and quantification of DNA fragments generated by differential PCR amplifications. Recently, this application of PCR, where two fragments correspond to two different genes, has been used more often for the quantitation of gene amplification, specially of oncogene amplification in human tumors (6,9–11,14). In combination with reverse transcription, this technique has been used for quantitation of mRNA reverse transcription PCR (RT-PCR) (5,21). The conventional techniques for the analysis of PCR products typically involve the use of agarose gel electrophoresis and densitometry of a negative photograph or of a black and white print. RT-PCR products have also been analyzed by densitometric scanning of silver-stained polyacrylamide gels (19) and by using a CCD video camera imaging system (13). However, the present data show that HPLC is superior in sensitivity to laser densitometry, densitometric scanning (white light source) and CCD imaging (Table 1). This may be due to limited linear response of the films used for photography of ethidium bromide-stained agarose gels. Also, the type of film used for black and white prints seems to determine the apparent degree of gene amplification. Interestingly, a widely used high-speed film (Polaroid type 57) in combination with scanning (Bio Image System 50) proved to be least sensitive compared to HPLC. However, compared to classical Southern blot hybridization, PCR and subsequent product analysis by gel electrophoresis are major steps towards routine analysis of oncogene amplification in general. Unfortunately, ethidium bromide staining produces a high background fluorescence signal, which limits both the sensitivity and dynamic range. The intensity of the fluorescent signal depends on the length of the DNA fragment. Moreover, nonlinear film sensitivity may be also a limiting factor. Therefore, the use of these methods may result in underestimation of the amount of PCR fragments. In the case of differential PCR this means underestimation of the degree of gene amplification.

Besides accuracy, the separation of differential PCR products by HPLC and quantitation by UV detection is an attractive alternative method to gel electrophoresis and densitometry of photographs. The method allows the analysis of about 10 to 15 samples per hour. A linear response has been reported over a broad DNA concentration range [2.4–240 ng (8); 0.86–220 ng (27)]. The sensitivity is much higher compared to ethidium bromide staining [0.2–0.4 ng vs. 1–5 ng (8)]. Since the analysis by HPLC of about 3% of the total reaction volume (50 μL) for differential PCR gave accurate results, a smaller reaction volume may be used, resulting in a reduced cost per reaction. Another major advantage of HPLC analysis is that the process may be automated by the use of an autosampler.

Therefore, it seems to be feasible to perform a retrospective study of gene amplification in large series of paraffin-embedded pathology specimens in a resonable time period. Moreover, the routine analysis of oncogene amplification in many clinical tumor samples, demonstrated in the example of the *c-erbB-2* oncogene, by differential PCR and subsequent separation and quantification of the PCR products by HPLC can be achieved in a single day. Because of the high sensitivity, this method may be also used for the analysis of fine needle aspirate biopsies, leading to possible preoperative treatment decisions in cancer patients.

REFERENCES

1. **Berns, E.M.J.J., J.G.M. Klijn, W.L.J. van Putten, I.L. van Staveren, H. Portengen and J.A. Foekens.** 1992. C-myc amplification is a better prognostic factor than HER2/neu amplification in primary breast cancer. Cancer Res. *52*:1107-1113.
2. **Berns, E.M.J.J., J.G.M. Klijn, I.L. van Staveren, H. Portengen, E. Noordegraaf and J.A. Foekens.** 1992. Prevalence of amplification of the oncogenes c-myc, HER-2/neu, and int-2 in one thousand human breast tumours: correlation with steroid receptors. Eur. J. Cancer. *28*:697-700.
3. **Borg, A., B. Baldetorp, M. Fernö, D. Killander, H. Olsson and H. Sigurdsson.** 1991. ERBB2 amplification in breast cancer with a high rate of proliferation. Oncogene *6*:137-143.
4. **Borg, A., A.K. Tandon, H. Sigurdsson, G.M. Clark, M. Fern, S.A.W. Fuqua, D.M. Killander and W.L. McGuire.** 1990. HER-2/neu amplification predicts poor survival in node-positive breast cancer. Cancer Res. *50*:4332-4337.
5. **Ferre, F.** 1992. Quantitative or semi-quantitative PCR: reality versus myth. PCR Methods Appl. *2*:1-9.
6. **Frye, R.A., C.C. Benz and E. Liu.** 1989. Detection of amplified oncogenes by differential polymerase chain reaction. Oncogene *4*:1153-1157.
7. **Gibbs, R.A.** 1990. DNA amplification by the polymerase chain reaction. Anal. Chem. *26*:1202-1214.
8. **Katz, E.D. and M.W. Dong.** 1990. Rapid analysis and purification of polymerase chain reaction products by high-performance liquid chromatography. BioTechniques *8*:549-554.
9. **Kury, F.D., R. Zeillinger, K.F. Czerwenka, U. Pruckner, Ch. Schneeberger, P. Speiser, G. Sliutz, W. Knogler and E. Kubista.** 1992. Detection of c-*erbB-2* oncogene expression and amplification in human breast cancer: evaluation of an immunohistochemical assay, an enzyme immunoassay and a differential PCR procedure, p. 443-448. *In* R. Klapdor (Ed.), Tumor Associated Antigens, Oncogens, Receptors, Cytokines in Tumor Diagnosis and Therapy at the Beginning of the Nineties. W. Zuckschwert, Munich.
10. **Liu, E., A. Thor, M. Barcos, B.-M Ljung and Ch. Benz.** 1992. The HER2 (c-erbB-2) oncogene is frequently amplified in in situ carcinomas of the breast. Oncogene *7*:1027-1032.
11. **Malka, O., Y. Pollack, D. Benharroch, Y. Cohen, R. Levy, I. Yanai-Inbar and J. Gopas.** 1991. Breast biopsy nuclear pellets are a convenient source of DNA for routine determination of HER-2/neu gene amplification. Breast Cancer Res. Treat. *19*:57-61.
12. **Merion, M., W. Warren, C. Stacey and M.E. Dwyer.** 1988. High resolution purification of DNA restriction fragments using HPLC. BioTechniques *6*:246-251.
13. **Nakayama, H., H. Yokoi and J. Fujita.** 1992. Quantification of mRNA by non-radioactive RT-PCR and CCD imaging system. Nucleic Acids Res. *20*:4939.
14. **Neubauer, A., B. Neubauer, M. He, P. Effert, D. Iglehart, R.A. Frey and E. Liu.** 1992. Analysis of gene amplification in archival tissue by differential polymerase chain reaction. Oncogene *7*:1019-1026.
15. **Ro, J., S.M. North, G.E. Gallick, G.N. Hortobagyi, J.U. Gutterman and M. Blick.** 1988. Amplified and overexpressed epidermal growth factor receptor gene in uncultured primary human breast carcinoma. Cancer Res. *48*:161-164.
16. **Saiki, R.K.** 1989. Principles and applications for DNA amplification. Chapter 1. *In* H.A. Erlich (Ed.), PCR Technology. Stockton Press, New York.
17. **Saiki, R.K., D.H. Gelfand, S. Stoffel, S.J. Scharf, R. Higuchi, G.T. Horn, K.B. Mullis and H.A. Erlich.** 1988. Primer-directed enzymatic amplification of DNA with a thermostable DNA polymerase. Science *239*:487-491.
18. **Saiki, R.K., S. Scharf, F. Faloona, K.B. Mullis, G.T. Horn, H.A. Erlich and N. Arnheim.** 1985. Enzymatic amplification of beta-globin genomic sequence and restriction site analysis for diagnosis of sickle cell anemia. Science *230*:1350-1354.
19. **Sarkar, F.H., Y-W. Li, D.E. Ball and J.D. Crissman.** 1992. Comparative method for detection of

RNA-PCR-amplified signals. BioTechniques *12*:23-26.

20. **Schuuring, E., E. Verhoeven, W.J. Mooi and R.J.A.M. Michalides.** 1992. Identification and cloning of two overexpressed genes, U21B32/PRAD1 and EMS1, within the amplified chromosome 11q13 region in human carcinomas. Oncogene *7*:355-362.

21. **Siebert, P.D. and J.W. Larrick.** 1992. Competitive PCR. Nature *359*:557-558.

22. **Slamon, D.J., G.M. Clark, S.G. Wong, W.J. Levin, A. Ullrich and W.L. McGuire.** 1987. Human breast cancer: correlation of relapse and survival with amplification of the HER-2/neu oncogene. Science *235*:177-182.

23. **Slamon, D.J., W. Godolphin, L.A. Jones, J.A. Holt, S.G. Wong, D.E. Keith, W.L. Levin, S.G. Stuart, J. Udove, A. Ullrich and M.F. Press.** 1989. Studies of the Her-2/neu protooncogene in human breast and ovarian cancer. Science *244*:707-712.

24. **Szepetowski, P., A. Courseaux, G.F. Carle, C. Theillet and P. Gaudray.** 1992. Amplification of 11q13 DNA sequences in human breast cancer: D11S97 identifies a region tightly linked to BCL1 which can be amplified separately. Oncogene *7*:751-756.

25. **Tsuda, H., S. Hirohashi, Y. Shimosato, T. Hirota, S. Tsugane, H. Yamamoto, N. Miyajima, K. Toyohashi, T. Yamamoto, J. Yokota, T. Yoshida, H. Sakamoto, M. Terada and T. Sugimura.** 1989. Correlation between long-term survival in breast cancer patients and amplification of two putative oncogene-coamplification units: hst/int-2 and c-erbB-2/ear1. Cancer Res. *49*:3104-3108.

26. **Warren, W. and J. Doniger.** 1991. HPLC purification of polymerase chain reaction products for direct sequencing. BioTechniques *10*:216-220.

27. **Warren, W., T. Wheat and P. Knudsen.** 1991. Rapid analysis and quantitation of PCR products by high-performance liquid chromatography. BioTechniques *11*:250-255.

28. **Zeillinger, R., F.D. Kury, K. Czerwenka, E. Kubista, G. Sliutz, W. Knogler, J. Huber, C.C. Zielinski, G. Reiner, R. Jakesz, A. Staffen, A. Reiner, F. Wrba and J. Spona.** 1989. HER-2 amplification, steroid receptors and epidermal growth factor receptors in primary breast cancer. Oncogene *4*:109-114.

150

Update to:

Rapid Quantitative Analysis of Differential PCR Products by High-Performance Liquid Chromatography

R. Zeillinger

University of Vienna Medical School, Vienna, Austria

Since the original publication of our paper, the combined use of differential PCR and HPLC for the detection of gene amplification (7) or quantification of viral DNA (1,3) has been described several times. Since cDNA may serve as template in PCR as well as genomic DNA does, PCR has also become an efficient tool in the study of gene expression. It has been shown that the combination of RT-PCR and HPLC provides a rapid and sensitive method for quantifying the expression of a variety of mRNAs (2,4–6). Furthermore, it has been shown that the sensitivity of the PCR product detection can be enhanced by the use of a fluorescein-labeled primer and a fluorescence monitor instead of UV detection (3).

REFERENCES

1. **Chan, A., J. Zhao M. and Krajden.** 1994. Polymerase chain reaction kinetics when using a positive internal control target to quantitatively detect cytomegalovirus target sequences. J. Virol. Methods *48*:223-236.
2. **Gaus, H., G.B. Lipford, H. Wagner K. and Heeg.** 1993. Quantitative analysis of lymphokine mRNA expression by a nonradioactive method using PCR and anion exchange chromatography. J. Immunol. Methods *158*:229-236.
3. **Oefner, P.J., C.G. Huber, E. Puchhammer-Stoeckl, F. Umlauft, K. Grunewald, G.K. Bonn and C. Kunz.** 1994. High-performance liquid chromatography for routine analysis of hepatitis C virus cDNA/PCR products. BioTechniques *16*:898-899,902-908.
4. **Rochlitz, C.F., E. de Kant, A. Neubauer, I. Heide, R. Bohmer, J. Oertel, D. Huhn and R. Herrmann.** 1992. PCR-determined expression of the MDR1 gene in chronic lymphocytic leukemia. Ann. Hematol. *65*:241-246.
5. **van Hille, B., A. Lohri, J. Reuter and R. Herrmann.** 1995. Nonradioactive quantification of mdr1 mRNA by polymerase chain reaction amplification coupled with HPLC. Clin. Chem. *41*:1087-1093.
6. **Wie, J., E.A. Walton, A. Milici and J.J. Buccafusco.** 1994. M1-m5 muscarinic receptor distribution in rat CNS by RT-PCR and HPLC. J. Neurochem. *63*:815-821.
7. **Yoshimura, S., A. Shishikura, S. Koido, S. Ushigome, H. Suemizu, Y. Taniguchi and T. Moriuchi.** 1992. Quantification of the c-myc gene in gastric carcinomas by the triplex polymerase chain reaction and high performance liquid chromatography. Nucleic Acids Symp. Ser. *27*:153-154.

CHAPTER 12

Quantitation of Metallothionein mRNA by RT-PCR and Chemiluminescence

Kathy Jessen-Eller[1], Enrico Picozza[2] and Joseph F. Crivello[1,3]

[1]University of Connecticut, Groton, CT, [2]Perkin-Elmer, Wilton, CT, and [3]University of Connecticut, Storrs, CT, USA

ABSTRACT

A general procedure has been developed for the determination of mRNA expression by reverse transcription polymerase chain reaction (RT-PCR), over a wide concentration range, with quick quantitation of amplified products by luminescence. The discriminating power of this approach is the specific hybridization of PCR product to ruthenium-labeled oligonucleotide probe(s). This method is sensitive enough to detect increases in the formation of PCR product by the 6th cycle. The accumulation of PCR product was successfully modeled with a recursive relationship. This procedure was capable of accurately determining starting template copies over a 9-log dynamic range, with a sensitivity limit of 10^2 copies. Inclusion of an mRNA internal standard (identical to amplified template except for a 6-bp deletion) corrected variabilities in the reverse transcriptase as well as PCR, allowing for the expression of data as mRNA copy number/µg total tissue RNA. This procedure was used to detect changes in levels of winter flounder (Pleuronectes americanus) liver metallothionein mRNA. Liver metallothionein mRNA levels ranged from 1.0×10^6 copies/µg total tissue RNA in control samples to 1.0×10^9 copies/µg total tissue RNA in samples treated with Cd (a known metallothionein mRNA inducer).

INTRODUCTION

Due to its extraordinary sensitivity, the polymerase chain reaction (PCR) is the method of choice for detection of nucleic acids at very low levels in biological samples (1,3). A number of procedures have been reported for the detection of both DNA and mRNA (after a reverse transcriptase step) by PCR (3,6,8,12,13,18,19,21,24,26). These reports clearly demonstrate the difficulties associated with quantitative PCR, even under the best conditions, in both the reverse transcriptase step (RT-PCR) as well as the PCR step. In addition, methods employed for the detection of PCR products, e.g., ethidium bromide and ^{32}P-labeled probes, are either semiquantitative or laborious. Therefore, there is a growing need for rapid, precise and accurate detection of PCR

(Reprinted from BioTechniques 17:962-973, 1994)

products that does not rely on radioactive isotopes or harmful organic chemicals.

Recently a new DNA detection system, based on luminescence, has been developed (9). This method involves specific hybridization of DNA templates to a ruthenium-labeled probe with subsequent detection of template mass by luminescence. This method is superior to those that detect DNA by radioisotopic means, in terms of its ease of use, safety and speed of analysis and quantitation, but is equivalent in terms of sensitivity.

At the heart of this technology is an oligonucleotide probe that is coupled to TBR [tris (2,2′-bipyridine) ruthenium II chelate]. TBR emits light at 620 nm in an electrochemical cell under appropriate redox conditions, allowing for direct and specific quantitation of DNA sequences. This detection system can also be used for the quantitation of PCR product (9).

As mentioned previously, quantitative PCR has been shown to have limitations, partly because the final PCR product is derived from the exponential amplification of starting template, and hence, minor differences in amplification efficiencies can result in large differences in product yield, especially if starting template concentration is low. Some researchers have observed a linear relationship between starting template and PCR product within the exponential range of amplification, but this is dependent on sample preparation, thermal cycler performance, reaction parameters and inhibitors (especially critical in RT-PCR). To overcome sample-to-sample variation in both the reverse transcriptase and PCR steps, procedures based on the use of internal standards have been developed (1,3,8,12). These internal standards generally share sequence homology (with additions or deletions) and utilize the same amplification primers to minimize variability. Previously, amplified PCR product was distinguished from internal standards by size differences [i.e., on polyacrylamide gel electrophoresis (PAGE) gels (26)], by presence or absence of a restriction enzyme site (12) or other means (13,17). Here we describe a method for the rapid construction of mRNA internal standards, as well as a protocol for directly quantitating PCR product formation by luminescence and the means by which this information can be used to determine starting template copy number. The use of mRNA internal standards allows for corrections in the variability in the RT step as well as PCR and allows the starting template copy number to be expressed as mRNA/mg total tissue RNA. Though this protocol is demonstrated with liver metallothionein (MT) mRNA from *Pleuronectes americanus*, it is generally applicable to any mRNA of interest.

MATERIALS AND METHODS

Internal Standard Construction

An mRNA internal standard (MT^{IS}) was designed from the winter flounder (*P. americanus*) MT cDNA (a kind gift from Dr. K. Chan, Nova Scotia) by site-directed deletion of a 6-bp fragment (bases 72–77 of the coding region) (4,15). To delete this fragment, the following primers were synthesized: AAG AGC TGC TGC CCA TGC (bases 88–105) and GGT GGT GCA GCT GCA GTT (bases 81–64). They were then added to a reaction tube at a final concentration of 5 µM, which also contained 200 µM deoxyribonucleoside triphosphates (dNTPs), 10 U AmpliTaq® DNA Polymerase (Perkin-Elmer, Norwalk, CT, USA), 50 mM KCl, 10 mM Tris-HCl pH 8.3, 2 mM MgCl$_2$ and 2 ng of the MT cDNA in a pUC13 vector, in a total volume of 50 µL. The sample was amplified in a GeneAmp™ PCR System 9600 Thermal Cycler (Perkin-Elmer), for 35 cycles with the following temperature profile: 94°C for 30 s, 55°C for 30 s and 72°C for 45 s.

The product was resolved on a 1% agarose gel (SeaPlaque®, FMC Bio-Products, Rockland, ME, USA), isolated and purified (Magic™ PCR Prep DNA purification system; Promega, Madison, WI, USA) and, after polishing with Klenow fragment, blunt-ligated to itself using T4 DNA ligase. After ligation, the deleted plasmid was used to transform competent DH5α cells. The plasmid was amplified, purified and sequenced over the deleted region. Since pUC13 does not contain RNA polymerase primer sites, MT^{IS} was subcloned into pBluescript™ KS+ (Stratagene, La Jolla, CA, USA), which contains T$_3$ and T$_7$ RNA polymerase primer sites.

MT^{IS} was generated by digesting the plasmid containing MT^{IS} with *Hin*dIII, which cleaves the MT cDNA 3′ of the poly(A) tail and allows for the synthesis of the full-length MT^{IS} mRNA. Then 1 µg of the digested plasmid was added to a reaction mixture that contained 5 U T$_7$ RNA polymerase, 40 mM Tris-HCl pH 7.5, 6 mM MgCl$_2$, 2 mM spermidine, 10 mM NaCl, 0.5 mM dNTPs, 2.5 U RNase inhibitor and 10 µM dithiothreitol (DTT), in a total volume of 20 µL. The solution was incubated at 37°C for 60 min, at the end of which 10 U of RNase-free DNase were added, and the reaction was continued for an additional 15 min. The reaction was then extracted with phenol:chloroform (1:1), (twice), chloroform (24:1 with isoamyl alcohol, once), and MT^{IS} mRNA was precipitated by the addition of 0.1 vol of RNase-free sodium acetate (3 M, pH 4.8) and 2.5 vol of absolute ethanol (22). The production of MT^{IS} mRNA was determined in parallel reactions that contained ^{32}P-UTP (at a known specific activity). A typical reaction produced 750–800 pmol of MT^{IS} mRNA/h when 1 µg of linearized plasmid was used as template.

Reverse Transcriptase Protocol

Total RNA was isolated from the skeletal muscle and liver of winter flounder by a guanidinium isothiocyanate protocol (5). Twenty-five micrograms of total RNA were used as a template for reverse transcription under the following conditions: 2.5 U r*Tth* polymerase (Perkin-Elmer), 20 pmol oligonucleotide dT$_{12-18}$, 10 mM Tris-HCl pH 8.3, 50 mM KCl, 200 µM dNTPs and 1 mM $ZnCl_2$, with or without variable amounts of MT^{IS} mRNA. The samples were incubated in a 9600 System Thermal Cycler with the following temperature profile: 25°C for 10 min, 42°C for 10 min and 70°C for 2.5 min. The reaction was halted by cooling the tubes to 2°C and adding 80 µL of TE buffer (10 mM Tris-HCl, pH 8.0, 1 mM EDTA) for a total volume of 100 µL. To determine the efficiency of the RT step, a parallel experiment containing ^{32}P-dATP (at a known specific activity) was performed. The presence of oligo-dT_{12-18} in the reaction mixture resulted in at least a 20-fold increase in trichloroacetic acid (TCA)-precipitable counts.

Primer and Luminescence Probe Design

Primers for the amplification and luminescence probes were designed with the Oligo® software package (National Biosciences, Plymouth, MN, USA), from the published winter flounder MT cDNA sequence (4). Amplification primers WF1- (TGC AAC TGC GGA GGA TC; bases 34–51) and WF2+ (biotin-GCA CAC GCA GCC AGA GG, bases 147–131) were selected to give a 114-bp and 108-bp PCR product from MT^N (the normal undeleted cDNA for MT) and MT^{IS} cDNA. Biotin was added to WF2+ to allow the subsequent capture of PCR product with streptavidin beads (i.e., streptavidin bovine serum albumin [BSA] coupled to 4.5 µm tosyl-activated magnetic beads; Dynal, Great Neck, NY, USA). Sequences were also selected for the hybridization of luminescence probes to both MT^N and MT^{IS} cDNA sequences: pMT^N {Ru}-TGC ACC ACC TGC AAC AAG AGC, bases 73–93; and pMT^{IS} {Ru}-TGC ACC ACC AAG AGC AGC, bases 73–99.

PCR

PCR conditions were optimized for primer annealing temperature, primer concentration, $MgCl_2$ concentration, AmpliTaq DNA Polymerase concentration and incubation times at denaturing, annealing and extension temperatures for WF1- and WF2+ amplification of MT^{IS} and MT^{IS} cDNA sequences. Optimal PCR conditions were identical for both 94°C for 30 s, 55°C for 30 s and 72°C for 45 s, using 3 mM $MgCl_2$, 0.3 µM primers, 1–1.5 U AmpliTaq in a 100-µL reaction volume.

For experiments that determined the accumulation of PCR product by luminescence as a function of cycle number, the PCR mixture was divided into 10 equal aliquots. One tube was placed on ice and used as the 0 cycle sample while the rest were amplified under conditions described above. One tube was then removed after each of the following cycles: 12th, 15th, 18th, 21st, 24th, 30th, 33rd and 36th and kept on ice. In some experiments, samples were removed at each cycle between 15 and 27.

To determine the relationship between the starting template copy number of MT^N and MT^{IS} cDNAs and luminescence signal generated during PCR, a series of serial dilutions were prepared of both purified cDNAs as well as purified mRNAs. The mRNAs were reverse-transcribed and then amplified as described. The template range examined was from 10^2 to 10^{12} copies of both the purified cDNAs and mRNAs. Some experiments examined the co-amplification of both templates in the same reaction vial. For these experiments, a known amount of MT^N mRNA (or cDNA) was mixed with a similar amount of MT^{IS} mRNA (or cDNA), or a constant amount of MT^N mRNA (or cDNA) was mixed with variable amounts of MT^{IS} mRNA (or cDNA) at a 0.1, 1 and 10 molar ratio.

Luminescence Procedure

To quantitate PCR product using the QPCR™ System 5000 (Perkin-Elmer), the following procedure was employed: 1–10 μL PCR mixture were mixed with 40–49 μL buffer (10 mM Tris-HCl pH 8.3, 50 mM KCl) and 10 μL luminescence probe (1–10 pmol), in a final volume of 60 μL. This mixture was heated to 94°C for 5 min, followed by cooling to the empirically derived optimal annealing for probe hybridization to PCR product. After a 10-min incubation at this temperature, 30 μg of 4.5-μm magnetic beads coupled to streptavidin were added to each tube and the incubation continued for an additional 15–20 min. PCR product (captured by bead streptavidin to biotin-oligonucleotide), was collected by centrifugation. After centrifugation, 150 μL of fresh buffer (without the probe) were added, the beads were mixed by up-and-down pipetting and allowed to hybridize for an additional 10 min. Hybridization was terminated by adding the entire hybridization mixture to a 75- × 100-mm polypropylene tube containing 400 μL of luminescence assay buffer (9). The tubes were placed in the QPCR 5000 carousel and the luminescence of each sample determined.

To determine luminescence, each sample was vortex mixed in the carousel prior to aspiration onto a working electrode. The beads (which contain the PCR product hybridized to the TBR-labeled probe) were trapped on the electrode magnetically. When a voltage was applied to the electrode, it led to a series of reactions, ultimately resulting in the emission of light at 620 nm by the ruthenium complex, which was detected and quantified by a photomultiplier tube. The dynamic range of the instrument was approximately 4 orders of magnitude.

The total detection procedure was fast (<60 min for hybridization, washing and sampling), accurate (<10% variability from experiment to experiment and <4% variability within an experiment) and reliable (as monitored by standards).

Mathematical Analysis of Amplified PCR Products

The amplification of PCR products was expected to be exponential in nature, but deviating from exponential amplification at high template concentrations (i.e., 1.0×10^8 M template concentration or 1.0×10^{11} copies in a 0.1-mL PCR) (24). The accumulation of PCR product during and outside the exponential phase can be successfully modeled with the recursive relationship:

$$Cn+1 = Cn(1+f/(1+aCn/(P_0+C_0-Cn)))$$

where $Cn+1$ = PCR product at the $n+1$ cycle; Cn = PCR product at the nth cycle, f = amplification efficiency (between 0 and 1), C_0 = original starting copy number, P_0 = original primer concentration and a = annealing efficacy between primer(s) and complementary strand(s) (2).

To solve for starting copy number, this recursive relationship was implemented in a spreadsheet program (Excel™ Version 4.0; Microsoft, Redmond, WA, USA). The spreadsheet program compares the experimentally derived PCR product concentration(s) (derived from luminescence values, see below) to concentration values derived from the model with different C_0s. The spreadsheet program alters the C_0, while holding all other factors constant, by a series of iterations (up to 1000) to minimize the difference between the experimental values and predicted values [i.e., a reduction in Σ (root mean)^2]. Experimentally, P_0 is known, and f can be determined by the portion of the amplification that is exponential in nature, i.e., slope of the line describing those values which, when plotted on a log plot, are linear. The value a can vary between 1 and 5, but in our experience usually varies between 2 and 3. After a C_0 has been calculated with a at 2.0, C_0 is held constant and Σ(root mean)^2 minimized by varying a through successive iterations. Then finally, the model holds the new a value constant and varies C_0 again, to minimize the Σ(root mean)^2. Using this procedure, the starting template copy number can be determined in the PCR and through the use of internal standards, the starting mRNA template copy number.

Calculation of Probe Luminescence

In some experiments, where C_0 was known (see above), this recursive model could be used to determine the relationship between luminescence and PCR product concentration. A number of PCRs were carried out with a serial dilution of MT^N and MT^{IS} cDNAs. The luminescence values were then mathematically converted to product concentration by assuming each luminescence value was equal to 1.0×10^{-12} mol (derived by assuming that the maximal PCR product amount possible was 3.0×10^{-7} mol, i.e., the amount of primers added). These experimentally obtained values were then compared to values derived from the model with the same C_0, P_0, f (determined as described above) and $a = 2.0$. The luminescence values were changed until a plot of experimentally derived C_0s and known C_0s were the same, i.e., regression analysis gave a P value <0.01.

To substantiate this method for the calculation of probe luminosity, PCRs with differing amounts of template were carried out as described previously.

A small fraction of each reaction was analyzed by luminescence, and the rest of the PCR mixture was fractionated on a C-18 HPLC column with an isocratic 0.1 M TEAA (triethanolamine acetate pH 7.0):acetonitrile 90:10 eluent. PCR product was calculated by OD_{260} and probe luminosity values directly calculated. The probe luminosity values were the same when calculated by the two methods (see Results).

RESULTS

PCR Optimization

The design of the mRNA internal standard is shown in Figure 1, as well as the sequence site(s) of the luminescence probe hybridization(s). Deletion mutagenesis with the described deletion primers, resulted in the predicted 6-bp deletion (determined by sequence analysis—not shown). The deleted cDNA was labeled MT[IS] and used to make mRNA internal standard as described in Materials and Methods.

Winter Flounder metallothionein cDNA, from [15].

```
                         1                              Primer WF1+→
AACCAGACAACCGCTGAGAGACATGGATCCCTGCGAATGCTCCAAGACTGGAACCTGCAACT

41                   60          Luminescence Probes        100
GCGGAGGATCTTGCACCTGCAAGAACTGCAGCTGCACCACCTGCAACAAGAGCTGCTGCCCAT

           120            ←Primer WF2-              160
GCTGCCCATCCGCCTGCCCCAAGTGCGCCTCTGGCTGCGTGTGCAAAGGGAAGACATGCGACAC

           180            200            220
CACTTGCTGTCAGTGAGAGGAGCCTGATGCACTGTGGAGCCGTGCGCCCTACACTGA
```

Primers:

   TGC AAC TGC GGA GGA TCT (Primer WF1+, 34-51)

   biotin-GCA CAC GCA GCC AGA GG   (Primer WF2-, 147-131)

Luminescence Probes:

TBR-TGC ACC ACC TGC AAC AAG AGC (pMT[N], 73-93)

TBR-TGC ACC ACC AAG AGC AGC   (pMT[IS], 73-99)

Figure 1. Description of the relevant sequences in the winter flounder MT cDNA (15), which were deleted for the generation of the mRNA internal standard, as well as the sequences which bind the amplification primer(s) and luminescence probes.

In Figure 2, Panel A, the optimal conditions are shown for the amplification of 1.0×10^8 (or 1.6×10^{-14} mol) copies of MT^N cDNA. PCRs were carried out as described, in a total volume of 50 μL for 30 cycles. One-half of the PCR mixture was applied to a 3.5% NuSieve® 3:1 GTG® agarose gel (FMC BioProducts), fractionated and analyzed by ethidium bromide staining. These optimized conditions, 1 U AmpliTaq, 0.3 μM primers, 0.2 mM dNTPs, 3 mM MgCl$_2$, 10 mM Tris-HCl pH 8.3, 50 mM KCl, were then used for all subsequent PCRs.

In Figure 2, Panel B, the amplification of a range of MT^N or MT^{IS} cDNAs (1.0×10^{12} to 1.0×10^6 copies or 1.6×10^{-18} to 1.6×10^{-12} mol) is shown. In Figure 2, Panel C, the co-amplification of both cDNAs is shown. By this detection method, i.e., ethidium bromide staining, amplified product could be

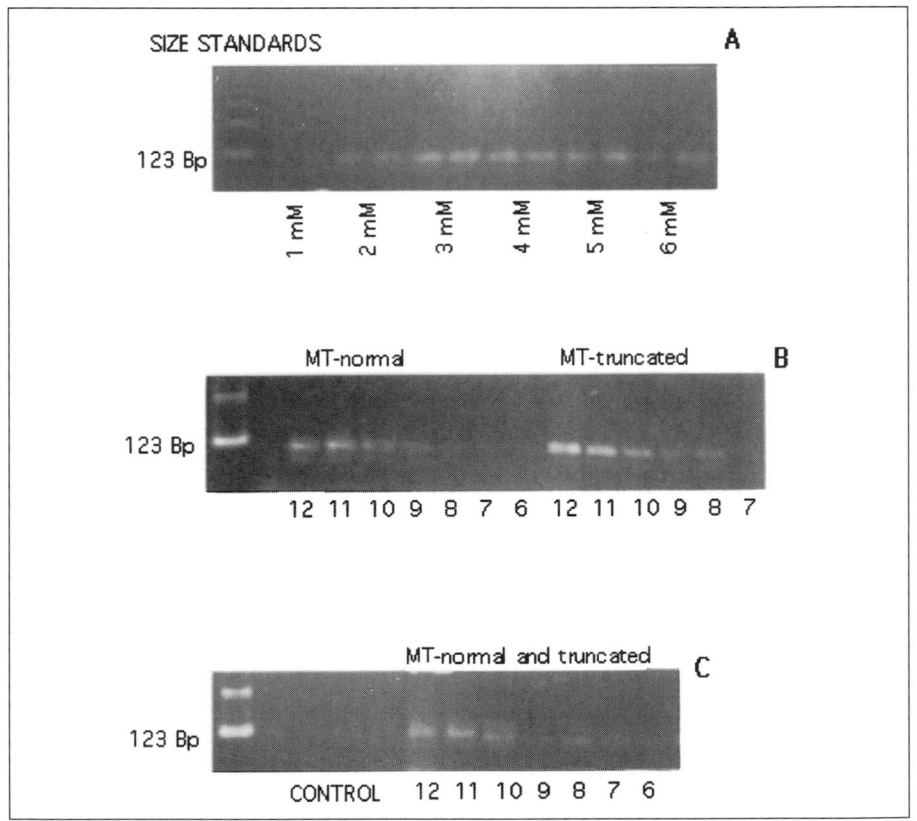

Figure 2. Analysis of PCR amplifications of MT^{IS} and MT^N cDNAs. Panel A: amplification of MT^N cDNA (0.16 fmol starting template content) as a function of MgCl$_2$ concentration. PCR methods are described in Materials and Methods. The PCR product was 114 bp and was fractionated on a 3.5% NuSieve 3:1 GTG agarose gel. PCR products were stained with ethidium bromide. In Panel B: amplification of MT^{IS} and MT^N cDNAs under optimal PCR conditions, as a function of starting template content. Starting template contents ranged from 1.0×10^6 to 1.0×10^{12} copies. In Panel C: coamplification of MT^{IS} and MT^N cDNAs at a range of starting template contents. In Panels B and C, the numerical values refer to the log of starting template content.

seen only in those reaction mixtures that started with $>1.0 \times 10^7$ copies of the target template. In addition, this procedure could not detect differences between MT^N or MT^{IS} amplified products.

Luminescence Probe Hybridization

Prior to the use of luminescence to detect PCR-amplified MT^N or MT^{IS} sequences, the specificity and optimal annealing temperature(s) for the two

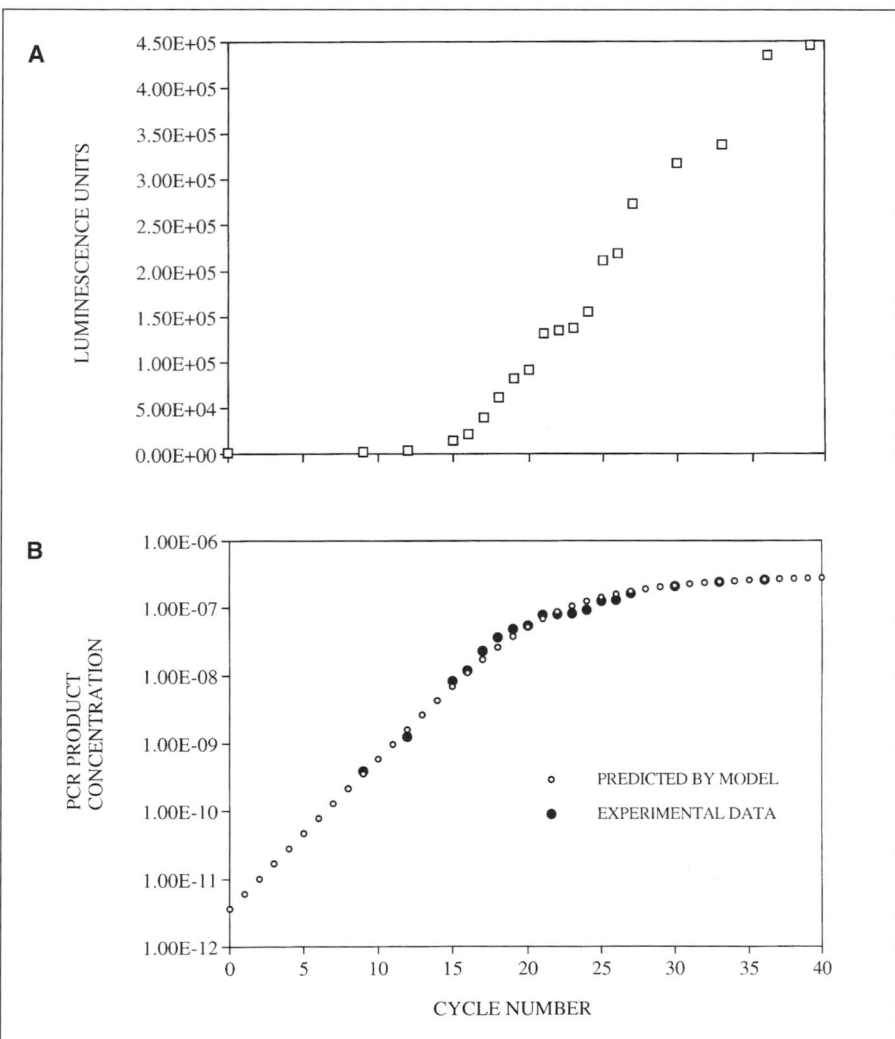

Figure 3. A) Amplification of MT^N cDNA (1.1×10^8 copies) by PCR as detection of product by luminescence. Luminescence values were corrected for the total PCR volume as well as the volume of sample that is used by the QPCR 500 instrument. B) These data from cycles 9–36 were converted to mass PCR product with 1 luminescence unit = 0.15×10^{-12} mol and fitted to the equation generated by the recursive model.

161

luminescence probes, pMT^N or pMT^{IS}, were determined. For these experiments, a 100-μL PCR mixture containing 1.6×10^{-14} mol starting template of either cDNA were amplified, and 2.5 μL of each reaction were mixed with both probes, over a range of concentrations and temperatures. Though not shown, hybridization was both concentration- and temperature-dependent, as expected. From these experiments, optimal conditions for pMT^N hybridiza-

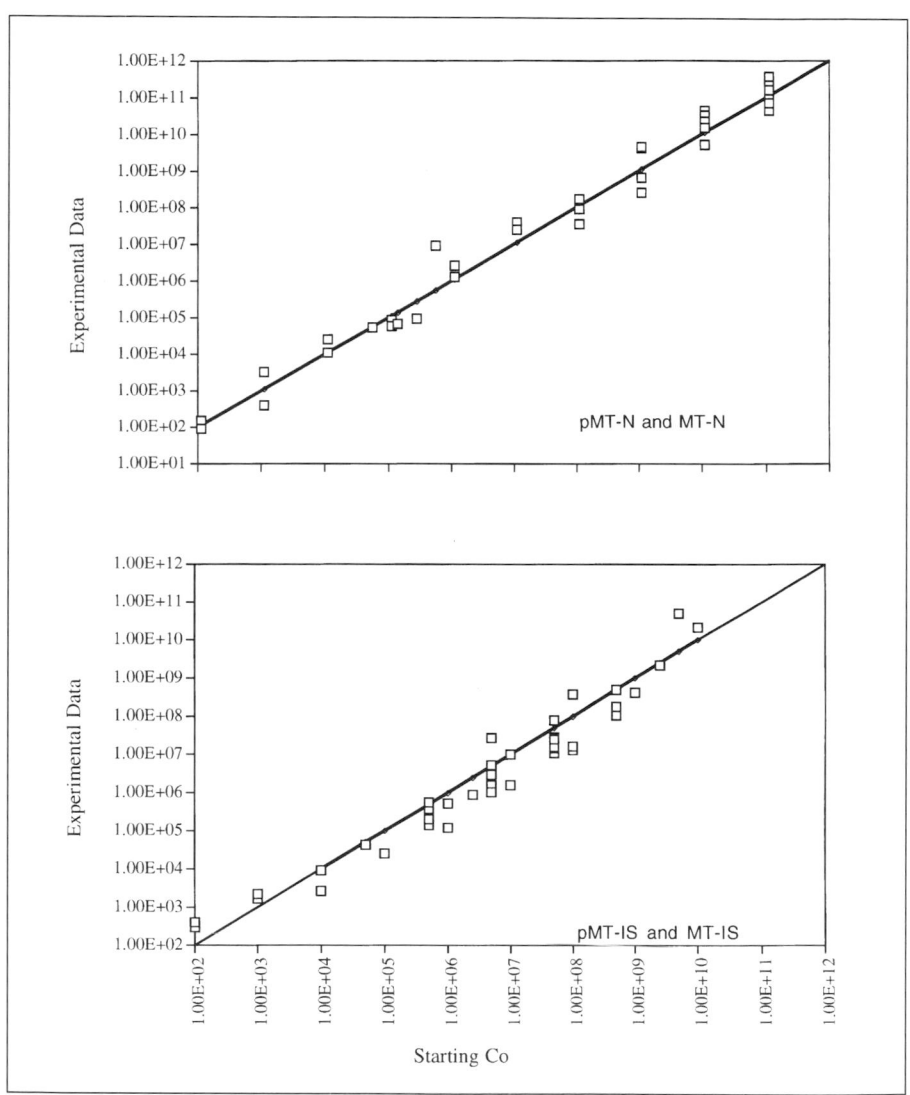

Figure 4. The relationship between the known starting template content and C_0s generated by the recursive model from the RT-PCR of a wide range of MT^{IS} and MT^N mRNAs. The heavy black line represents the expected values with the points generated by the recursive model from the experimental data.

tion were 4 pmol and a 60°C annealing temperature, defined operationally as the greatest signal with the best specificity (i.e., no cross-hybridization to MT^{IS}). The other probe, pMT^{IS}, hybridized to its target sequences in a concentration- and temperature-dependent manner, with optimal conditions—5 pmol and 65°C annealing temperature.

Quantitative PCR

Figure 3A shows a typical amplification of MT^{N} cDNA, with detection of product by luminescence. In this experiment, MT^{N} cDNA starting template was 1.1×10^{8} copies. The first cycle that gave luminescence signal greater than control (i.e., the luminescence of the 0 cycle) was the 9th, and the increase in signal was sigmoidal through the 36th cycle. This detection procedure was sensitive to an increase of 1.5×10^{-12} mol PCR product over control. In Figure 3B, the same data from Figure 3A were converted to PCR product concentration by the conversion factor 1 luminescence unit = 0.15×10^{-12} mol of MT^{N}. This conversion factor was obtained as described in Materials and Methods.

If examined more closely, the increase in luminescence was exponential through the 21st cycle, as demonstrated by the linear increase in PCR product on a log plot, after which the rate of increase slowed (Figure 3B). During the exponential portion of the amplification, the luminescence signal increased at the rate of 1.6-fold per cycle. In other experiments, with different starting copy numbers, the luminescence signal increased between 1.1-fold and 1.8-fold per cycle. There was an inverse relationship between starting template copy number and the fold increase per cycle, with the highest fold increases seen with very low starting template copy numbers (e.g., 1.8-fold to 1.9-fold per cycle with 1.0×10^{3} copies). It is clear from Figure 3B that the recursive model correctly predicted PCR product accumulation outside the exponential range. When used to solve for C_0, the model predicted that a value of 1.25×10^{8} starting copies was not significantly different than the correct value—1.1×10^{8} copies.

Figure 4 shows the generality of the recursive model to determine C_0s, when both MT^{N} or MT^{IS} cDNAs are used at different starting template concentrations (from 1.0×10^{2} to 1.0×10^{11} copies). In Figure 4, top panel, the recursive model was used to solve C_0 for a series of experiments in which known amounts of MT^{N} cDNA were amplified. The same conversion factor described above was used to convert luminescence signal to PCR product mass. In Figure 4 the heavy line represents the known copy numbers of MT^{N} cDNA in each sample, with the points being those values of C_0 generated by the model from the experimental data. Statistical analysis indicates that there exists a highly significant correlation ($P < 0.001$) between the expected and experimental values with $r = 0.991$.

In Figure 4, bottom panel, the recursive model was used to solve C_0 for a series of experiments in which known amounts of MT^{IS} cDNA were

Table 1. Assay Accuracy for Detection of Starting Template Concentration When MT^N and MT^{IS} cDNAs are Assayed Separately and Together

Sample	Amount Added		$C_0 \pm SD$
MT^N	1.12×10^{11}		$2.05 \times 10^{11} \pm 1.12 \times 10^{11}$
	1.12×10^{10}		$2.69 \times 10^{10} \pm 1.13 \times 10^{10}$
$+MT^{IS}$		$+ 1.0 \times 10^{11}$	$1.22 \times 10^{11} \pm 0.65 \times 10^{11}*$
		$+ 1.0 \times 10^{10}$	$1.58 \times 10^{10} \pm 1.10 \times 10^{10}*$
MT^{IS}	5.00×10^7		$1.73 \times 10^7 \pm 2.01 \times 10^7$
	5.00×10^6		$2.40 \times 10^6 \pm 2.02 \times 10^6$
$+MT^N$		$+ 1.12 \times 10^8$	$1.21 \times 10^7 \pm 2.72 \times 10^7*$
		$+ 1.12 \times 10^7$	$2.90 \times 10^6 \pm 1.80 \times 10^6*$

C_0 determined from recursive model.

* Not significantly different than sample without added other mRNA, with $P < .1$ (or greater) and $n = 5$

amplified. The conversion factor for pMT^{IS} was 0.12×10^{-12} mol/luminescence unit compared to 0.15×10^{-12} mol/U for pMT^N. Statistical analysis indicates that there was a highly significant correlation ($P < 0.001$) between the expected and experimental values with $r = 0.997$. These results clearly indicate that the dynamic range of this assay is over 8–9 log units and that the assay (and recursive model) can accurately detect starting template copy number over this wide range of concentrations.

mRNA Detection

The ability of the assay to indirectly determine MT^N or MT^{IS} mRNA was then examined. For these experiments, total RNA isolated from mammalian skeletal muscle (known to have no detectable PCR signal with these MT primers) was mixed with various amounts of either MT^N or MT^{IS} mRNA, generated from the relevant plasmids with T_7 RNA polymerase, alone or as a mixture of both. After the RT step, the solutions were used as templates for PCR analysis. When 1.12×10^{11} copies of MT^N mRNA were assayed a number of times, the experimental value(s) obtained by this procedure, i.e., $2.05 \times 10^{11} \pm 1.12 \times 10^{11}$ ($n = 5$, with mean $\pm$SD), was not significantly different than the known starting template content (Table 1). To do this analysis, it was assumed that there was a one-to-one conversion of MT^N mRNA to cDNA in the RT step. The same results were obtained when 1.12×10^{10} copies of MT^N mRNA were used or 5.0×10^7 and 5.0×10^6 copies of MT^{IS} mRNA. The variability was typical of the error seen when determining either mRNA by this method and was not substantially different over a wide range of mRNA concentrations. When calculating starting mRNA concentrations by this assay with the recursive model, three to five runs of the same sample, carried out on the same and different days, were averaged.

Table 2. The Effect(s) of Various Amounts of MT^{IS} mRNA on the Reverse Transcription and Amplification of MT^N mRNA

Sample #	MT^N mRNA (exp.:obs. - C_0)	MT^{IS} mRNA exp.:obs. - C_0)	MT^N mRNA (corrected)
1	$1.1 \times 10^9 : 1.0 \times 10^9$	$1.9 \times 10^8 : 1.66 \times 10^8$	1.15×10^8
2	$1.1 \times 10^9 : 0.34 \times 10^9$	$1.9 \times 10^9 : 0.95 \times 10^9$	0.68×10^9
3	$1.1 \times 10^9 : 0.44 \times 10^9$	$1.9 \times 10^{10} : 1.28 \times 10^{10}$	0.65×10^{10}
average	$0.61 \times 10^9 \pm 0.38 \times 10^9$*		$0.83 \times 10^9 \pm 0.28 \times 10^9$*

C_0 determined from recursive model.
* Not significantly different than expected with $P < 0.086$, with $n = 5$

The next series of experiments examined the effects of the presence of both mRNAs on the RT step and PCR analysis. In these experiments, the amounts of each mRNA was the same in a specific sample and covered a wide range (i.e., 10^6, 10^7, 10^{10} and 10^{11} copies). The samples were treated as described above, and the same PCR mixture was separately probed with pMT^N or pMT^{IS}. It is clear from Table 1 that co-RT and co-amplification of equal amounts of both templates do not affect subsequent determination of C_0. This suggests that the presence of equal amounts of both templates does not interfere with amplification of either, or subsequent detection by the two luminescence probes.

Experiments were then carried out to determine the effect of mixing variable amounts of MT^{IS} mRNA with a constant amount of MT^N mRNA on subsequent detection of both mRNAs by this procedure. In this experiment, 1.1×10^9 copies of MT^N mRNA were mixed with 1.9×10^7, 1.9×10^8, 1.9×10^9 and 1.9×10^{10} copies of MT^{IS} mRNA and the mixtures treated as described. Each PCR mixture was probed separately with either pMT^N or pMT^{IS}. It is clear in Table 2 that the amount of MT^N mRNA determined experimentally ($0.61 \times 10^9 \pm 0.38 \times 10^9$) was not significantly different than the amount added. The amount(s) of MT^{IS} mRNA determined experimentally was then used to correct MT^N mRNA values for variability in the RT and PCR steps. To do this, the MT^N mRNA values were corrected by the ratio: observed value for the internal standard/expected value for the internal standard, and the corrected values, $0.80 \times 10^9 \pm 0.24 \times 10^9$, were not significantly different than the expected values ($P < 0.086$, paired Student's t test). Most importantly, the error associated with the corrected values (ca. 25%–30%) was less than that associated with uncorrected values (ca. 40%–50%).

Finally, this assay was used to determine winter flounder MT^N mRNA contents in the livers of fish from control and from tanks treated with $CdCl_2$ at 50 ppb. It was determined that in control fish, MT liver mRNA values were approximately 1.0×10^6 copies per µg total RNA and that $CdCl_2$-treated samples had approximately 1.0×10^9 copies per µg total RNA. These biological levels fall within the middle portion of the dynamic range of this assay. When the copy number is converted to mass of mRNA [i.e., using the size of the full-length mRNA as reported by Chan et al. (4)], in control tissues, the mRNA for MT is 0.001%–0.004% of total mRNA (assuming mRNA is 1%–5% of total RNA). Treatment with $CdCl_2$ at 50 ppb increases this mRNA to 1%–4% of total RNA.

DISCUSSION

A new procedure is reported here for the determination of specific mRNA copy number in biological samples by RT-PCR, with mRNA internal standards, by direct determination of amplified products by luminescence. The use of a luminescence-based detection system for PCR product accumulation has several advantages over other detection systems, in that, it is rapid, accurate, reliable, sensitive and allows for direct determination of PCR product by nonradioactive or nonhazardous means.

The luminescence probes (pMT^N or pMT^{IS}) were shown to be specific for their targets (i.e., MT^N or MT^{IS} templates). The hybridization of the probes was concentration- and temperature-dependent, which was not unexpected, since annealing of the probe to the target occurs prior to assaying luminescence, with a large probe excess. Under the hybridization-wash conditions selected, both probes could readily discriminate the absence or presence of the 6-bp deletion, since there was no significant cross-hybridization. In other experiments, we have noted that these types of probes can detect 2-bp differences by the same hybridization-wash protocol.

The luminescence signal generated by the probes was very similar when equal amounts of the PCR products were present. The sensitivity of the assay was about 10 luminescence units above background, or a change in PCR product of 1.5×10^{-12} mol. It was possible to detect increases in PCR product in samples after the 6th cycle, when high starting copy numbers were used ($>1.0 \times 10^{10}$ copies), and in samples with much lower starting copy numbers (ca. 1.0×10^5), increases in product were detectable after the 15th cycle. In early cycles, increases in PCR product were exponential in nature, i.e., it could be modeled with the equation $\mathbf{y} = k10^{xs}$, where x = cycle number, s = slope, k = constant and $\mathbf{y}$ = PCR product at a given cycle. But in samples with high copy number ($>1.0 \times 10^9$ copies), there were very few cycles in which product accumulation was exponential, and in all samples, when total PCR product began to exceed 1.0×10^{-8} M, product accumulation was linear. Presumably, this is due to the increasingly effective self-hybridization of the double-stranded product compared with the hybridization of the primer to the product (23).

The recursive model was capable of successfully modeling increases in PCR product outside the exponential phase, and thus, the model increases in product when the starting template ranged over 9 log units. Furthermore, the model was readily adapted to a spreadsheet program for the rapid calculation of C_0 in unknown samples by curve-fitting. Others have reported the ability to detect starting template content and detectable PCR product over a 3–4 log range (3,7,10). This procedure is many orders of magnitude more sensitive and can readily detect starting template contents of 10^2 copies.

The accuracy of this assay (defined as the ability to correctly determine the amount of mRNA in a sample) was shown to be quite high over an 8–9 log range ($r = 0.991$ to 0.997). The precision (defined as the error associated with multiple sampling of the same sample) was on the order of 50% when not corrected with the internal standard. When corrected for variabilities in both the RT and PCR steps, errors of 20%–30% were common. The greatest variability in these measurements was found to be in the RT step, which has been reported by others (7), and is likely due to inhibitors co-purified with the RNA.

To effectively determine mRNA levels in biological samples, one must either assume that there is a one-to-one relationship between the starting template copy number in PCR and mRNA in the RNA pool or use internal standards to account for variability in these steps. For the internal standard to be effective, it must not interfere with the RT step for the target mRNA nor its amplification but must be efficiently transcribed and amplified to correct sample-to-sample variation. Ideally, the internal standard should be as close as possible in sequence to the target mRNA, to ensure the same efficiencies in both steps. Internal standards have been developed by others (3,8,12,13), but in this case, the mRNA internal standard is identical to the MT mRNA except for a 6-bp deletion. This has the advantage of allowing for the use of the same primers, presumably the same optimal conditions for the RT and PCR steps and the same efficacy of each step for the two templates. This was demonstrated by conventional agarose gel electrophoresis:ethidium bromide detection of PCR products. But conventional analysis was not able to differentiate the two products. Others have reported the use of internal standards with 30–50-bp deletions (3,8) to differentiate the different amplification products by these means. Presumably, PCR products with these relatively large differences in size (especially with small products of <300 bp) would not be amplified with the same efficacy.

Even though the templates were almost identical, both luminescence probes could readily discriminate between the two, even when both were present in equal or one at greater amounts. The presence of the internal standard at molar ratios of 0.1, 1 and 10 did not affect reverse transcription or amplification of the target template. Variability in the reverse transcription or amplification of the internal standard could then be assumed to reflect variability in both steps for the target template.

The use of internal standards did not affect the accuracy of the assay but the error associated with sampling. The typical error associated with determination of target mRNA without the internal standard was >50%, but when values were corrected for the variability demonstrated by the internal standard, errors decreased to about 20%–30%.

The assay was then used to measure liver MT mRNA levels in winter flounder. In control animals, levels were low, 1.0×10^6 copies per μg total RNA or 0.001%–0.004% of total mRNA, as expected. In $CdCl_2$-treated samples, levels were increased 1000-fold to 1.0×10^9 copies per μg total RNA or 1%–4% of total RNA. $CdCl_2$ has been previously reported to be a potent inducer of liver MT levels in winter flounder. Increases of 50-fold to 100-fold in MT protein levels after $CdCl_2$ treatment have been reported (11,14,16,20,25). The increases in MT mRNA levels reported here are consistent with increases of MT mRNA reported by others with Northern blot analysis under the same inducing conditions.

The RT-PCR procedure described here is a powerful and sensitive means to detect mRNA content in a wide range of biological samples, and the use of an internal standard allows for corrections in the RT and PCR steps. The detection of PCR product is rapid, sensitive and accurate and nonradioactive and nonhazardous. This procedure should be useful in many biological applications.

ACKNOWLEDGMENTS

The authors would like to thank the following people for their help in the preparation of this manuscript and for supplying reagents or advice for the experiments described in this paper: Dr. Lisa Kaplan (New England College, Henniker, VT), Dr. B. Chapple and Dr. A. Moiseff (University of Connecticut), Dr. J. DiCesare and Dr. J. Atwood (Perkin-Elmer). This work was supported in part by grants from the University of Connecticut Research Foundation and the Perkin-Elmer Corporation.

REFERENCES

1.**Arnheim, N. and H. Erlich.** 1992. Polymerase chain reaction strategy. Ann. Rev. Biochem. *61*:131-156.
2.**Atwood, J.** Patent pending. 1994.
3.**Bouaboula, M., P. Legoux, B. Pessegue, B. Delpech, X. Dumont, M. Piechaczyk et al.** 1992. Standardization of mRNA titration using a PCR method involving co-amplification with a multispecific internal control. J. Biol. Chem. *267*:21830-21838.
4.**Chan, K.M., W.S. Davidson, C.L. Hew and G.L. Fletcher.** 1989. Molecular cloning of metallothionein cDNA and analysis of metallothionein gene expression in winter flounder tissues. Can. J. Zool. *67*:2520-2527.
5.**Chomczynski, P.A. and N. Sacchi.** 1987. Single-step method of RNA isolation by acid guanidinium thiocyanate-phenol-chloroform extraction. Anal. Biochem. *162*:156-159.
6.**Clairborne, S. and J. McCombs.** 1992. Rapid production of DNA controls for PCR. BioTechniques *12*:372-374.
7.**Cone, R.W., A.C. Hobson and M.W. Huang.** 1992. Coamplified positive control detects the inhibition of PCR. J. Clin. Microbiol. *30*:3185-3189.
8.**Delidow, B.C., J.J. Peluso and B.A. White.** 1989. Quantitative measurements of mRNAs by PCR. Gene Anal. Tech. *6*:120-124.

9. Di Cesare, J., B. Grossman, E. Katz, E. Picozza, R. Ragusa and T. Woudenberg. 1993. A highly-sensitive electrochemiluminescence-based detection system for automated PCR product quantitation. BioTechniques *15*:152-157.

10. Diviacco, S., P. Norio, L. Zentilin, S. Menzo, M. Clementi, G. Biamonti et al. 1992. A novel procedure for quantitative PCR by coamplification of competitive templates. Gene *112*:313-320.

11. George, S.G. 1990. Biochemical and cytological assessments of metal toxicity in marine animals, p. 124-136. *In* S.G. George (Ed.), CRC Press, Boca Raton.

12. Gilliland, G., S. Perrin, K. Blanchard and H.F. Bunn. 1990. Analysis of cytokine mRNA and DNA: detection and quantitation by competitive PCR. Proc. Natl. Acad. Sci. 87:2725-2729.

13. Henco, K. and M. Heibey. 1990. Quantitative PCR: the determination of template copy numbers by temperature gradient gel electrophoresis (TGGE). Nucleic Acids Res. *18*:6733-6734.

14. Holwerda, D.A., F.A. Knecht, F. Hemelraad and P.R. Veenhof. 1989. Cd kinetics in freshwater clams. Uptake of Cd by the excised gill of *Anodota anatina*. Bull. Environ. Contam. Toxicol. *42*:382-388.

15. Imai, Y., Y. Matsushima, T. Sugimura and M. Terada. 1991. A simple method for generating a deletion by PCR. Nucleic Acids Res. *19*:2785.

16. McCarthy, J.F. and L.R. Shugart. 1990. Stress proteins: potential as multitiered biomarkers, p.165-199. J.F. McCarthy and L.R. Shugart (Eds.), Biomarkers of Environmental Contamination. Lewis Publishing, New York.

17. Mulder, J., M. McKinnney, C. Christopherson, J. Sninsky, L. Greenfield and S. Kwok. 1994. Rapid and simple PCR assay for quantitation of human immunodeficiency virus type I RNA in plasma —application of acute retroviral infection. J. Clin. Microbiol. *32*:292-300.

18. Nedelman, J., P. Heagerty and C. Lawrence. 1992. Quantitative PCR: procedures and precisions. Bull. Math. Biol. *54*:477-502.

19. Nedelman, J., P. Heagerty and C. Lawrence. 1992. Quantitative PCR with internal controls. Cabios *8*:65-70.

20. Poulsen, E., H.U. Riisgard and F. Mohlenberg. 1982. Accumulation of Cd and bioenergetics in the mussel, *Mytilus edulis*. Mar. Biol. *68*:25-29.

21. Rappolee, D.A., D. Wang, D. Mark and Z. Werb. 1989. Novel method for studying mRNA phenotypes in single or small number of cells. J. Cell. Biochem. *39*:1-11.

22. Sambrook, J., E.F. Fritsch and T. Maniatis. 1989. Molecular Cloning: A Laboratory Manual. Cold Spring Harbor Laboratory Press, Cold Spring Harbor, NY.

23. Sardelli, A.D. 1993. Plateau effect—Understanding PCR limitations. Amplifications *9*:1-3.

24. Singer-Sam, J., M.O. Robinson, A.R. Belive, M.I. Simon and A.D. Riggs. 1990. Measurement by quantitative PCR of changes in HPRT, PGK-1, PGK-2, APRT, MTase and Zfy gene transcripts during mouse spermatogenesis. Nucleic Acids Res. *18*:1255-1259.

25. Viarengo, A., M.N. Moore, G. Mancinelli, A. Mazzucotelli, R.K. Pipe and S.V. Farrar. 1987. Metallothioneins and lysosomes in metal toxicity and accumulation in marine mussels: the effects of Cd in the presence and absence of phenanthrene. Marine Biol. *94*:251-257.

26. Wang, A.M., M.V. Doyle and D.F. Mark. 1989. Quantitation of mRNA by the polymerase chain reaction. Proc. Natl. Acad. Sci. USA 86:9717-9721.

Update to:

Quantitation of Metallothionein mRNA by RT-PCR and Chemiluminescence

Joseph F. Crivello

University of Connecticut, Groton, CT, USA

Quantitative polymerase chain reaction (QPCR) in association with cDNA synthesis (i.e., reverse transcription PCR [RT-PCR]) has rapidly supplanted other methods for quantitating mRNA sequences, such as Northern blots and RNase protection assays. In the past few years, beginning with the work of Becker-Andre and Hahlbrook (2), Gilliland et al. (3), and Wang et al. (10), QPCR, which uses an internal competitively amplified standard, has become the method of choice for the rapid, sensitive, accurate quantitation of specific DNA or RNA templates for as little as 10 copies.

There is general agreement on the conditions required for successful QPCR (9,11):

- optimized methods for the isolation of nucleic acids
- optimized conditions for reverse transcription
- optimization of PCR conditions
- suitable internal standards that use the same primers and have the same amplification efficiency
- the ability to detect PCR product formation in the non-exponential phase

The critical differences arise in the detection and analysis of PCR products. Optimally, PCR product is detected in "real time" to decrease additional manipulations, which not only take more time but increase the possibility of error. Until recently, PCR product formation could be determined by conventional gel electrophoresis with ethidium bromide staining, capillary gel electrophoresis, HPLC, fluorescence-labeled deoxyribonucleoside triphosphates (dNTPs) coupled with automated laser fluorometers, biotinylated primers coupled with streptavidin coupled to solid supports, incorporated radioactivity, digoxigenin incorporation and finally immunoenzymatic detection.

This manuscript describes early work that exploited chemiluminescence for quantitation of PCR product. This protocol still involved post-PCR manipulation, i.e., capture of biotinylated PCR product on streptavidin-coated magnetic beads, and a dedicated instrument to determine the TBR (tris(2,2'-bipyridine)ruthenium II chelate) reporter group. This approach had the advantage of tremendous sensitivity (detection of $7.4 \times 10{-}11$ mol of PCR product above background), very high throughput, and a sophisticated software package (developed by John Atwood, Perkin-Elmer), which could accurately determine starting template copy number after PCR had entered the non-exponential phase. This approach could determine starting template copy number over 10 orders of magnitude (from 10 copies to 10^{11}) with 90% accuracy (see Reference 9).

Recently, a true real-time QPCR system was developed (4); this approach exploits the TaqMan™ technology from Perkin-Elmer. QPCR is carried out with three primers, two for amplification and a third that hybridizes internally in the amplified region. This internal primer has two fluorescent tags, a reporter fluorescent dye (such as FAM, 6-carboxyfluorescein), which has its emission spectra quenched by a second fluorescent dye (such as TAMRA, 6-carboxy-teramethyl rhodamine). The 5′ nuclease activity of *Taq* polymerase (5) degrades the internal probe, releasing the quenching of the fluoroescent reporter dye, resulting in an increased fluorescent emission (which in the case of FAM is 418 nm) (1,6,7,8). The fluorescence emission of the PCR tube can be determined during PCR, making this a true real-time determination of PCR product formation. This approach is not as sensitive as chemiluminescence (the limit of detection is about 10000 copies) but it has very high throughput and allows for the detection of target template amplification and internal standard at the same time (by use of different fluorescent tags).

QPCR technologies are becoming increasingly sophisticated and, in time, will become a standard of modern molecular biology.

REFERENCES

1.**Bassler, H.A., S.J. Flood, K.J. Livak, J. Marmaro, R. Knorr and C.A. Batt.** 1995. Use of a fluorogenic probe in a PCR-based assay for the detection of Listeria monocytogenes. App. Environ. Microbiol. *61*:3724-3728.
2.**Becker-Andre, M. and K. Hahlbrock.** 1989. Absolute mRNA quantification using the polymerase chain reaction (PCR). A novel approach by a PCR aided transcript titration assay (PATTY). Nucleic Acids Res. *22*:9437-9446.
3.**Gilliland, G., S. Perrin, K. Blanchard and H.F. Bunn.** 1990. Analysis of Eytokine mRNA and DNA: detection and quantitation by competitive polymerase chain reaction. Proc. Natl. Acad. Sci. USA *87*:2725-2729.
4.**Heid, C.A., J. Stevens, K.J. Livak and P.M. Williams.** 1996. Real time quantitative PCR. Genome Methods *6*:986-994.
5.**Holland, P.M., R.D. Abramson, R. Watson and D.H. Gelfand.** 1991. Detection of specific polymerase chain reaction product by utilizing the 5′A3′ exonuclease activity of Thermus aquaticus DNA polymerase. Proc. Natl. Acad. Sci. USA *88*:7276-7280.
6.**Lee, L.G., C.R. Connell and W. Bloch.** 1993. Allelic discrimination by nick-translation PCR with fluorogenic probes. Nucleic Acids Res. *21*:3761-3766.
7.**Livak, K.J., S.J. Flood, J. Marmaro, W. Giusti and K. Deetz.** 1995. Oligonucleotides with fluorescent dyes at opposite ends provide a quenched probe system useful for detecting PCR product and nucleic acid hybridization. PCR Methods Appl. *4*:357-362.
8.**Livak, K.J., J. Marmaro and J.A. Todd.** 1995. Towards fully automated genome-wide polymorphism screening. Nat. Genet. *9*:341-342.
9.**Raemaekers, L.** 1995. A commentary on the practical applications of competitive PCR. Genome Res. *5*:91-94.
10.**Sunderman, F.W., M.C. Plowman, O. Slaisova, S. Grbac-Ivankovic, L. Foglia and J.F. Crivello.** 1995. Effects of Teratogenic concentrations of Zn+2, Cd+2, Ni+2, Co+2, and Cu+2 in FETAX assays

on metallothionein (MT) and MT-mRNA contents of *Xenopus* laevis embryos. Pharmacol. Toxicol. *76*:178-184.

11. **Wang, A.M., M.V. Doyle and D.F. Mark.** 1989. Quantitation of mRNA by the polymerase chain reaction. Proc. Natl. Acad. Sci. USA *86*:9717-9721.

12. **Zimmerman, K. and J.W. Mannhalter.** 1996. Technical aspects of quantitative competitive PCR. BioTechniques *21*:268-279.

CHAPTER 13

Quantitation of PCR Products with Chemiluminescence

C.S. Martin, L. Butler and I. Bronstein

Tropix, Bedford, MA, USA

ABSTRACT

Quantitative PCR and reverse transcription PCR (RT-PCR) are widely used in biomedical, industrial and other research applications to determine the number of RNA or DNA molecules of a specific type and/or sequence in a sample of interest. We have developed an assay system to accurately quantitate PCR products that utilizes solid-phase capture and an enzyme-linked chemiluminescent detection method. The entire assay is performed in a single tube or microplate well. Biotinylated PCR products are quantitated by capture onto a streptavidin-coated surface, followed by hybridization of an internal fluorescein-labeled oligonucleotide probe and subsequent detection with an anti-fluorescein-alkaline phosphatase conjugate and CSPD® chemiluminescent substrate. Light signal is measured in a luminometer. The assay sensitivity enables accurate quantitation of target DNA because the measurement is performed on product generated during the exponential phase of amplification. The broad dynamic range of the assay, which is greater than three orders of magnitude of PCR product concentration, simplifies the determination of the number of amplification cycles necessary for accurate quantitation of target molecules. The PCR-Light™ system is an ultrasensitive, non-isotopic and rapid assay for PCR product detection that also has general application to solution hybridization assays and other quantitation methods.

INTRODUCTION

The development of amplification techniques utilizing the polymerase chain reaction (PCR) (16) has enabled the detection of extremely small quantities of DNA or RNA. PCR methods are now widely used to determine the presence of specific nucleic acid sequences. More recently, methods have been developed for quantitating nucleic acids with PCR. Quantitative PCR is based on establishing a relationship between the initial concentration of target DNA or RNA and the amount of PCR product generated (7,10,19).

(Reprinted from BioTechniques 18:908-912, 1995)

During PCR amplification, an exponential accumulation of product DNA occurs. Assuming the reaction efficiency for each amplification cycle remains constant, the concentration of product DNA following PCR is directly proportional to the amount of initial target DNA. With increasing cycle number, the rate of accumulation of product eventually becomes non-exponential and reaches a "plateau". At "plateau", the relationship between the initial number of molecules and the DNA product no longer exists; thus, accurate quantitation is not possible (10). In practice, quantitative PCR determination is performed by selecting a number of amplification cycles so that the rate of product accumulation is exponential, and a linear relationship between the initial number of molecules and the amplified target is established. The concentration of target DNA in an unknown sample is determined from a standard curve generated by amplification of known quantities of target DNA. Since reaction efficiency may vary between samples because of temperature gradients, incomplete denaturation and product strand reannealing (12), it is useful to co-amplify a known control target DNA of similar size and known quantity in the same reaction vessel.

Quantitative PCR is used to determine viral copy number or level of infection in cell cultures, blood and tissues (10). The level of bacterial

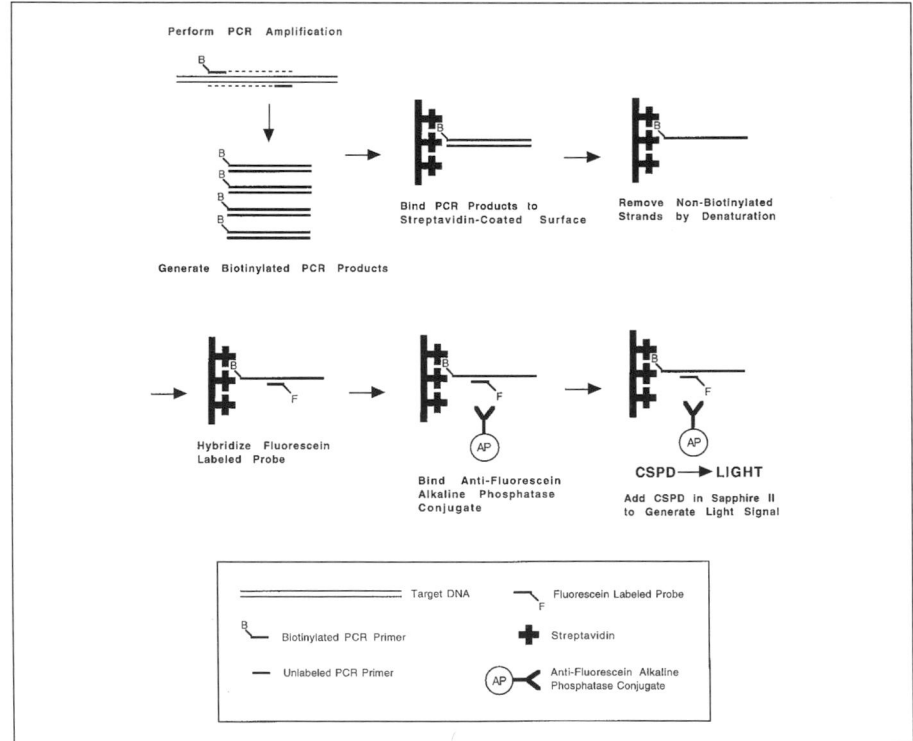

Figure 1. PCR-Light assay system.

contamination in food samples may also be quantitated with PCR at a much faster rate compared with culture techniques (5). Accurate RNA quantitation to measure cellular gene expression can be accomplished with reverse transcription PCR (RT-PCR), in which RNA message is converted to complementary DNA sequence by reverse transcription and is subsequently PCR-amplified (1,2,11,15,18).

PCR quantitation is dependent on the ability to accurately determine the concentration of PCR products after amplification. Widely used methods for post-PCR product detection include agarose and polyacrylamide gel electrophoresis. With these techniques, specific PCR products are identified by their mobility and subsequently quantified by a variety of methods, which are considered semiquantitative at best. Products separated on an agarose gel are visualized by ethidium bromide staining. More sensitive quantitation is achieved when PCR products are transferred to a membrane and detected by hybridization with radiolabeled probes (7). Relative quantitation is achieved by film densitometry or with a phosphor screen imaging device. Alternatively, radiolabeled nucleotides or primers may be incorporated during the amplification process. Products are subsequently separated by gel electrophoresis, excised and quantitated by scintillation counting. These methods are laborious, require hazardous radioactive compounds, exhibit poor sensitivity and assay precision. Recently, solid-phase capture and ligand-binding assays for the detection of nucleic acids have been developed, including radioisotopic (17) and non-isotopic detection methods (1,6,8,13,14). These assays are performed in a single tube or a microplate well, exhibit increased assay precision and are less labor-intensive than gel electrophoresis methods.

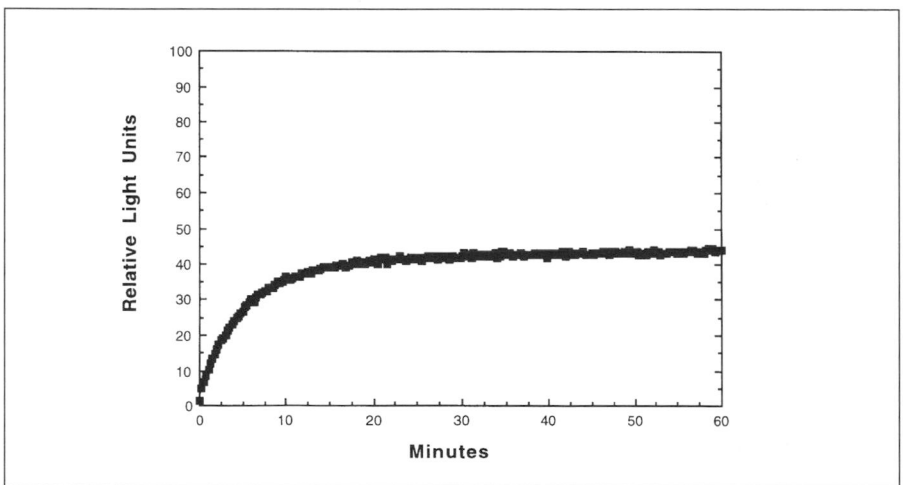

Figure 2. Light emission kinetics of CSPD with Sapphire II. 7.35×10^{-8} mg of purified alkaline phosphatase was incubated with 0.4 mM CSPD in 0.1 M diethanolamine, pH 10.0, 1 mM $MgCl_2$, 10% Sapphire II. Light signal was measured in a Turner Model 20e luminometer (Turner Designs, Mountain View, CA, USA).

We have developed the PCR-Light™ system, an extremely sensitive, non-radioactive assay for detection of PCR products. The PCR-Light system utilizes rapid hybridization of an oligonucleotide probe followed by an enzyme-linked chemiluminescent detection method (Figure 1). The assay is performed in streptavidin-coated microplates, or with streptavidin-coated beads or microparticles, and can be completed in three hours. The assay sensitivity enables detection of low levels of product generated during the early exponential phase of amplification, permitting accurate quantitation over a wide range of initial target concentrations. The PCR-Light assay employs CSPD®, a direct chemiluminescent substrate for alkaline phosphatase (4). Upon enzymatic dephosphorylation, CSPD decomposes and produces a prolonged, constant emission of light with a maximum at 463 nm in Sapphire II™ enhancer. Sapphire II, a water-soluble macromolecular compound, significantly increases the signal-to-noise ratio of the chemiluminescent assay. The light signal from alkaline phosphatase-activated CSPD is produced as a "glow", and maximum intensity is reached in 10–20 min following substrate addition (Figure 2). The rate of light emission of the PCR-Light assay enables the use of luminometers without injection capabilities and facilitates automation.

MATERIALS AND METHODS

PCR-Light reagents, including CSPD, Sapphire II, I-Block™, Fluorx-AP™ anti-fluorescein-alkaline phosphatase conjugate, and preformulated hybridization and wash buffers, are available from Tropix (Bedford, MA, USA). Streptavidin-coated plates were obtained from Xenopore

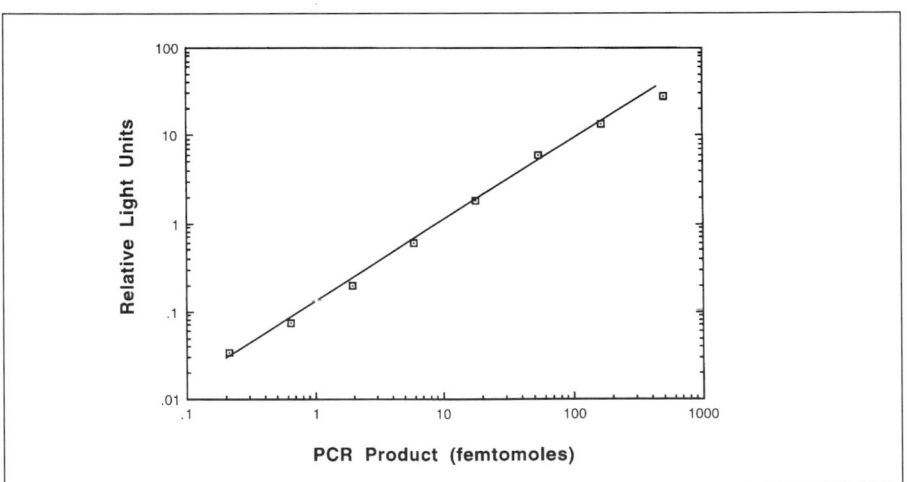

Figure 3. Sensitivity and linearity of the PCR-Light system. Serial dilutions (500 fmol-0.2 fmol) of a 246-bp biotinylated PCR product were assayed in triplicate with the PCR-Light system using an 18-nucleotide 5′-fluorescein-labeled probe and streptavidin-coated microplate strips.

(Saddlebrook, NJ, USA). Purified alkaline phosphatase was obtained from Biozyme (Catalog No. ALPI-12G; San Diego, CA, USA). PCR amplification was performed using GeneAmp™ reagents and a DNA Thermal Cycler 480 (Perkin-Elmer, Norwalk, CT, USA). A "hot-start" method using Ampli-Wax™ PCR Gem 100 (Perkin-Elmer) was utilized. Primers were synthesized on an Applied Biosystems Model 392 DNA Synthesizer (currently Perkin-Elmer/Applied Biosystems Division [PE/ABI], Foster City, CA, USA). Biotinylation of oligonucleotides was accomplished using Aminolink 2™ (PE/ABI) and Biotin-XX-NHS (Glen Research, Sterling, VA, USA). An 18-nucleotide 5′-fluorescein-labeled oligonucleotide probe was obtained from Genosys (The Woodlands, TX, USA). Primer and probe sequences for a 246-bp PCR product from the human interleukin-4 (IL-4) gene were from Alard et al. (1). Plasmid pcD-hIL-4 containing the IL-4 cDNA was obtained from ATCC (Rockville, MD, USA). The 246-bp PCR product was purified by gel chromatography (FPLC®; Pharmacia Biotech, Piscataway, NJ, USA) and quantified by absorbance spectrophotometry.

The PCR-Light assay was performed in streptavidin-coated microplates according to the following protocol. All steps were performed at room temperature unless noted. Streptavidin-coated wells were rehydrated with 2× phosphate-buffered saline (PBS) for 10 min. PCR product was diluted with binding buffer (0.5 M NaCl, 10 mM Tris-HCl pH 7.5, 1 mM EDTA) and subsequently added to the wells and incubated for 60 min. Non-biotinylated DNA strands were removed by a 10-min incubation with 0.25 mM NaOH. Wells were rinsed with hybridization buffer (0.9 M NaCl, 50 mM NaPO$_4$ pH 7.7, 5 mM EDTA, 10% 50× Denhardt's solution [1% each of Ficoll® Type 400, polyvinylpyrrolidone, and bovine serum albumin (BSA) Fraction V]). Fluorescein-labeled probe was diluted in hybridization buffer (0.5 pmol/100

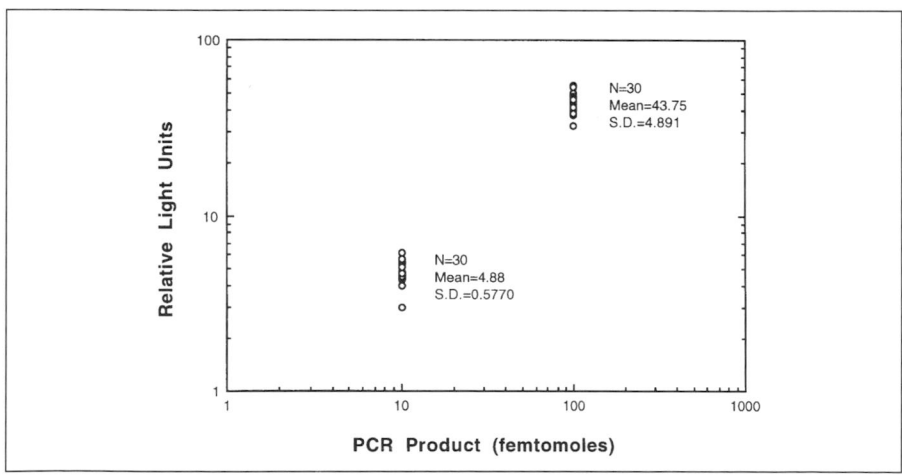

Figure 4. PCR-Light assay precision. Biotinylated 246-bp PCR product (100 and 10 fmol) was assayed 30 times with the PCR-Light system in streptavidin-coated microplates.

177

μL), added to the wells, and incubated for 15 min at 55°C. Wells were subsequently washed twice at room temperature with wash buffer I (0.9 M NaCl, 50 mM $NaPO_4$ pH 7.7, 5 mM EDTA), once at room temperature and twice at 55°C for 5 min with wash buffer II (3 M tetramethylammonium chloride (TMAC), 50 mM Tris-HCl pH 8.0, 2 mM EDTA, 0.025% Triton® X-100), and then incubated with blocking buffer (0.2% I-Block, 0.1% Tween 20, PBS) for 10 min. Fluorx-AP anti-fluorescein alkaline phosphatase conjugate was diluted in blocking buffer, added to the wells and incubated for 60 min. Wells were washed three times with detection wash buffer (0.1% Tween 20, PBS), twice with assay buffer (0.1 M diethanolamine, 1 mM $MgCl_2$, pH 10.0) and incubated for 10 min with 0.4 mM CSPD and 10% Sapphire II diluted in assay buffer. Light signal was measured on a microplate luminometer (ML 2250; Dynatech Laboratories, Chantilly, VA, USA).

RESULTS

The PCR-Light assay is accomplished by detecting the concentration of probe hybridized to an immobilized PCR product (Figure 1). Samples containing target molecules are PCR-amplified with one 5′-biotin-labeled primer and one unlabeled primer. Double-stranded biotinylated PCR product is bound to the streptavidin-coated solid support. The PCR product is denatured and the non-biotinylated strands are removed. A fluorescein-labeled oligonucleotide probe, which is complementary to sequences internal to the primer sequences, is hybridized to the captured strand. Stringency washes are

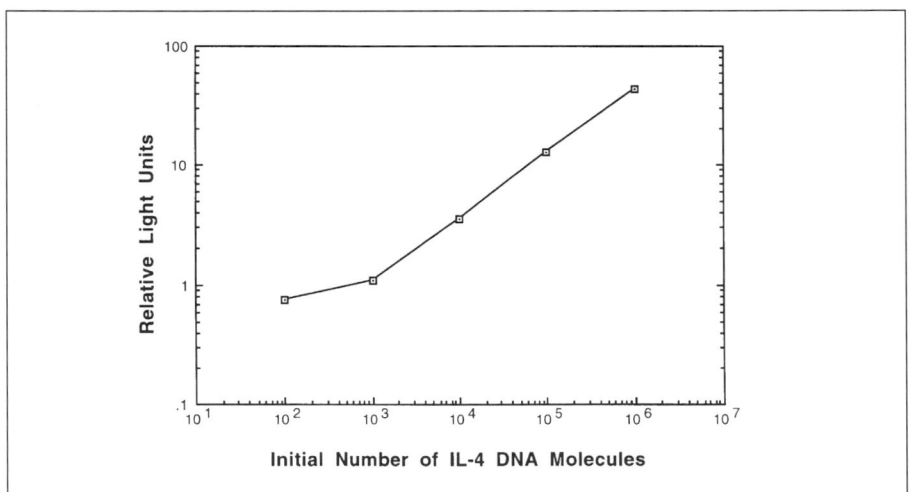

Figure 5. Quantitative PCR assay for IL-4 DNA. Dilutions of a linearized plasmid containing the cDNA clone of the human IL-4 gene were PCR-amplified with 1 μg of human K562 DNA. Amplification conditions: 94°C for 2 min, then 30 cycles of 94°C for 1 min, 55°C for 1 min and 72°C for 2 min. Each amplification reaction was assayed in triplicate with the PCR-Light system in streptavidin-coated microplate strips.

performed in 3.0 M TMAC buffer, which eliminates melting point temperature differences between oligonucleotide probes resulting from G/C or A/T base composition (9). The wells or beads are incubated with anti-fluorescein-alkaline phosphatase conjugate and then CSPD chemiluminescent substrate. Following a 10-min incubation, the light emission is measured in a luminometer.

The PCR-Light assay is extremely sensitive and exhibits a broad linear detection range greater than three orders of magnitude of PCR product concentration. A 246-bp PCR product was serially diluted from 500 to 0.2 fmol and assayed with PCR-Light (Figure 3). The assay was performed in triplicate for each dilution. Within this range, the chemiluminescent signal is directly proportional to the concentration of PCR product. The high sensitivity enables detection of a broad range of initial target molecule concentrations. Assay sensitivity for a particular PCR product is inversely proportional to the PCR product size (8) and dependent on the efficiency of probe hybridization.

PCR-Light exhibits high assay precision compared with gel and membrane-based PCR quantitation methods. The entire assay is performed in a single reaction vessel that leads to greater precision due to the absence of transfer manipulation steps. Typical precision achieved with the PCR-Light system is shown in Figure 4. Two samples containing 100 and 10 fmol of the 246-bp IL-4 PCR product were assayed 30 times with PCR-Light. The average relative light units and standard deviations for the 100- and 10-fmol samples are 43.75 ± 4.891 and 4.88 ± 0.5770, respectively.

PCR amplification was performed on a dilution series of a linearized IL-4 cDNA plasmid (10^2–10^6 copies) in genomic DNA. The reaction products were assayed with the PCR-Light system, following 30 amplification cycles (Figure 5). The concentration of PCR product assayed for each dilution was within the dynamic range of the PCR-Light system. The broad dynamic range simplifies the determination of the number of cycles necessary to obtain the measurement in the exponential phase of amplification. With the PCR-Light system, analysis of only one set of reactions is necessary. In comparison, less sensitive detection systems require analysis of results from two different experiments with different numbers of amplification cycles in order to obtain data from the same range of target DNA concentrations.

DISCUSSION

The PCR-Light system is an accurate, efficient method for PCR product quantitation. The broad dynamic range and high sensitivity of the PCR-Light assay enable quantitation during the exponential phase of amplification. The probe hybridization step eliminates detection of PCR artifacts. The assay is performed in a single reaction vessel, thus sources of error associated with transfer manipulations are eliminated. Chemiluminescent detection with

CSPD substrate provides a rapid, sensitive and non-isotopic quantitative detection system.

The PCR-Light components and protocol ensure high sensitivity and reproducibility of results. The benefits of the chemiluminescent detection system are well known and include a wide dynamic range and high sensitivity compared with colorimetric detection systems (Elizabeth Bonney and Olivier Lantz, personal communication). The anti-fluorescein-alkaline phosphatase conjugate exhibits high affinity and low nonspecific binding (data not shown). The hybridization procedure is designed for optimized performance with widely varying probes and target DNA sequences. The broad dynamic range of the assay simplifies the process of determining the number of amplification cycles necessary for accurate quantitation. To ensure that amplification artifacts are not detected as specific products, hybridization of a probe to an internal sequence is performed. An alternative method that utilizes two labeled amplification primers, one with biotin and one with fluorescein, followed by detection with anti-fluorescein-alkaline phosphatase and CSPD, resulted in the detection of primer dimers present in some PCRs (data not shown). The enzyme-linked chemiluminescent-detection technique may be used for other solid-phase capture assay formats, including methods utilizing sandwich hybridization protocols where an unlabeled PCR product is captured with support-bound probe followed by hybridization of a reporter probe (6,13).

Further improvements to the PCR-Light assay may be possible. Assay sensitivity is limited by nonspecific binding of the probe and anti-fluorescein conjugate that generates background signal 25-fold to 50-fold higher than substrate background from CSPD and Sapphire II alone. Reduction in nonspecific binding may be possible using improved blocking conditions or streptavidin-coated surfaces. Alternatively, this system can be adapted for use with an alkaline phosphatase-conjugated oligonucleotide probe assay system (5,14) and detected with CSPD chemiluminescent substrate. Benefits of alkaline phosphatase-coupled probes include a reduction of the number of assay steps as well as potentially lowering the detection limit of the assay by eliminating background signal generated by the antibody enzyme conjugate. The detection limit for purified alkaline phosphatase approaches 0.01 amol (3), which may be the theoretical detection limit for this assay.

In summary, PCR-Light is a highly accurate assay for quantitative PCR which is a tool for various applications including viral quantitation, measurement of gene expression, and therapeutic and environmental monitoring. The design of the assay is well suited to automation in the microplate, tube and other formats.

ACKNOWLEDGMENTS

We wish to thank Olivier Lantz in Villejuif, France, and Elizabeth Bonney at NIH for helpful discussions and John C. Voyta and Corinne E.M. Olesen at Tropix for assistance.

REFERENCES

1. **Alard, P., O. Lantz, M. Sebagh, C.F. Calvo, D. Weill, G. Chavanel, A. Senik and B. Charpentier.** 1993. A versatile ELISA-PCR assay for mRNA quantitation from a few cells. BioTechniques *15*:730-737.
2. **Becker-André, M. and K. Hahlbrock.** 1989. Absolute mRNA quantification using the polymerase chain reaction (PCR). A novel approach by a PCR aided transcript titration assay (PATTY). Nucleic Acids Res. *17*:9437-9446.
3. **Bronstein, I. B. Edwards and J. C. Voyta.** 1989. 1,2-Dioxetanes: novel chemiluminescent enzyme substrates. Applications to immunoassays. J. Biolumin. Chemilumin. *4*:99-111.
4. **Bronstein, I., R.R. Juo, J.C. Voyta and B. Edwards.** 1991. Novel chemiluminescent adamantyl 1,2-dioxetane enzyme substrates, p. 73-82. *In* P. Stanley and L.J. Kricka (Eds.), Bioluminescence and Chemiluminescence: Current Status. John Wiley, Chichester, U.K.
5. **Cano, R.J., S.R. Rasmussen, G.S. Fraga and J.C. Palomares.** 1993. Fluorescent detection-polymerase chain reaction (FD-PCR) assay on microwell plates as a screening test for salmonellas in foods. J. Appl. Bacteriol. *75*:247-253.
6. **Carcillo, J.A., R.A. Parise and Marjorie Romkes-Sparks.** 1994. Comparison of the enzyme-linked oligonucleotide sorbent assay to the ^{32}P-labeled PCR/Southern blotting technique in quantitative analysis of human and rat mRNA. PCR Methods Appl. *3*:292-297.
7. **Coen, D.** 1989. The polymerase chain reaction, p. 15.3.1-15.3.8. *In* F.M. Ausubel, R. Brent, R.E. Kingston, D.D. Moore, J.A. Smith, J.G. Seidman and K. Struhl (Eds.), Current Protocols in Molecular Biology. John Wiley & Sons, NY.
8. **DiCesare, J., B. Grossman, E. Katz, E. Picozza, R. Ragusa and T. Woudenberg.** 1993. A high-sensitivity electrochemiluminescence-based detection system for automated PCR product quantitation. BioTechniques *15*:152-157.
9. **DiLella, A.G. and S.L.C. Woo.** 1987. Hybridization of genomic DNA to oligonucleotide probes in the presence of tetramethylammonium chloride. Methods Enzymol. *152*:447-451.
10. **Ferre, F.** 1992. Quantitative or semi-quantitative PCR: Reality versus myth, p. 1–9. PCR Methods and Applications. Cold Spring Harbor Laboratory Press, Cold Spring Harbor, NY.
11. **Gause, W.C. and J. Adamovicz.** 1994. The use of the PCR to quantitate gene expression. PCR Methods Appl. *3*:S123-S135.
12. **Gilliland, G., S. Perrin and H.F. Bunn.** 1990. Competitive PCR for quantitation of mRNA, p. 60-69. *In* M.A. Innis, D. Gelfand, J.J. Sninsky and T.J. White (Eds.), PCR Protocols: A Guide to Methods and Applications. Academic Press, San Diego.
13. **Inouye, S. and R. Hondo.** 1990. Microplate hybridization of amplified viral DNA segment. J. Clin. Microbiol. *128*:1469-1472.
14. **Ishii, J.K. and S.S. Ghosh.** 1993. Bead-based sandwich hybridization characteristics of oligonucleotide-alkaline phosphatase conjugates and their potential for quantitating target RNA sequences. Bioconjug. Chem. *4*:34-41.
15. **Lantz, O. and A. Bendelac.** 1994. An invariant T cell receptor α chain is used by a unique subset of major histocompatibility complex class I-specific CD4+ and CD4-8- T cells in mice and humans. J. Exp. Med. *180*:1097-1106.
16. **Saiki, K., D.H. Gelfand, S. Stoffel, S.J. Scharf, R. Higuchi, G.T. Horn, K.B. Mullis and H.A. Erlich.** 1988. Primer-directed enzymatic amplification of DNA with a thermostable DNA polymerase. Science *239*:487-491.
17. **Syvanen, A., M. Bengstrom, J. Tenhunen and H. Soderlund.** 1990. Quantification of polymerase chain reaction products by affinity-based hybrid collection. Nucleic Acids Res. *16*:11327-11338.
18. **Wang, A.M., M.V. Doyle and D.F. Mark.** 1989. Quantitation of mRNA by the polymerase chain reaction. Proc. Natl. Acad. Sci. USA *86*:9717-9721.
19. **Wang, A.M. and D.F. Mark.** 1990. Quantitative PCR. *In* M.A. Innis, D.H. Gelfand, J.J. Sninsky and T.J. White (Eds.), PCR Protocols: A Guide to Methods and Applications. Academic Press, San Diego.

181

Update to:

Quantitation of PCR Products with Chemiluminescence

Chris S. Martin
Tropix, Bedford, MA, USA

We have not modified the procedure described in the original article. The additional reference (1) provides an example of a kinetic PCR approach to quantitation utilizing the chemiluminescent detection assay described in the manuscript.

REFERENCE

1.**Umlauf, S, B. Beverly, O. Lantz and R.H. Schwartz.** 1995. Regulation of interleukin 2 gene expression by CD28 costimulation in mouse T-cell clones: both nuclear and cytoplasmic RNAs are regulated with complex kinetics. Mol. Cell. Biol. *15*:3197-3205.

CHAPTER 14

Optimized Chemiluminescent Detection of DNA Amplified in the Exponential Phase of PCR

X. Su, T.F. Sullivan, S. Bursztajn and S.A. Berman

McLean Hospital and Harvard Medical School, Belmont, MA, USA

ABSTRACT

We describe a procedure for quantitative detection of nucleic acids by coupling PCR with improved chemiluminescent detection techniques. After performing PCR in the exponential phase in the presence of a trace amount of digoxigenin-11-dUTP, the amplified products are transferred to a positively charged nylon membrane. The membrane is cleaned with a modified method involving sodium dodecyl sulfate and ethanol washing steps to ensure low backgrounds in chemiluminescent detection. The membrane is exposed to two stacked x-ray films to increase the dynamic response in a film exposure. This sensitive and quantitative procedure is useful for molecular biology studies.

INTRODUCTION

Quantifying limited amounts of specific nucleic acid sequences is often required in molecular biology research. Detecting picogram levels can be accomplished by radioactive (5) or nonradioactive labeling and appropriate detection techniques (3,10). Methods based on the polymerase chain reaction (PCR) (9), i.e., competitive PCR methods (2,4), can measure femtograms or even smaller amounts of specific nucleic acid species. When a suitable competitor DNA is not available or there is a need to cross-check questionable competitive PCR results, one may wish to perform PCR in the exponential phase and generate a linear standard curve (6,8,12). In this case, the detection sensitivity is lowered. To overcome this problem, nucleotide analogs can be included in the PCR so that the products may be detected in a more sensitive way, such as by chemiluminescent detection (1). Chemiluminescent detection, however, is often associated with high backgrounds (11) as recorded on x-ray film. In this short report, we describe techniques to improve the chemiluminescent detection as well as a procedure that couples the exponential

(Reprinted from BioTechniques 17:734-736, 1994)

PCR with the improved detection method for sensitive and quantitative analysis of nucleic acids.

MATERIALS AND METHODS

PCR was performed in a volume of 10 µL containing 2 µM digoxigenin-11-dUTP (dig-11-dUTP; Boehringer Mannheim, Indianapolis, IN, USA), 100 µM deoxyribonucleoside triphosphate (dNTP), 10 nM of each primer, as well as other reagents necessary for PCR (Instruction for GeneAmp® DNA Amplification Kit; Perkin-Elmer, Norwalk, CT, USA). To ensure equal experimental conditions among a set of reactions, PCR reagents, except for DNA, were mixed to form a master solution. A 9–µL aliquot was retrieved and added to each DNA sample (1 µL, concentration varies) before starting a PCR program, which was 94°C for 2 min, followed by 20 or 25 cycles of 94°C for 30 s, 55°C for 30 s and 72°C for 1 min. PCR was performed in the TwinBlock™ System (Ericomp, San Diego, CA, USA). As part of an RNA quantification study, the DNA template used in this study was cDNA-generated from in vitro transcribed cRNA of the δ subunit of the acetylcholine receptor (7). The cRNA was made from the δ subunit cDNA cloned in pBluescript® (Stratagene, La Jolla, CA, USA), according to Stratagene's protocol. PCR primers used in this study were AGAGTGACCCAGAGACTGAA and GATCATAGAGCTGCTCCTCGCT. The PCR product size was 741 bp.

DNA Blotting and Membrane Cleaning

To analyze PCR product specifically, amplified DNAs were separated in an agarose gel (12 × 14 cm, with 2 × 17 wells) in TBE buffer (50 mM Tris-boric acid, pH 8.3, 10 mM EDTA) with no ethidium bromide. The gel was trimmed to a 2-cm slice containing a row of bands. DNAs in two of such gel slices could be blotted, in a solution containing 0.4 M NaOH and 1.5 M NaCl, to a 4- × 12-cm strip of positively charged nylon membrane (Boehringer Mannheim). Using such a membrane strip could save reagents in later steps. After transfer, the membranes were cleaned by the following sodium

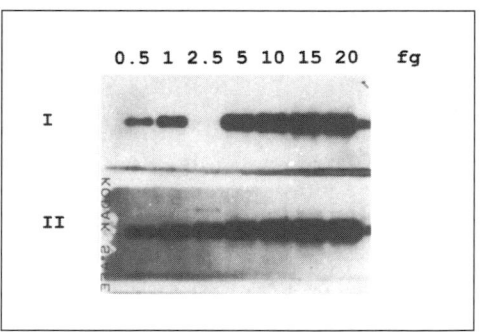

Figure 1. Effects of washing procedures. PCR template was cDNA corresponding to cRNA of 0.5–20 fg. Half of the DNA in each 10-µL sample was transferred to a membrane strip and subjected to chemiluminescent detection under identical conditions except for the washing steps. The membrane (I) washed with the SDS-ethanol method produced clean signals. The membrane (II) washed with no SDS and ethanol generated high backgrounds. One band (1, 2.5) was missing due to a failed PCR. The SDS-ethanol washing method appeared to reduce the signals slightly.

dodecyl sulfate (SDS)-ethanol method: the membrane was soaked in 1% SDS (45°–55°C) for 10 min with gentle shaking, rinsed with deionized water several times and washed in 70% ethanol for 5 min, followed by washing with deionized water.

Chemiluminescent Detection

Chemiluminescent detection was carried out with the Genius™ kit following the manufacturer's instructions (Boehringer Mannheim) with some modifications. For a membrane strip of 4 × 12 cm, about 200 μL of Lumi-Phos™ 530 were added on a sheet of polyvinyl chloride (PVC) (Fisher Scientific, Pittsburgh, PA, USA). The membrane strip was laid on the solution and turned a few times during a 1-min period so that the substrate was evenly distributed over the membrane. The membrane was then covered with a clean sheet of PVC and excess liquid was removed by gentle wiping of the wrapped membranes with a piece of Kimwipe® tissue. Two x-ray films (XAR-5; Eastman Kodak, Rochester, NY, USA) were stacked and placed between the membrane and a piece of flattened aluminum foil in an autoradiography cassette. Several exposures could be made (a few minutes to 30 min) by changing the exposure positions of the same films. The recorded signals were analyzed by the Bio Image™ Analysis System (Millipore, Bedford, MA, USA).

RESULTS

Chemiluminescent detection is sensitive and nonradioactive; however, we have sometimes encountered background problems with the detection method. We noticed that the SDS-ethanol washing steps could result in better signals than the washing procedure recommended by Boehringer Mannheim (Genius kit instructions). To illustrate the results, we performed PCR with dig-11-dUTP for 25 cycles, and the amplified DNAs were transferred to membranes under identical conditions. The membranes were then washed separately by the two procedures before exposing them to the anti-digoxigenin antibody (Boehringer Mannheim). Figure 1 is an example demonstrating that the membrane washed with the SDS-ethanol steps had a clean background, but the membrane washed by the standard method could sometimes leave high or uneven backgrounds.

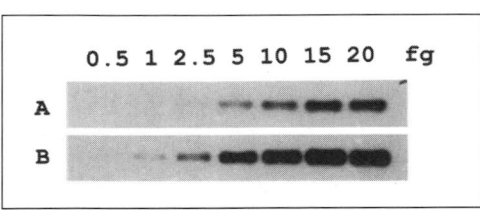

Figure 2. Chemiluminescent signals recorded by stacked films. PCR was performed for 20 cycles with cDNA corresponding to indicated amounts of cRNA. After DNA blotting and membrane cleaning, two (stacked) x-ray films A and B were exposed for about 154 min (A-B-membrane).

With 25 cycles of PCR, DNAs were barely detectable by ethidium bromide staining, but the recorded signals were no longer in the linear range likely because the PCR was performed beyond the exponential phase. To obtain a quantitative linear relation between the input template and the detected signals, we made two modifications. For DNA in femtogram range, we performed PCR for no more than 20 cycles, ensuring that the amplification would remain exponential (at most a 1×10^6-fold of amplification). We used two stacked films in each exposure to improve the film's dynamic range because the film farthest from the membrane would receive fewer photons and have less chance of being exposed beyond the nonlinear response region. Figure 2 illustrates the result of an experiment using this stacked film method. After PCR and other necessary steps, two x-ray films were exposed (film A—film B—membrane). As expected, the bands in film A were weaker than those in film B. In film B, the signal for cDNA generated from 0.5 fg of cRNA was just visible, but the signal for 5 fg of cRNA was in the nonlinear response range (>1 OD). In film A, signals for template in the lower range were not detected, but those for 5–20 fg were in the linear response region of the x-ray film.

To demonstrate the quantitative relations of recorded signals, four sets of replicated reactions were performed and processed as for Figure 2. In Figure 3, the lower curve (A) and upper curve (B) were derived from film A and film B, respectively. Curve A was linear from 5 fg to about 20 fg of cRNA. Curve B only showed linearity from 0.5 to 5 fg of cRNA. The signals started to plateau when the values reached 200 (about 1 OD unit). Apparently, the signal values from film A were linear in the upper range of the template and those from film B in the lower range. Based on linear regression analysis, a signal value of film B beyond the linear range (Y_B) could be calculated from the value of film A within the linear range (Y_A): $Y_B = 3.75Y_A + 14.96$. For our data, the number 14.96 was insignificant compared with actual signal values, and, thus, the linear range of curve B could simply be extended 3.75-fold, which is the slope ratio of line b over line a. More generally, the constant (e.g., 14.96 in this example) could be used in the calculation.

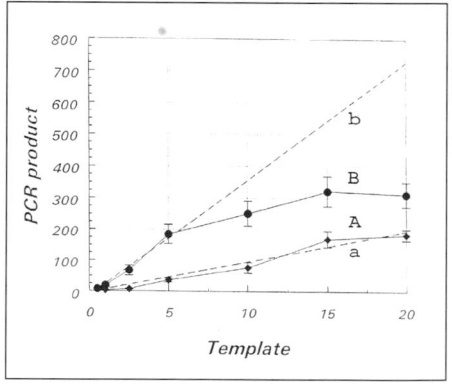

Figure 3. Analysis of x-ray film recorded signals. Each data point was determined from 4 replicated bands obtained under the same conditions. Curve A represents data from film A and curve B from film B (see Figure 2). Lines a and b were made from the linear ranges of the two curves. A and B., respectively. (a: $Y = -7.16 + 9.99X$, $r^2 = 0.96$; b: $Y = -11.90 + 37.47X$, $r^2 = 0.98$). PCR product values were in arbitrary units and were equivalent to the integrated values of the bands. The template was shown in fg of cRNA from which the cDNA was generated.

DISCUSSION

We have described a procedure for obtaining quantitative data for specific DNA or RNA sequences in femtogram ranges. The coupling of the two enzymatic reactions (PCR and chemiluminescent detection) produced higher detection sensitivity than that obtained by exponential PCR alone. We used a low concentration of dig-11-dUTP not only to reduce the cost but also to avoid the possibility that local availability of the Lumi-Phos 530 substrate could be a limiting factor during the alkaline phosphatase reaction. However, dig-11-dUTP concentration of lower than indicated might not be sufficient to produce clear signals (data not shown).

We improved the detection procedure by generating clean signals and expanding the dynamic range of film exposure. We consistently produced signals with lower noise because SDS could remove stains and dirt while ethanol removed SDS. It appeared that background could be a problem if SDS was not removed completely. The stacked film exposure method was a simple way to record broad linearity from a set of samples in a single exposure. Therefore, a linear standard curve can be readily generated from known amounts of template and their signal values. This procedure is thus useful for DNA and RNA quantification.

ACKNOWLEDGMENT

This work was supported partly by NIH grant 24377 and a grant from M.D.A.

REFERENCES

1. **An, S.F., D. Franklin and K.A. Fleming.** 1992. Generation of digoxigenin-labelled double-stranded and single-stranded probes using the polymerase chain reaction. Mol. Cell. Probes 6:193-200.
2. **Becker-Andre, M. and K. Hahlbrock.** 1989. Absolute mRNA quantification using the polymerase chain reaction (PCR). A novel approach by a PCR aided transcript titration assay (PATTY). Nucleic Acids Res. 17:9437-9446.
3. **Bronstein, I., J.C. Voyta, K.G. Lazzari, O. Murphy, B. Edwards and L.J. Kricka.** 1990. Rapid and sensitive detection of DNA in Southern blots with chemiluminescence. BioTechniques 8:310-314.
4. **Gilliland, G., S. Perrin, K. Blanchard and H.F. Bunn.** 1990. Analysis of cytokine mRNA and DNA: Detection and quantification by competitive polymerase chain reaction. Proc. Natl. Acad. Sci. USA 87:2725-2729.
5. **Lee, J.J. and N.A. Costlow.** 1987. A molecular titration assay to measure transcript prevalence levels, p. 633-648. In S.L. Berger and A.R. Kimmel (Eds.), Methods in Enzymology, Vol. 152. Academic Press, New York.
6. **Murphy, L.D., C.E. Herzog, J.B. Rudick, A.T. Fojo and S.E. Bates.** 1990. Use of the polymerase chain reaction in the quantitation of mdr-1 gene expression. Biochemistry 29:10351-10356.
7. **Nef, P., A. Mauron, R. Stalder, C. Alliod and M. Ballivet.** 1984. Structure, linkage, and sequence of the two genes encoding the δ and γ subunits of the nicotinic acetylcholine receptor. Proc. Natl. Acad. Sci. USA 81:7975-7979.
8. **Noonan, K.E., C. Beck, T.A. Holzmayer, J.E. Chin, J.S. Wunder, I.L. Andrulis, A.F. Gazdar, C.L. Willman, B. Griffith, D.D. Von Hoff and I.B. Roninson.** 1990. Quantitative analysis of MDR1 (multidrug resistance) gene expression in human tumors by polymerase chain reaction. Proc. Natl. Acad. Sci. USA 87:7160-7164.
9. **Saiki, R.K., D.H. Gelfand, B. Stoffel, S.J. Scharf, R. Higuchi, G.T. Horn, K.B. Mullis and H.A. Erlich.** 1988. Primer directed enzymatic amplification of DNA with thermostable DNA polymerase. Science 239:487-489.

10.**Schaap, A.P., M.D. Sandison and R.S. Handley.** 1987. Chemical and enzymatic triggering of 1,2-dioxetanes. 3. Alkaline phosphatase catalyzed chemiluminescence from aryl phosphate-substituted dioxetane. Tetrahedron Lett. *28*:1159.

11.**Steck, T.R.** 1994. Improved chemiluminescent detection of DNA. BioTechniques *16*:406-407.

12.**Wang, A.M., M.V. Doyle and D.F. Mark.** 1989. Quantitation of mRNA by the polymerase chain reaction. Proc. Natl. Acad. Sci. USA *86*:9717-9721.

Update to:
Optimized Chemiluminescent Detection of DNA Amplified in the Exponential Phase of PCR

S. Bursztajn and S.A. Berman[1]

McLean Hospital and Harvard Medical School, Belmont, MA and
[1]Louisiana State University Medical Center, Shreveport, LA, USA

The reverse transcription polymerase chain reaction (RT-PCR) has gained wide use in biomedical, industrial and forensic work. Our published report on the quantitative detection of nucleic acids using chemiluminescence has many advantages and has been widely utilized. The assay is sensitive, does not require the use of radioactivity, and, because the measurements are performed on products generated during the exponential phase of PCR, it allows for accurate quantitation. Since its publication, numerous other reports using quantitative RT-PCR and chemiluminescence have appeared (3–7). The non-radioactive features of the chemiluminescent assays coupled with their high sensitivity contributes to their use as diagnostic tools in a number of pathological conditions (1,2,4,5). Because the efficiencies of RT-PCR are affected by the length, GC content and base order of the substrate molecules, as well as the reaction conditions, an RNA quantitation strategy using RT-PCR must consider both the sequence-dependent and independent factors in each of the enzymatic reactions.

We have designed an in vitro transcribed target-specific cRNA containing the full or near full-length coding region as a specific quantitative standard to minimize structural differences from the target. The employment of such a standard requires the separation of the standard from the target sequences. Together with the described chemiluminescent detection procedure, we are able to precisely measure the absolute quantities of several mRNAs in the same sample. This amplification method is especially useful for measuring the amount of mRNA in a sample expressing multiple receptor subunits, or different protein isoforms that are coded by different genes.

REFERENCES

1. **Fujii, T., S. Oguni, M. Kikuchi, N. Kanai and K. Saito.** 1995. p53 mutation in carcinomas arising in ovarian cystic teratomas. Pathol. Int. 45:649-654.
2. **Hicks, K.E., S. Bear, B.J. Cohen, and J.P. Clewley.** 1995. A simple and sensitive DNA hybridization assay for the routine diagnosis of human parvovirus B19 infection. J. Clin. Microbiol. 33:2473-2475.

3. **Martin, C.S., L. Butler and I. Bronstein.** 1995. Quantitation of PCR products with chemilumines-cence. BioTechniques *18*:908-913.
4. **Mekus, F., T. Dork, T. Deufel, N. Morral and B. Tummler.** 1995. Analysis of microsatellites by direct blotting electrophoresis and chemiluminescence detection. Electrophoresis *16*:1886-1888.
5. **Nanba, E. Y. Kohno, A. Matsuda, M. Yano, C. Sato, K. Hashimoto, T. Koeda, K. Yoshino, M. Kimura and Y. Maeoko.** 1995. Nonradioactive DNA diagnosis for the fragile X syndrome in mentally retarded Japanese males. Brain Dev. *17*:317-321.
6. **Stevens J., F.S. Yu, P.M. Hassoun and J.J. Lanzillo.** 1996. Quantification of polymerase chain reac-tion products: enzyme immunoassay based systems for digoxigenin-and biotin-labelled products that quantify either total or specific amplicons. Mol. Cell. Probes *10*:31-41.
7. **Yu, H., J.G. Bruno, T.C. Cheng, J.J. Calomiris, M.T. Goode and D.L. Gatto-Menking.** 1995. A comparative study of PCR product detection and quantitation by electro-chemiluminescence and fluo-rescence. J. Biolumin. Chemilumin. *10*:239-245.

CHAPTER 15

Rapid Colorimetric Quantification of PCR-Amplified DNA

Joakim Lundeberg, Johan Wahlberg and Mathias Uhlén

Royal Institute of Technology, Stockholm, Sweden

ABSTRACT

A diagnostic system for rapid colorimetric quantification of the initial amount of DNA template amplified by the polymerase chain reaction is described. The method is based on co-amplification of target DNA with a cloned DNA fragment, in which a lac *operator sequence has been introduced by in vitro mutagenesis. The in vitro-amplified material is immobilized on magnetic beads using the biotin-streptavidin system, and the ratio of target DNA to cloned mutated DNA can be determined using a fusion protein consisting of the* Escherichia coli LacI *repressor and β-galactosidase. This method for quantitative detection of immobilized amplified nucleic acids is well adapted for rapid automated or semi-automated assays. Here, we show that it can be used to detect and quantify* Plasmodium falciparum *genomic DNA in clinical samples.*

INTRODUCTION

For many analytical biotechniques, the polymerase chain reaction (PCR) has many advantages as compared to conventional immunological and biochemical methods; i.e., the generality and sensitivity of the assay (3,11). However, the fact that most PCR assays are qualitative, limits their use to applications where only the presence or absence of a specific DNA molecule is to be determined. A great need exists for simple, accurate and non-isotopic quantitative PCR assays where the initial amount of the template DNA or RNA can be determined. Such quantification would be important for many clinical and research applications and for standardization of assays between laboratories. Recently, several systems have been described either based on co-amplification, followed by restriction cleavage and size analysis on agarose gels (1,5), or by DNA/RNA hybridizations, followed by isotopic or enzymatic detection (2,9). However, since these methods require time-consuming manual operations (1,5) or careful titrations of the PCR efficiencies (2,9), a great need exists for novel, quantitative PCR assays suitable for routine work.

(Reprinted from BioTechniques 10:68-75, 1991)

We have recently described a qualitative solid-phase approach to detection of immobilized, amplified nucleic acids, designated DIANA (8,10), which has been used for colorimetric detection of in vitro-amplified DNA. The assay is based on the use of a biotinylated PCR primer, which is used to capture the in vitro-amplified material on streptavidin-coated magnetic beads. The other PCR primer contains a *lac* operator "handle" sequence, which allows a colorimetric detection of the captured DNA using a LacI repressor-β-galactosidase fusion protein (10). The detection of the PCR products has thus been adapted to a format similar to the enzyme-linked immunosorbent assay (ELISA), which has found extensive use both in clinical and research applications.

Here, we show that this approach to detect PCR-amplified DNA can be modified to allow simple and rapid quantification of the initial concentration of target DNA in a sample. A protocol involving competitive titration has been developed in which a cloned version of the target sequence, which has been changed by in vitro site-specific mutagenesis to contain the *lac* operator sequence, is used for detection. It is noteworthy that this quantitative assay involves analysis of the relative yield of amplified material after completed PCR, and therefore it is rather insensitive to variations in the PCR between different samples. Thus, sample inhibition is not a major concern and the assay does not need a certain number of temperature cycles to be performed. Here, we show that this method can be used for colorimetric quantification of malaria-specific DNA in clinical samples.

MATERIALS AND METHODS

Blood Samples

Plasmodium falciparum parasites from patient blood samples were prepared as described by Zolg et al. (13). One microliter from the lysis mixture corresponds to 0.2 µL whole blood. A microscopical examination to determine the number of parasites per µL blood was performed as described earlier (6).

Cloning and In Vitro Mutagenesis

In vitro amplification was performed using a Techne Programmable Dri-Block PHC-2 (Techne, Cambridge, UK) with primers synthesized by phosphoramidite chemistry on an automated DNA synthesis machine (Gene Assembler® Plus DNA Synthesizer; Pharmacia LKB Biotechnology, Uppsala, Sweden) as described by Wahlberg et al. (10). The PCR cloning and in vitro mutagenesis were performed by subcloning the upstream fragment of the Pf155/RESA gene generated by PCR with primers RIT52 and RIT54. It was cleaved with *Bam*HI and *Eco*RI and ligated into pUC8 (12) digested with the same enzymes. The resulting construction was digested with *Bam*HI and

*Hin*dIII and ligated with the downstream fragment generated by PCR with RIT53 and RIT55, digested with *Bam*HI and *Hin*dIII. The mutant clone was verified by DNA sequencing as described by Hultman et al. (7).

In Vitro Amplification

The co-amplification was run in two consecutive steps: first 35 cycles with RIT52 and RIT53, followed by a 1/100 dilution and a second amplification for 25 cycles with RIT48 and RIT61 for the quantitative DIANA assay. The PCR buffer used contained 20 mM TAPS (3-[tris(hydroxymethyl)methy-lamino]-1-propanesulfonic acid) (pH 9.3), 8 mM MgCl$_2$, 50 mM KCl and 0.1% Tween 20®. The sample and a decreasing amount of competitor DNA, 1 µL each, were put into PCR tubes with 10 µL PCR buffer. The tubes were subsequently heated to 99°C for 5 min to lyse the samples and then rapidly cooled to 0°C. To the lysed mixture, 40 µL of a PCR buffer with 0.2 mM of each deoxyribonucleoside triphosphate (dNTP), 0.2 µM of each primer and 1.0 unit of AmpliTaq® DNA Polymerase (Perkin-Elmer, Stockholm, Sweden) were added. The reaction mixture was covered with a layer of light mineral oil (Sigma Chemical, St. Louis, MO, USA). One temperature cycle on the PCR was as follows: denaturation of template 96°C, 0.5 min; annealing of primers 59°C, 1 min; extension of primers 72°C, 1 min. The size of the PCR product was determined by agarose gel electrophoresis.

Colorimetric Analysis

The PCR mixture was immobilized on 300 µg of magnetic beads with co-valently coupled streptavidin, Dynabeads M280-Streptavidin (Dynal AS, Oslo, Norway), and incubated for 20 min at room temperature. The beads with the immobilized DNA were mixed with 200 µL of fusion protein (LacI-β-galactosidase, 0.2 mg/mL) and 200 µg of sonicated herring sperm DNA in a microcentrifuge tube and incubated for 20 min. The beads were washed four times with 250 µL TST buffer (0.1 M Tris-HCl, pH 7.5, 0.15 M NaCl, 0.1% Tween 20) containing 10 mM β-mercaptoethanol. The substrate, *o*-nitrophenyl-β-D-galactoside (ONPG), was added (200 µL) and the change of absorbances at room temperature was measured. The reaction was terminated by taking one-half of the reaction mixture to a microtiter plate with β-galactosidase stop solution (Pharmacia LKB Biotechnology). The result is presented as the change of absorbance per min at 405 nm.

RESULTS

The Basic Concept

The qualitative DIANA, described by Wahlberg et al. (10), which allows for colorimetric detection of in vitro-amplified DNA by immobilization to a solid support, has been modified to allow quantification of the initial amount of target DNA in the sample. The principle of the quantitative PCR assay is outlined in Figure 1. A competitive system is used where identical amounts of the target template are "spiked" with decreasing amounts of known competitor plasmid DNA identical to the target DNA to be analyzed, except for a

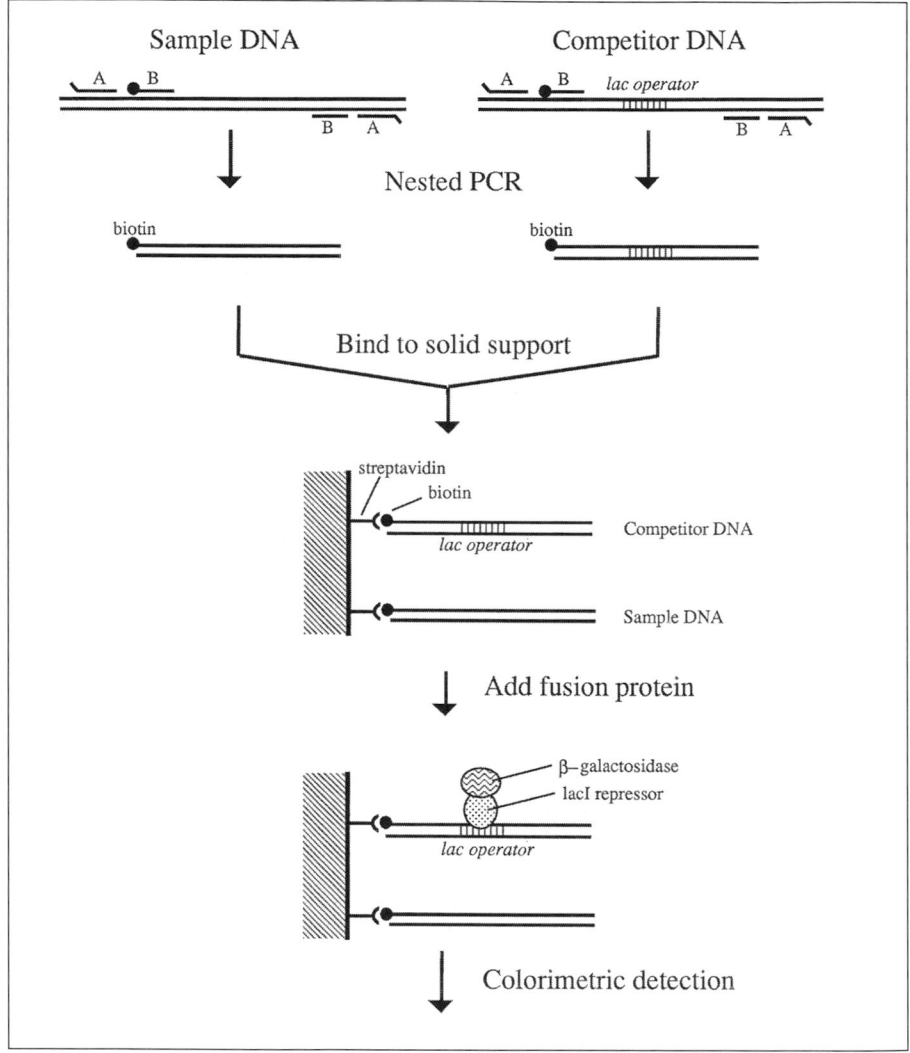

Figure 1. Schematic drawing showing the principle of the quantitative DIANA, in which a solid support is used to capture the biotinylated DNA fragment obtained after a nested PCR, with outer primers A and inner primers B. A LacI repressor-β-galactosidase fusion protein is used for the subsequent colorimetric detection of a *lac* operator sequence which is introduced into a cloned version of the target sequence. A known amount of this cloned DNA is co-amplified with the target DNA. The detection is based on the specific interaction of the LacI-β-galactosidase fusion protein with the *lac* operator sequence and the subsequent hydrolysis of the substrate *o*-nitrophenyl-β-D-galactoside (ONPG) to give the colorimetric response.

21-base pair *lac* operator sequence introduced into the middle of the cloned fragment by in vitro mutagenesis (Figure 1). The fragments produced by the PCR are captured on a solid phase by using labeled primers. When competitor DNA is used as template, it can be detected by a LacI repressor-β-galactosidase fusion protein, while the fragment produced using the target template DNA does not contain a *lac* operator sequence and is therefore not detected. The amount of *lac* operator captured on the solid phase can subsequently be determined by the colorimetric assay. A high signal in the competitive assay reflects a low ratio of target DNA to competitor DNA in the sample. Note, only the complete reactions are analyzed and there is therefore no need to determine the amount of amplified material during the PCR.

Cloning and Mutagenesis of a Malaria-Specific Chromosomal DNA Fragment

As a model system, the quantitative DIANA assay was used to determine the absolute amounts of the parasite *P. falciparum* in clinical human blood samples. The Pf155 gene was chosen as amplification of this gene and has been shown to accurately reflect the presence of parasite in clinical material (8). The primers used to clone and assay the Pf155 gene are shown in Figure 2. The gene fragment used as competitor was cloned in two parts by in vitro amplification of the chromosomal gene (see Materials and Methods for details). The DNA fragment was assembled by cloning the two PCR fragments into a plasmid vector. A 21-base pair *lac* operator sequence was simultaneously introduced into the cloned fragment, replacing 21 nucleotides from 1519 to 1539 of the gene (Figure 2). The in vitro mutagenesis was performed

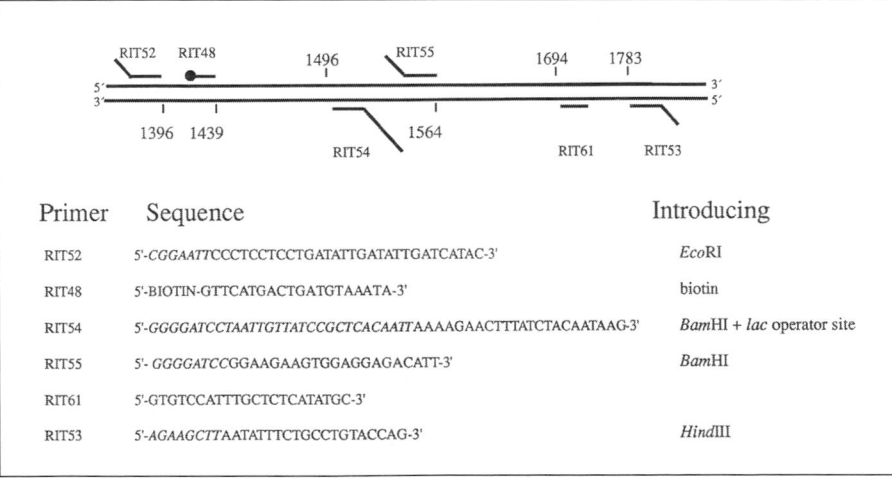

Primer	Sequence	Introducing
RIT52	5'-*CGGAATT*CCCTCCTCCTGATATTGATATTGATCATAC-3'	*Eco*RI
RIT48	5'-BIOTIN-GTTCATGACTGATGTAAATA-3'	biotin
RIT54	5'-*GGGGATCC*TAATTGTTATCCGCTCACAATT*AAAAGAACTTTATCTACAATAAG*-3'	*Bam*HI + *lac* operator site
RIT55	5'-*GGGGATCC*GGAAGAAGTGGAGGAGACATT-3'	*Bam*HI
RIT61	5'-GTGTCCATTTGCTCTCATATGC-3'	
RIT53	5'-*AGAAGCTT*AATATTTCTGCCTGTACCAG-3'	*Hind*III

Figure 2. The structure of the Pf155/RESA target gene of *P. falciparum* and the primers used for the detection, quantification and cloning. The numbers refer to the sequence published by Favaloro et al. (4). The nucleotides shown in italics correspond to sequences introduced by in vitro mutagenesis.

to keep the size and the GC content of the native and the mutated gene constant. In addition, the sequences where the primers anneal are identical to minimize the difference in the amplification efficiencies between target and competitor template DNA.

Efficiency of the Co-Amplification Procedure

Since the total amplification efficiency is a product of the techniques used for sample preparation (including cell lysis), as well as structural features of both the target and the competitor plasmid DNA, the relative amplification efficiency must initially be determined by using a reference sample containing known amounts of the DNA. Here, we used a clinical sample containing 500 *P. falciparium* parasites (Figure 3) as template in the quantitative DIANA. A 1:1.5 serial dilution of the competitor DNA containing the *lac* operator sequence was performed, starting with approximately 16 000 competitor plasmid molecules in the first step (dilution 1). Note that since the parasite is haploid in its asexual stage, there is only 1 target chromosomal region per parasite.

A two-step nested primer PCR was performed to keep the background signal low (10), and the cloned competitor DNA was added in serial dilutions before the first PCR cycle to samples containing the same amount of the initial target DNA. After the two-step PCR procedure, biotin-containing material was captured by the magnetic beads. The Lac repressor fusion protein was added to the beads and the amount of immobilized *lac* operator containing DNA was determined by adding the chromogenic substrate. Results presented in Figure 3 show that dilutions 1 through 6 give a high colorimetric response, while a low activity is observed for dilutions 9 and 10, which demonstrates that little competitor DNA has been co-amplified in the latter two samples. An intermediate response is obtained for dilutions 7 and 8 containing approximately 1400 and 950 competitor molecules, respectively.

These results show that the quantitative DIANA assay gives a titration curve with the expected sigmoid shape. Interestingly, the inflection point of the curve corresponds to approximately 1000 competitor molecules, suggesting that the plasmid

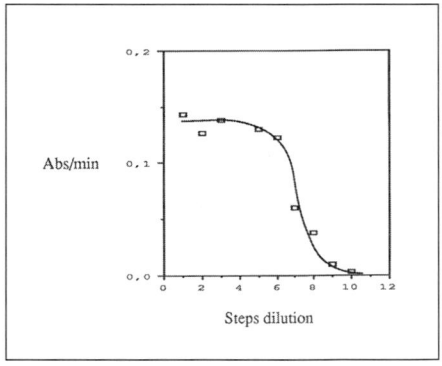

Figure 3. The results of the quantitative DIANA for a human clinical sample containing 500 *P. falciparum* parasites as determined by microscopical examination. A 1:1.5 serial dilution of the competitor DNA containing the *lac* operator sequence was performed, starting with approximately 16 000 competitor plasmid molecules in the first step (dilution 1). A two-step nested primer PCR was performed using the outer primers RIT52 and RIT53 for an initial 35 cycles and then, after dilution of the amplified material, an additional 25 cycles with the inner primers RIT48 and RIT61. See text for details.

target DNA is amplified with an approximately twofold lower efficiency as compared to the genomic target DNA. Thus, the initial cycles of the PCR are relatively inefficient for the competitor DNA, in the form of a supercoiled covalently closed plasmid molecule, as compared to the target DNA, which is in the form of linear chromosomal molecules.

Quantitative DIANA Assay to Detect Malaria DNA in Clinical Samples

The procedure described above was subsequently used to quantitate malaria parasites in clinical human blood samples obtained from The Gambia (Africa). A standard quantification using microscopical examination was performed in parallel (6). To allow quantification over a relatively broad range (20–50 000 parasites/sample) without more than eight serial dilutions, the assay was modifed by performing 1:3 dilutions. The results obtained with four independent and representative samples are shown in Figure 4. The qualitative DIANA shows that samples A, B and C contain parasites, while sample D is negative (data not shown). The quantitative DIANA for sample A shows a cut-off point between dilutions 3 and 4, which represent 5550 and 1850 parasites, respectively. This result is in good agreement with the amount malaria parasites (2600) determined by the microscopical method. Sample B has a cut-off point between dilutions 4 and 5, which represent 1850 and 620 molecules, respectively; this is also in good agreement with the 900 parasites detected with the microscopical method. Sample C, which contains 40 parasites, has a cut-off point at dilutions 7 and 8, containing competitor molecules corresponding to 70 and 20 parasites, respectively. This shows that as few as 40 parasites can be detected with the assay. In contrast, all dilutions for sample D show a high colorimetric response, which confirms the sample does not contain any parasite.

Note that the absolute values for the individual positive signals vary considerably, probably because of the variations of the efficiencies during the latter part of the PCR procedure. However, this variation does not limit the assay, since only the inflection point must be determined for each sample. Thus a high signal-to-noise ratio is sufficient. In Figure 4A, B and C, the signal in each positive point is indeed high while the background is low.

DISCUSSION

The quantitative DIANA assay described here provides a rapid method to detect and quantitate PCR-amplified DNA after immobilization to a solid support. The fact that the detection of PCR products is converted to a colorimetric method similar to standard ELISA makes the DIANA concept suitable for routine work, both in research and for clinical applications. Hence, it is possible to use plate readers and other instruments developed and available for immunological assays, such as ELISA. The assay can also be combined with an initial reverse transcriptase reaction to allow colorimetric quantitation of specific RNA in crude samples (Lundeberg and Uhlén, unpublished). Thus, diagnosis of viral, bacterial and protozoan pathogens can be performed as well as more research-oriented quantification, such as assays to determine amplification of specific genes or transcription assays to investigate gene regulation in vivo.

The use of a solid support makes it possible to avoid electrophoresis, precipitations, centrifugations, filtrations and other manual operations not well suited for routine analytical methods. The two-step nested primer approach minimizes the amplification of nonspecific target DNA (10) and no size assays (1) or hybridizations of RNA/DNA molecules (2) are needed. The support used here was monodispersed magnetic beads with covalently linked

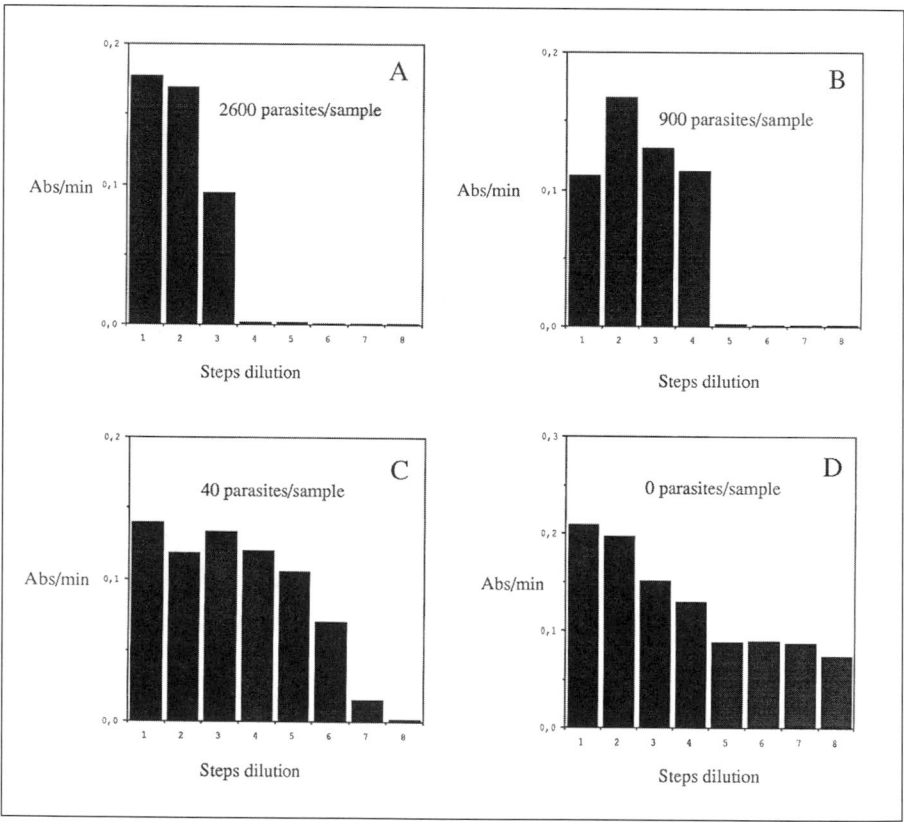

Figure 4. The results of the quantitative DIANA on four human clinical samples containing *P. falciparum* parasites. A lysis mixture corresponding to 0.2 μL whole blood of patients from The Gambia were used and the amount of parasites was independently determined by microscopical determination. In each case, a 1:3 serial dilution of the competitor DNA was performed with 100 000 plasmid molecules added to dilution 1, corresponding to 50 000 target molecules because of the two-fold lower efficiency for plasmid DNA as compared to genomic DNA (Figure 3). The amount of parasites per sample determined by the microscopical method is indicated.

streptavidin. However, other solid supports such as the walls of microtiter wells or plastic dip sticks can also be used. The assay technique is very simple and rapid, thus making it easy to process a large number of samples either by manual or automated procedures.

The method described here can be used for sensitive quantification of *P. falciparum* in small sample volumes. This is important since standard microscopical examination is very laborious, and thereby rapid, nonisotopic methods to detect and quantitate malaria parasites are greatly needed (6). The assay described in Figure 4A–D was designed to detect and quantitate between 20–50 000 parasites per sample, which makes it possible to diagnose the patients with both acute (high titer) or chronic infection in a single experiment. Obviously, calculations similar to those used for standard competitive ELISA will make it possible to obtain more precise quantification of the initial target DNA of the sample and not just a rough cut-off point between two dilutions as described here.

For the diagnosis of the malaria parasite, a qualitative DIANA (10) might be included to show the presence or absence of target DNA in the sample. This simple assay can therefore be a good supplement to the quantitative DIANA. Further improvements of this system might be achieved by a standard sample preparation that avoids differences in inhibition within the initial DNA samples. For many applications, it might also be important to perform a qualitative DNA determination involving a host-specific gene, such as actin or β-globin, to demonstrate that the in vitro amplification has not been inhibited due to the presence of polymerase inhibitors in the sample. Such a combination of the qualitative and the quantitative DIANA provides a powerful system for the diagnosis of pathogens in clinical samples.

ACKNOWLEDGMENTS

We are grateful to Drs. E. Hornes, T. Hultman, S. Ståhl and S. Bergh for useful discussions and Drs. M. Holmberg and B.M. Greenwood for providing the clinical samples. This work was supported by the Swedish Board for Technical Development.

REFERENCES

1.**Becker-André, M. and K. Hahlbrock.** 1990. Absolute mRNA quantification using the polymerase chain reaction (PCR). A novel approach by a PCR aided transcript titration assay (PATTY). Nucleic Acids Res. *17*:9437-9446.
2.**Coutlée, F., B. Yang, L. Bobo, K. Mayur, R. Yolken and R. Viscidi.** 1990. Enzyme immuno assay for detection of hybrids between PCR-amplified HIV-1 DNA and a RNA Probe: PCR-EIA. AIDS Res. Hum. Retroviruses 6:775-784.
3.**Erlich, H.A., D.H. Gelfand and R.K. Saiki.** 1988. Specific DNA amplification. Nature *331*:461-462.
4.**Favaloro, J.M., R.L. Coppel, L.M. Corcoran, S.J. Foote, G.V. Brown, R.F. Anders and D.J. Kemp.** 1986. Structure of the RESA gene of *Plasmodium falciparum*. Nucleic Acids Res. *14*:8265-8268.
5.**Gilliland, G., S. Perrin and H.F. Bunn.** 1990. Competitive PCR for quantitation of mRNA, p. 60-69.

PCR Protocols. Academic Press, San Diego.

6. **Holmberg, M., A.B. Vaidya, F.C. Shenton, R.W. Snow, B.M. Greenwood, H. Wigzell and U. Pettersson.** 1990. A comparison of two DNA probes; one specific for *Plasmodium falciparum* and one with wider reactivity in the diagnosis of malaria. Trans. R. Soc. Trop. Med. Hyg. *84*:202-205.

7. **Hultman, T., S. Ståhl, E. Hornes and M. Uhlén.** 1989. Direct solid phase sequencing of genomic and plasmid DNA using magnetic beads as solid support. Nucleic Acids Res. *17*:4937-4946.

8. **Lundeberg, J., J. Wahlberg, M. Holmberg, U. Pettersson and M. Uhlén.** 1990. Rapid colorimetric detection of *in vitro* amplified DNA sequences. DNA Cell Biol. *9*:289-292.

9. **Syvänen, A.C., M. Bengtström, J. Tenhunen and H. Söderlund.** 1988. Quantification of polymerase chain reaction products by affinity-based hybrid collection. Nucleic Acids Res. *16*:11327-11338.

10. **Wahlberg, J., J. Lundeberg, T. Hultman and M. Uhlén.** 1990. General colorimetric method for DNA diagnostics allowing direct solid-phase genomic sequencing of the positive samples. Proc. Natl. Acad. Sci. USA *87*:6569-6573.

11. **White, T.J., N. Arnheim and H.A. Erlich.** 1989. The polymerase chain reaction. Trends Genet. *5*:185-189.

12. **Viera, J. and J. Messing.** 1982. The pUC plasmids, an M13mp7-derived system for insertion mutagenesis and sequencing with synthetic universal primers. Gene *19*:259-268.

13. **Zolg, W., E. Scott and M. Wendlinger.** 1988. High salt lysates: a simple method to store blood samples without refrigeration for subsequent use with DNA probes. Am. J. Trop. Med. Hyg. *39*:33-40.

CHAPTER 16

A Simplified Method for Determination of Specific DNA or RNA Copy Number Using Quantitative PCR and an Automatic DNA Sequencer

Catherine Porcher, Marie-Claire Malinge, Christiane Picat and Bernard Grandchamp

Laboratoire de Génétique Moléculaire, Faculté Xavier Bichat, Paris, France

ABSTRACT

Quantification of specific RNA or DNA molecules that are present in minute amounts in biological samples has previously been performed using PCR in the presence of an internal standard. We have adapted this concept by introducing several modifications that facilitate the quantification of the products and obviate the need for radioisotopes. After amplification, individual products are separated on sequencing gels and directly quantified using a fluorescent automated DNA sequencer. We describe two applications of this approach: the quantitation of minute amounts of bcr-abl *hybrid mRNA from malignant cells and the determination of gene copy number in cells stably transfected with a plasmid bearing a chloramphenicol acetyltransferase gene.*

INTRODUCTION

Sensitive and accurate quantitation of specific DNA or RNA sequences in biological samples is a general problem in medical and biological research. Due to its powerful amplification capacity, the PCR (6) has been widely used to detect specific nucleic acid molecules present in limited amounts either because few cells expressing the sequence are available or because the sequence of interest is present in low copy number. During the PCR process, an unknown initial number of target sequences are used as a template from which a large quantity of specific product can be obtained. Although the amount of product formed is easy to determine, it is difficult to deduce the initial copy number of the target molecule because the efficiency of the PCR is largely unknown. This problem has recently been addressed by coamplifying the

(Reprinted from BioTechniques 13:106-113, 1992)

molecule of interest with an internal standard bearing identical sites for attachment of the primers (1,4,10). Accordingly, the ratio of unknown product to the internal standard is expected to remain constant during the amplification process, or at least during the exponential phase of the PCR. In order to quantify the relative amount of PCR products, they are separated from each other, either directly according to their size (4,10) or after cutting specifically one of the two species using a restriction enzyme (1, 4). The amounts of the individual products are then usually calculated from their radioactivity (1,4,10). We have adapted the concept of competititive PCR by introducing several modifications to previously described techniques which facilitate the quantification of the products and which obviate the need for radioisotopes: 1) an internal standard is used which is very similar to the target molecule but slightly different in size (less than 5%), 2) PCR products are labeled during the PCR using a fluorescent primer, and 3) individual products are separated on sequencing gels and quantified using a fluorescent automated DNA sequencer. We describe two applications of this approach: the quantitation of minute amounts of *bcr-abl* hybrid mRNA and the determination of gene copy number in cells stably transfected with a plasmid bearing a chloramphenicol acetyltransferase (*CAT*) gene.

MATERIALS AND METHODS

Construction of the Internal Standards

Standard for *bcr-abl* cDNA. RNA and DNA were prepared as previously described (5). One microgram of RNA from the human leukemic cell line K562 was reverse-transcribed in two separate reactions using primers 2 and 3AS, respectively (Figure 1). Each reaction was carried out for 30 min at 42°C in 20 µL of a mixture containing 50 mmol/L KCl, 20 mmol/L Tris-HCl, pH 8.4, 2.5 mmol/L $MgCl_2$, 1 mmol/L dATP, 1 mmol/L dCTP, 1 mmol/L dGTP, 1 mmol/L dTTP, 20 pmol of primer and 200 units of Moloney murine leukemia virus (MMLV) reverse transcriptase (GIBCO BRL/Life Technologies, Cergy Pontoise, France). Two partially overlapping fragments were then amplified using oligonucleotides 1-3AS and 3-2 as primer pairs, respectively. After an initial denaturation for 3 min at 95°C, 30 cycles of PCR were performed; each cycle consisted of 15 s at 94°C, 30 s at 56°C and 30 s at 72°C. The two PCR fragments were purified on a polyacrylamide gel and allowed to diffuse from the gel into 200 µL of distilled water. Then 5 µL of each solution were directly added to 100 µL of a reaction mixture containing 50 mmol/L KCl, 20 mmol/L Tris-HCl, pH 8.4, 2.5 mmol/L $MgCl_2$, 0.2 mmol/L dATP, 0.2 mmol/L dCTP, 0.2 mmol/L dGTP, 0.2 mmol/L dTTP and 2 units of *Taq* DNA polymerase. After an initial denaturation for 3 min at 94°C and hybridization at 65°C for 2 min, partially single-stranded heteroduplexes were elongated for 2 min at 72°C and 5 cycles of PCR were performed for 15 s at 94°C, 2 min at 65°C and 2 min at 72°C; then primers 1 and 2 were added (50 pmol each), and the PCR was allowed to process for 25 additional cycles (15 s at 94°C, 30 s at 56°C, 30 s at 72°C). The reaction mixture was phenol-extracted, PCR products were ethanol-precipitated, digested with *Eco*RI and *Hind*III, ligated into the corresponding sites of the plasmid PGEM3Z and cloned in *E. coli* JM109. The recombinant plasmid was purified, linearized with *Eco*RI and quantified by spectrophotometry before being used as an internal standard.

Standard for the *CAT* gene. The *CAT* gene containing plasmid pblCAT8+ (8) was digested at a unique *Nco*I site located within the *CAT* gene, blunt-ended by filling in with the Klenow fragment of DNA polymerase I, ligated and cloned in *E. coli* JM109. The resulting plasmid contained a 4-base addition relative to the initial sequence and was used as an internal standard.

Separation and Quantification of PCR Products

PCR products are loaded onto an Automated Laser Fluorescent DNA Sequencer (A.L.F.™; Pharmacia LKB Biotechnology, Uppsala, Sweden) and submitted to electrophoresis. After the end of the electrophoresis, the registered signals are plotted on a HP plotter (Hewlett Packard, San Diego, CA, USA) and the areas of the peaks are calculated using the integration software Smart™ Manager (Pharmacia LKB Biotechnology) after conversion of the A.L.F. file. Subsequently, the number of molecules in the initial sample is calculated from the ratio of individual PCR products and from the amount of internal standard as follows:

n(target) = N(target)/N(internal standard) × n(internal standard)

where N represents the number of molecules after amplification and n the initial number of molecules.

Oligonucleotides

Oligonucleotides were synthesized with an Applied Biosystem 381A synthesizer. Fluorescent oligonucleotides were

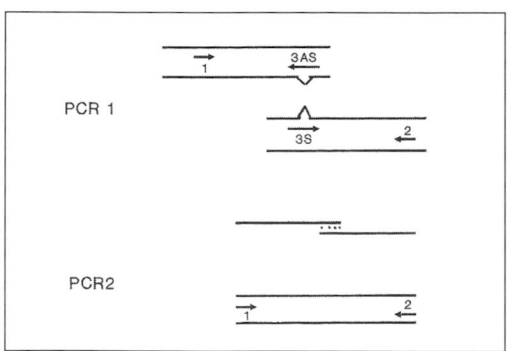

Figure 1. Construction of internal standard for bcr-abl. Two reactions of reverse transcription and PCR were carried out as described in the methods section. The sequence of the oligonucleotides is as follows:

1	5′ tccgggagcagcagaagaagt 3′
2	5′ cggaattcacaccattccccattgtg 3′
3S	5′ cttggagttccaacgagctcagtccctgaggc 3′
3AS	5′ gcctcagggtctgagctcgttggaactccaag 3′

Primer 1 corresponds to a sequence in exon 1 of the *bcr* region. Primer 2 is situated in exon 3 of the *abl* gene and contains a 3-base overhang at its 5′ extremity creating an *Eco*RI site. Primer 3 and 3AS are complementary to each other and correspond to two sequences of 15 bases each in exon 2 of the *abl* gene that are separated by an 8-base interval in the original sequence and thus introduce an 8-base deletion with respect to the natural sequence.

Quantitative PCR with an Automated DNA Sequencer

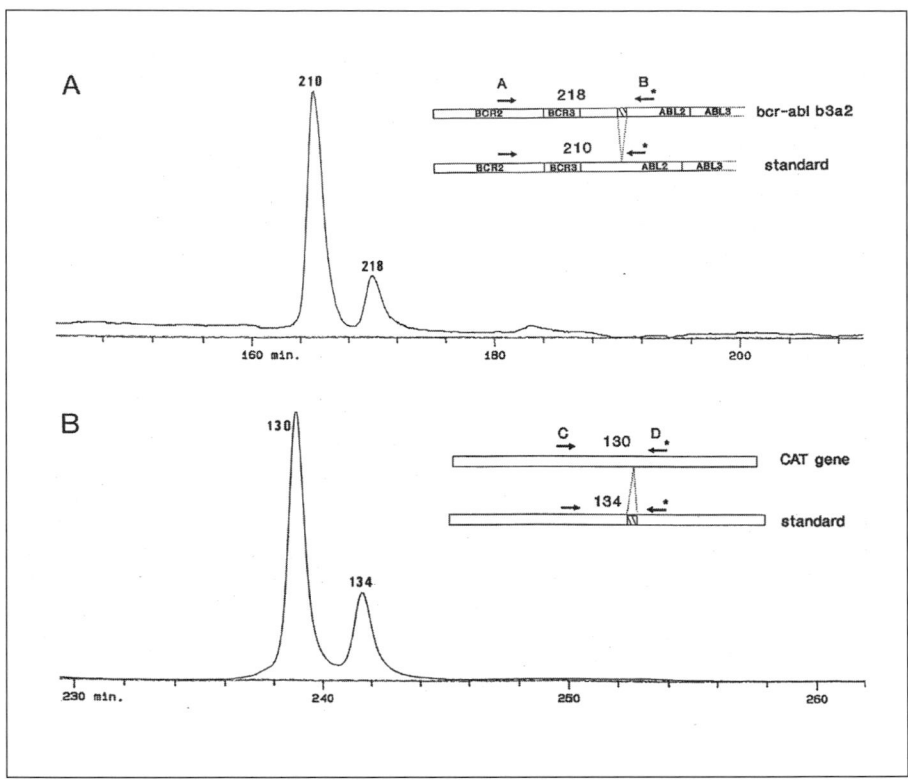

Figure 2. Separation and quantitation of PCR products after coamplification of a target molecule with an internal standard. (A) *Bcr-abl* cDNA: 10 pg of total RNA from the cell line K562 were mixed with 100 ng of RNA from a normal human lymphoblastoid cell line and submitted to RT-PCR. The reverse-transcription was carried out for 30 min at 42°C in 20 µL of a mixture containing 50 mmol/L KCl, 20 mmol/L Tris-HCl, pH 8.4, 2.5 mmol/L MgCl₂, 1 mmol/L dATP, 1 mmol/L dCTP, 1 mmol/L dGTP, 1 mmol/L dTTP, 20 pmol of a primer 5′ TTCACACCATTCCCCATTGTG 3′ and 200 units of MMLV reverse transcriptase; 5 µL of cDNA were submitted to PCR in the presence of 300 molecules of internal standard. The PCR was carried out in 50 µL of a reaction mixture containing 50 mmol/L KCl, 20 mmol/L Tris-HCl, pH 8.4, 2.5 mmol/L MgCl₂, 0.2 mmol/L dATP, 0.2 mmol/L dCTP, 0.2 mmol/L dGTP, 0.2 mmol/L dTTP and 25 pmol of each primer A and B. After an initial denaturation of 5 min at 95°C, 2 units of *Taq* DNA polymerase were added while the reaction mixture was maintained at 80°C. The tube was transferred to a Perkin-Elmer DNA thermal cycler (Saint Quentin, France) prewarmed at 94°C, and 50 cycles of PCR were performed, each cycle consisted of 1 min at 94°C and 1 min at 70°C. The primer sequences were as follows:

  -Upstream primer: A 5′ CGTGTGTGAAACTCCAGACTGTCCACAGCA 3′
  -Downstream primer: B 5′ AGCGAGAAGTTTTCCTTGGAGTTCCAACG 3′

The downstream primer was labeled with fluorescein. PCR products were diluted 1:20 with water, 2 µL of the dilution were mixed with an equal volume of denaturing solution (95% formamide, 1% dextran blue) and loaded onto a sequencing gel. The sizes of PCR products are 218 bp and 210 bp as indicated.

(B) CAT gene from transfected cells: DNA was prepared from stably transfected cells, then coamplified with 20 pg (3 × 10⁶ molecules) of internal standard. The sequence of the amplimers were:

  -Upstream primer: C 5′ CGTCTCAGCCAATCCCTGGG 3′
  -Downstream primer: D 5′ CAGCGGCATCAGCACCTTGT 3′

Primer D was fluorescently labeled. The PCR was carried out in the same reaction mixture as above in the presence of 25 pmol, each of primers C and D for 30 cycles. Each cycle consisted of 10 s at 92°C, 40 s at 49°C and 20 s at 72°C. Products derived from the genomic DNA (130 bp) and from the internal standard (134 bp) were resolved on a sequencing gel.

204

5′ labeled using the fluorescein phosphoramidite Fluoreprime (A.L.F.). After deprotection, the primers were purified on a 15% polyacrylamide gel, allowed to elute from the gel by diffusion, ethanol-precipitated and quantitated by spectrophotometry at 260 nm.

RESULTS

Separation and Detection of PCR Products Using an Automatic DNA Sequencer

PCR products in the range of 100 to 400 bases that differ from each other by 4 to 8 bases are very easy to separate and to detect using an automatic DNA sequencer provided they are labeled during the PCR by introduction of a fluorescent primer (Figure 2).

Linear Relationship Between the Amplified Signal and the Amount of PCR Product

The DNA sequencer used (A.L.F.) detects molecules passing through a fixed laser beam. The laser excites the fluorescent dye simultaneously in all the lanes by penetrating the gel perpendicular to band migration. The emitted light is detected by 40 separate photodiodes. Signals are automatically collected, digitized and sent to the computer for storage.

Because the intensity of the laser beam is not equivalent throughout the gel and the emitted light is detected by separate photodetectors, signals are not directly comparable from lane to lane, but the amount of individual products migrating in the same lane can be accurately compared. We first verified that the ratio of the fluorescent signals precisely reflected the relative amounts of individual products loaded onto the gel in each lane. A series of tubes were set up, each containing 20 attomol of fluorescent internal standard for *bcr-abl* and a variable amount of fluorescent *bcr-abl* cDNA from the K562 cell line. As shown in Figure 3, this experiment demonstrated that there is a linear relationship between the recorded signal and the amount of product loaded onto the gel over a 100-fold range.

Comparative Efficiency of Amplification Between the Internal Standard and the Target of Interest

This question was addressed in two model systems: *bcr-abl* cDNA and CAT gene DNA. The results presented for the quantitation of *bcr-abl* cDNA are representative of those obtained with both systems.

Individual solutions containing fluorescent PCR products from *bcr-abl* cDNA and the corresponding internal standard were mixed, and an aliquot of this mixture was loaded on A.L.F. to determine the ratio of the two species. Serial dilutions from 10^{-1} to 10^{-8} were then prepared, and 5 μL of each dilution were submitted to 25 cycles of PCR in a final volume of 50 μL. One aliquot from each reaction was electrophoresed into agarose gel and stained with ethidium bromide, and another aliquot from the same reaction was analyzed using A.L.F. From the ethidium bromide-stained gel, it was estimated that for initial dilutions from 10^{-1} to 10^{-6} a similar amount of PCR products was obtained; this suggests that reaction rates had declined, so the amount of amplified product had reached a plateau. On the contrary, from the initial dilution 10^{-7}, less product accumulated, and for the highest dilution, no signal was detectable by ethidium bromide staining (not shown), suggesting that in this latter case, the PCR was still in its exponential phase. Analysis of the products accumulated during the PCR shows that the ratio of the amplified *bcr-abl* cDNA to the product from the internal standard is independent of the initial dilution and that this ratio is very similar before and after amplification (Figure 4A). Comparable results were obtained when the same initial dilution (10^{-7}) was amplified through different numbers of cycles (data not shown). These results demonstrate that *bcr-abl* cDNA and the internal standard are amplified with a similar efficiency throughout the amplification process, not only during the exponential phase of the PCR.

We also examined if the initial ratio between the amount of internal

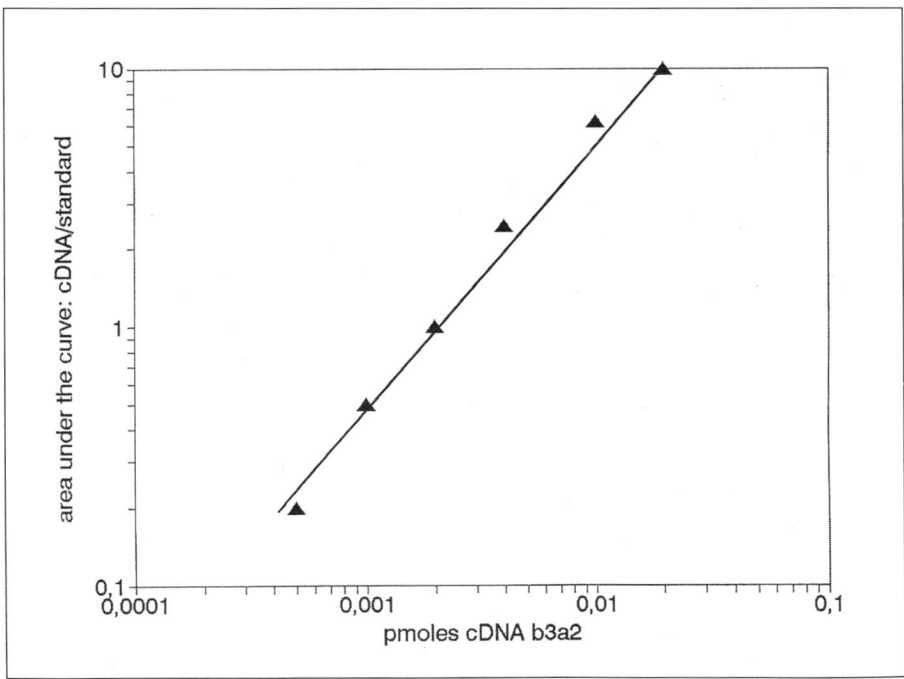

Figure 3. cDNA from K562 cells and DNA from the internal standard were amplified separately in the presence of a fluorescently labeled primer. Serial dilutions of the amplified cDNA were prepared and mixed with a constant amount of product from the internal standard. Each sample was analyzed with the DNA sequencer. The ratio of the fluorescent signals was calculated from the areas under the peaks and plotted against the amount of cDNA product loaded onto the gel.

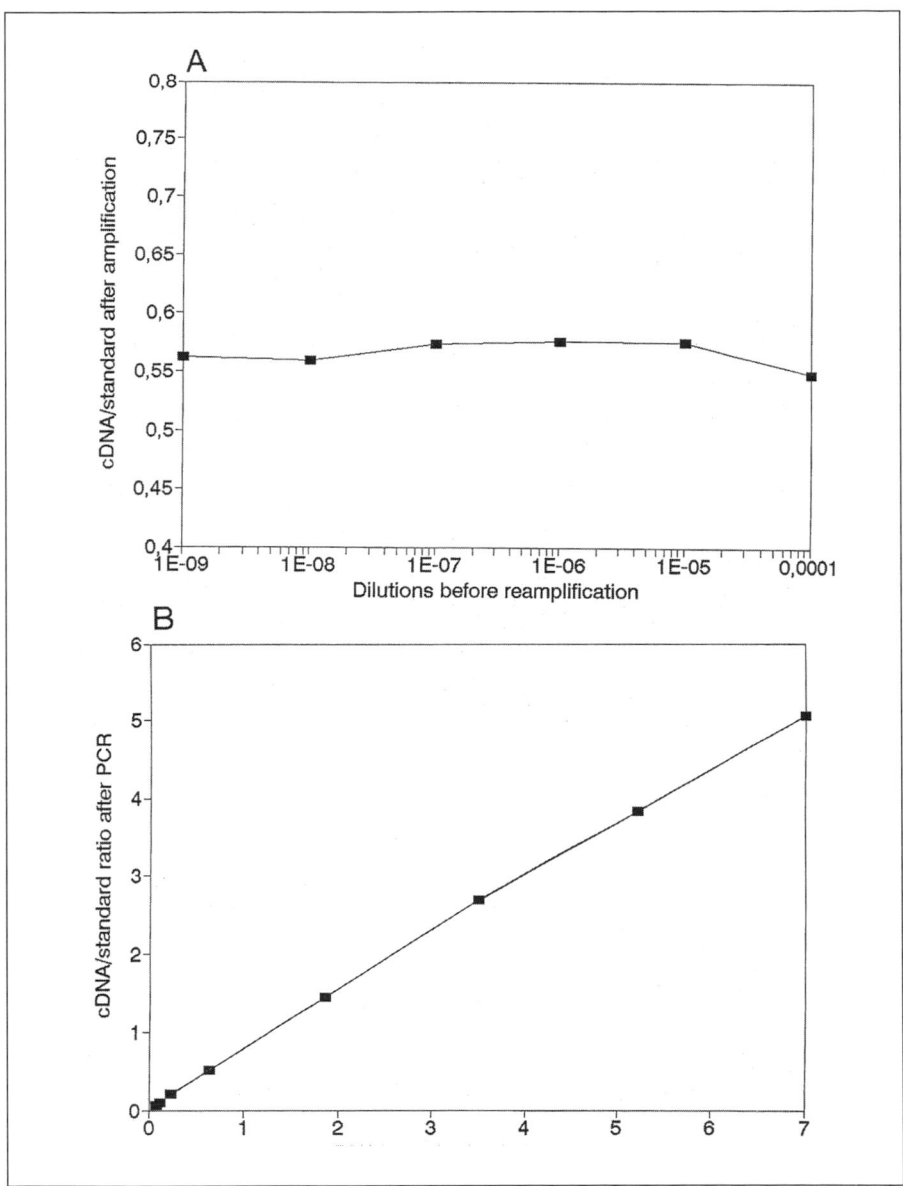

Figure 4. (A) cDNA from K562 cells was amplified in the presence of the internal standard. The ratio of the PCR products was determined to be 0.5 as described in the Materials and Methods section. Serial dilutions of the first amplification mixture were prepared and an aliquot from each dilution was reamplified for 30 cycles under the same experimental conditions, and the products from the second amplification were analyzed on the DNA sequencer. The ratio of individual products following the second amplification is plotted as a function of the dilutions. (B) cDNA from K562 cells and DNA from the internal standard were amplified separately in the presence of a fluorescently labeled primer and mixed in different proportions. An aliquot from each sample was analyzed on the DNA sequencer and the ratio of the two products was determined. A fixed dilution (10^{-7}) from each sample was then reamplified for 30 cycles; the ratio of cDNA/standard was determined and plotted against the corresponding ratio prior to amplification.

207

standard and the target molecule had any effect on the results of the quantitation. Accordingly, different proportions of internal standard to target were performed ranging from 1:7 to 7:1; each mixture was diluted by 10^{-7} and amplified for 30 cycles. After PCR, an aliquot was analyzed again, and the postamplification ratio of the two individual PCR products was determined. As shown in Figure 4B, the values of the final ratios were close to those of the initial ratios although consistently higher, suggesting that the amplification of the standard was slightly more efficient than that of the cDNA.

Quantitation of *bcr-abl* cDNA

Serial dilutions of *bcr-abl* RNA from K562 cells were prepared in a solution containing RNA extracted from normal circulating nucleated cells. One hundred nanograms from each dilution were reverse-transcribed, submitted to 50 cycles of PCR in the presence of appropriate dilutions of the internal standard and the products were analyzed on A.L.F. Figure 5 shows that there

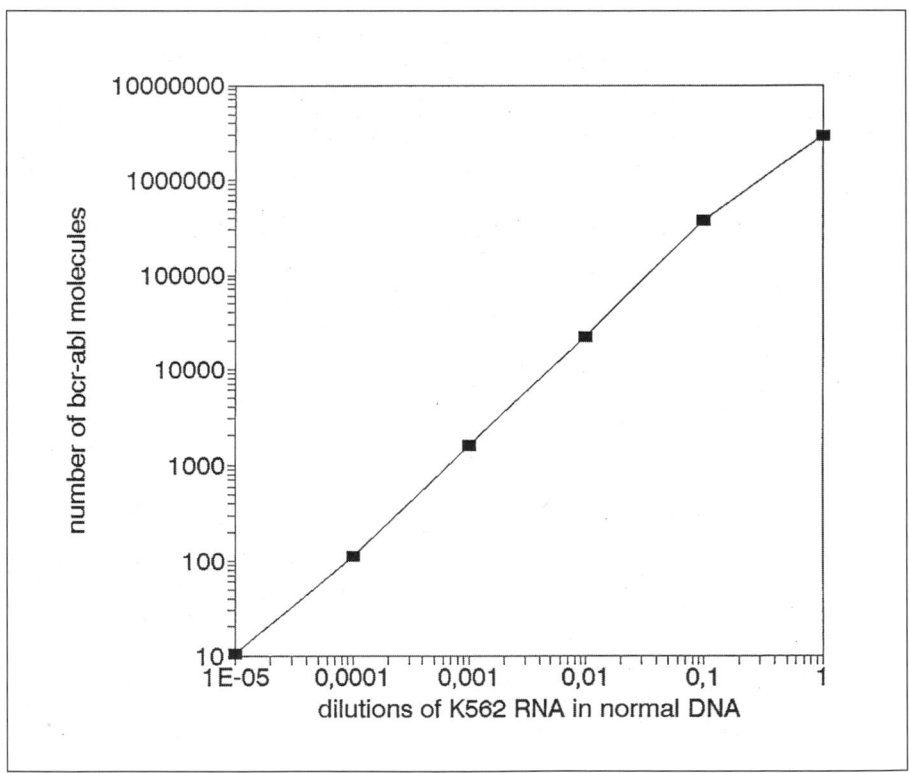

Figure 5. Serial dilutions of *bcr-abl* RNA from K562 cells (0.5 mg/mL) were made in a solution containing the same concentration of RNA extracted from normal circulating nucleated cells. One hundred nanograms from each dilution were reverse-transcribed and submitted to 50 cycles of PCR in the presence of various dilutions of the internal standard (from 30 to 3×10^6 molecules). The PCR products were analyzed on A.L.F., and the number of cDNA molecules corresponding to the *bcr-abl* RNA was calculated.

is a linear relationship between the amount of input K562 RNA and the calculated number of *bcr-abl* cDNA molecules.

Quantitation of the Copy Number of a Transfected Gene

Determination of the gene copy number is necessary to interpret the results of expression studies either in stably transfected cells or in transgenic animals. We have used our quantitative PCR assay for this purpose. As a model system, we cotransfected mouse erythroleukemic cells (MEL) with the plasmid pBLCAT8+ containing the CAT gene and plasmid pSV2neo containing a neomycine resistance gene (7). After selection in the presence of G418, DNA was extracted from transfected MEL cells, mixed in various proportions with DNA from untransfected cells, and an aliquot from each sample, corresponding to 10^5 genomes, was amplified in the presence of 3×10^6 molecules of internal standard. As shown in Figure 6, the ratio of individual PCR products is a direct linear function of the amount of transfected cells in the assay. The average copy number of the *CAT* gene in the population of transfected cells used in this experiment was deduced to be 75 copies per cell, assuming that 1 µg of mouse DNA represents 3×10^5 genomes.

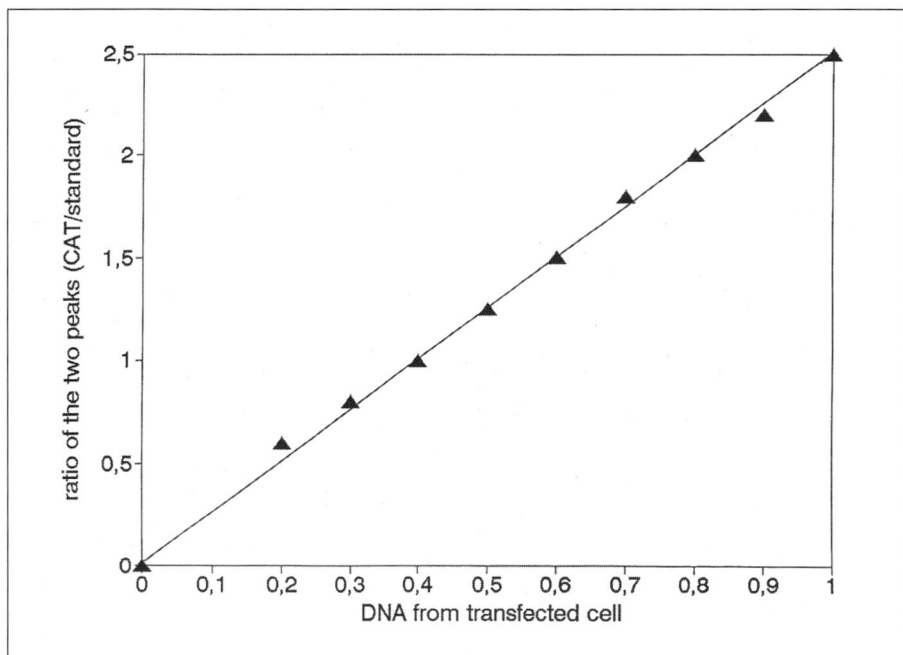

Figure 6. DNA (100 µg/mL) from a cell line stably transfected with a CAT gene was mixed with a solution containing the same concentration of DNA from untransfected cells. The proportion of DNA from the transfected cells ranged from 0 to 1. An aliquot from each sample, corresponding to 300 ng of DNA, was amplified in the presence of 20 pg of internal standard (3×10^6 molecules). The ratio of the signal of PCR products from the genomic DNA to that from the internal standard is expressed as a function of the fraction of DNA from transfected cells present in the samples.

DISCUSSION

Different strategies for quantitating DNA or RNA levels using PCR have been previously described (1–4,10). Most of these methods have addressed this problem by coamplifying the molecule of interest with an internal standard (1,4,10). We have taken this approach and modified it in such a way that 1) the radioactive labeling is replaced by a fluorescent labeling, 2) PCR products are separated from each other on a denaturing gel, and 3) the use of an automatic sequencer allows the separation and the quantitation of individual PCR products at the same time.

The reliability of using PCR in the presence of an internal standard for quantification purposes depends on the standard and the molecule of interest having the same efficiency of amplification. Different types of internal standards have been used to fulfill this requirement. One approach has been to use a synthetic standard using the same primers as the target but without any homology in the internal sequence (10). However, the efficiency of the PCR is only the same for both types of molecules during the exponential phase of the PCR. Consequently, in order to obtain reliable results, it is necessary to first titrate the specific target molecule to find the range of concentration that gives exponential amplification over a defined range of cycles. Another approach is to use internal standards that are very closely related to the sequence of interest by creating a small deletion or insertion or by modifying a restriction site (1,4,9,10). This ensures that the efficiency of amplification of the internal standard remains identical with that of the sequence of interest after the exponential phase of the PCR. With such standards, formation of heteroduplexes occurs during the late phase of the PCR from annealing between heterologous strands of mutant and wild-type templates. Because product strand reannealing can be a major factor limiting the amplification, the formation of heteroduplexes may be important for the plateau effect to occur simultaneously in the amplification of both templates. However, the formation of heteroduplexes may complicate the quantification of individual PCR products, since hybrid molecules are not cleaved when a restriction enzyme is used to distinguish the two types of molecules after PCR.

Several advantages of our system over previously described methods should be emphasized: The technique is simple since the labeling that is introduced in one of the primers is stable for more than one year when a stock solution is kept at -20°C. Direct detection of the emitted fluorescence is linear over a much wider range than scanning of autoradiograms and easier than counting the radioactivity of gel fragments. In addition, this approach overcomes one of the major obstacles that complicates the use of internal standards which are structurally very similar to the sequence of interest, that of heteroduplex formation (1,4). Indeed, single-stranded molecules are separated in a sequencing gel. The use of an automatic DNA sequencer allows one to quantify as many as 40 individual samples on a single gel. Therefore, this approach is particularly suitable for the analysis of large series of samples and is applicable for clinical purposes such as the quantitation of various DNA or RNA molecules in infectious diseases, leukemias or for evaluation of gene dosage. It is also a very valuable tool for the study of gene expression in stably transfected cells or in transgenic animals. In many cases, the internal sequence can be the endogenous gene itself, provided the transfected gene is labeled by inserting or deleting a few base pairs (Catherine Porcher, unpublished results). In these cases, both the copy number and the expression level of a transfected gene can be quickly determined relative to that of the endogenous gene or by the method described here.

ACKNOWLEDGMENTS

This work was supported by INSERM, CJF8904 and by the Association Française pour les Myopathies. We are grateful to Drs. A. Hance and C. Beaumont for helpful discussions and review of the manuscript. The first and second authors had an equal contribution to the work. We thank E. Thibaud for helping us with the integration software.

REFERENCES

1. **Becker-André, M. and K. Hahlbrock.** 1989. Absolute mRNA quantification using the polymerase chain reaction. A novel approach by a PCR aided transcription assay (PATTY). Nucleic Acids Res. *17*:9437-9446.
2. **Chelly, J., J.C. Kaplan, P. Maire, S. Gautron and A. Kahn.** 1988. Transcription of the dystrophin gene in human muscle and nonmuscle tissue. Nature *333*:858-860.
3. **Gaudette, F. and W.R. Crain.** 1991. A simple method for quantifying specific mRNAs in small numbers of early mouse embryos. Nucleic Acids Res. *19*:1879-1884.
4. **Gilliland, G., S. Perrin, K. Blanchard and H.F. Bunn.** 1990. Analysis of cytokine mRNA and DNA: detection and quantitation by competitive polymerase chain reaction. Proc. Natl. Acad. Sci. USA *87*:2725-2729.
5. **Grandchamp, B., C. Picat, F. de Rooij, C. Beaumont, P. Wilson, J.C. Deybach and Y. Nordmann.** 1989. A point mutation in exon 12 of the porphobilinogen deaminase gene results in exon skipping and is responsible for acute intermittent porphyria. Nucleic Acids Res. *17*:6637-6649.
6. **Mullis, K.B. and F.A. Faloona.** 1987. Specific synthesis of DNA in vitro via a polymerase catalyzed chain reaction. Methods Enzymol. *155*:335-350.
7. **Porcher, C., G. Pitiot, M. Plumb, S. Lowe, H. de Verneuil and B. Grandchamp.** 1991. Characterization of hypersensitive sites, protein-binding motifs, and regulatory elements in both promotors of the mouse porphobilinogen deaminase gene. J. Biol. Chem. *266*:10562-10569.
8. **Seiler-Tuyns, A., P. Walker, E. Martinez, A.M. Merillat, F. Givel and W. Wahli.** 1986. Identification of estrogen-responsive DNA sequences by transient expression experiments in a human breast cancer cell line. Nucleic Acids Res. *14*:8755-8770.
9. **Singer-Sam, J., J. Le Bon, R.L. Tanguay and A.D. Riggs.** 1990. A quantitative HpaII-PCR assay to measure methylation of DNA from a small number of cells. Nucleic Acids Res. *18*:687.
10. **Wang, A., M.V. Doyle and D.F. Mark.** 1989. Quantitation of mRNA by the polymerase chain reaction. Proc. Natl. Acad. Sci. USA *86*:9717-9721.

Update to:

A Simplified Method for Determination of Specific DNA or RNA Copy Number Using Quantitative PCR and an Automatic DNA Sequencer

Bernard Grandchamp

INSERM U409, Faculté Xavier Bichat, Paris, France

The following references illustrate the use of our quantitative PCR assay for different purposes: studies of gene expression, RNA quantitation and determination of gene copy number in acute leukemias.

REFERENCES

1. **Cave, H., B. Gerard, E. Martin, C. Guidal, I. Devaux, J. Weissenbach, J. Elion, E. Vilmer and B. Grandchamp.** 1995. Loss of heterozygosity in the chromosomal region 12p12-13 is very common in childhood acute lymphoblastic leukemia and permits the precise localization of a tumor-suppressor gene distinct from p27kip1. Blood *86*:3869-3875.
2. **Guidal-Giroux, C., B. Gerard, H. Cave, M. Duval, P. Rohrlich, J. Elion, E. Vilmer and B. Grandchamp.** 1996. Deletion mapping indicates that MTS1 is the target of frequent deletions at chromosome 9p21 in paediatric acute lymphoblastic leukaemias. Br. J. Haematol. *92*:410-419.
3. **Porcher, C., C. Picat, D. Daegelens, C.Beaumont and B. Grandchamp.** 1995. Functional analysis of DNase-I hypersensitive sites at the mouse porphobilinogen deaminase gene locus. J. Biol. Chem. *270*:17368-17374.
4. **Raja, K.B., B. Gerard, A. McKie, R.J. Simpson, T.J. Peters, B. Grandchamp and C. Beaumont.** 1996. Duodenal expression of NFE2 in mouse models of altered iron metabolism. Br. J. Haematol. *91*:483-489.

CHAPTER 17

Random Primer p(dN)$_6$-Digoxigenin Labeling for Quantitation of mRNA by Q-RT-PCR and ELISA

William Lear, Michael McDonnell, Sonya Kashyap and Poppo H. Boer

University of Ottawa Heart Institute, Ottawa, Ontario, Canada

ABSTRACT

The ability to accurately measure mRNA levels in samples of total RNA is essential for studies on control of gene expression. The mRNAs from the housekeeping gene for phosphoglycerate kinase (PGK-1) can serve as a quality control for RNA samples. We describe an enzyme-linked immunosorbent assay (ELISA) method for mRNA determination by Q-RT-PCR, a quantitative reverse transcriptase-mediated PCR assay with competitive internal standards. After PCR, two biotinylated capture primers, one specific for PGK-1 cDNA and another one for internal standard, are annealed in separate assays so that each can attach DNA to a streptavidin-coated microplate. The captured DNA is either internally labeled with digoxigenin (DIG) or is "developed" after annealing with DIG-labeled primers. Bound DNA is then quantitated by adding DIG-specific antibody with attached alkaline phosphatase and measuring phosphatase activity with a chromogenic substrate and a plate reader. We compared different capturing methods and various primers labeled with DIG at their 3' ends. We determined that amplified PGK-1 DNA specifically captured with biotinylated primers was efficiently assayed with random p(dN)$_6$-DIG.

INTRODUCTION

The development of reverse transcriptase-mediated polymerase chain reactions (RT-PCR) has greatly advanced studies on control of gene expression (5,6), and this leads to greater insight into diseases, including those of the heart. Because of the exponential nature of PCR, introduction of a competitive internal DNA standard has proven to be a great advantage for reliable mRNA quantitation by Q-RT-PCR (5). The PCR primers can become incorporated into nonspecific amplification products (6), such as low-molecular-weight "primer-dimer" DNA and heterogeneous products of various size not easily seen on agarose gels. Therefore, the quantitation of specific amplification products in PCR assays should make use of nested primers. Recently,

(Reprinted from BioTechniques 18:78-83, 1995)

Table 1. PGK-1 PCR Primer Sequences

PB3 ←: 5′-<u>A</u>CCATCCAGCCAGCAGGTAT	941–960 (forward primer, exon 8)
PB5 →: 5′-GTGAAGGGGAAGCGGGTCGT	62–81 (reverse primer, exon 1)
PB1 ←: 5′-ACCTTGTTCCCAGAAGCATC	422–441 (reverse primer, exon 4)
PB4 ←: 5′-<u>A</u>CCCTCCCAAGATAGCCAGG	646–665 (forward primer, exon 6)
PB2 ←: 5′-GA<u>AA</u>ACCTCCGCTTTCATGT	380–399 (reverse primer, exon 4)
*MM3 →: 5′-C<u>C</u>CGTTGT<u>C</u>CTTATGAGCCACC	192–212 (reverse primer, exon 2)
*MM4 ←: 5′-AGATAGCCAGGAAGGGTCGC	637–656 (forward primer, exon 6)

Arrows indicate primer orientation, * indicates biotin; coordinates are from the RNPGKXL/MMXPGK sequences from rat and mouse, respectively.

The cloned PGK-1 cDNA originated from the mouse and shares 92% nucleotide identity with the counterpart in rat, and there are no sequence differences in the primer binding sites for the rat PCR templates, except for underlined nucleotides.

immuno-PCR and enzyme-linked immunosorbent assay (ELISA) approaches have been adapted to RT-PCR, and they successfully combine the high throughput nature of microplates with the sensitivity of PCR (1,8,9). In such post-PCR applications, end-labeled primers are used for the specific capture and subsequent detection of amplified DNA. For example, a biotinylated oligonucleotide can hold the PCR product in a streptavidin-coated microplate, enabling subsequent reactions on captured DNA. This DNA may be labeled during amplification with radioactive nucleotides or modified nucleotides containing covalently-linked groups such as digoxigenin (DIG) that can be recognized by enzyme-linked antibodies. Unlabeled PCR products that were specifically captured may be annealed to developing or "revealing" primers (1) carrying a terminal DIG label. In either case, quantitation is allowed by incubation with an enzyme-linked antibody to DIG and subsequent reaction with a chromogenic substrate. In this study, we have focused on mRNAs from the housekeeping gene for phosphoglycerate kinase-1 (PGK-1), which we use as a quality control (external standard) for Q-RT-PCR on cardiac RNA preparations (2,3).

MATERIALS AND METHODS

RNA was extracted (4) with TRIzol™ (Promega-Fisher, Ottawa, Ontario, Canada), and all nucleic acids were quantitated by UV absorbance. Reverse transcription of 2 µg total RNA was performed with 0.2 µg p(dT)$_{15}$ primer and 8 U of avian myeloblastosis virus (AMV) reverse transcriptase per 25 µL assay. The cDNA corresponding to 2–20 ng total cardiac RNA was used in 30-µL PCRs (2), with 1.5 U of *Taq* DNA polymerase (Promega-Fisher), supplied PCR buffer (10 mM Tris-HCl, pH 9.0, 50 mM KCl, 1% vol/vol Triton® X-100), MgCl$_2$ at 2.5 mM and deoxyribonucleotides (Pharmacia Biotech,

Baie d'Urfe, Quebec, Canada) at 0.25 mM. A Coy I Model 60 Thermocycler (Coy, Grass Lake, MI, USA; obtained from Diamed, Ottawa, Ontario, Canada) was used with settings of 94°C for 1 min, 55°C for 1 min and 72°C for 1.5 min using 30 cycles. Figure 1 shows the PCR primers used and the data bank coordinates; their sequences are listed in Table 1. Primers were designed with the help of computer programs like Oligo (National Biosciences, Plymouth, MN, USA) or PC/GENE (Intelligenetics, Mountain View, CA, USA). Primers, including the 5'-biotinylated ones for capturing, were synthesized by General Synthesis Diagnostics (GSD-Inc., Toronto, Ontario, Canada). A laminar flow cabinet was used to assemble PCR assays for "hot start," and PCR samples of 10 µL were electrophoresed in agarose gels (in TBE: 89 mM Tris, 89 mM boric acid, 2 mM EDTA, pH 8.0) and stained with ethidium bromide, as described before (3). For quantitation of amplified DNA, Polaroid T57 or T55 film (Cambridge, MA, USA) was used, and the negatives were scanned with a laser densitometer (UltroScan™, Pharmacia Biotech). For quantitation of PGK-1 mRNA by Q-RT-PCR, an internal standard, ΔPGK-1, was created by removing an internal $ApaI$-$BglII$ region of 497 bp from the cloned cDNA (7).

Microplates (MaxiSorp 96-well; Nunc-BRL, Ottawa, Ontario, Canada) were coated with 50 µL streptavidin (Sigma Chemical, St. Louis, MO, USA) at 100 µg/mL in 0.1× phosphate-buffered saline (PBS) (1× PBS: 137 mM NaCl, 2.7 mM KCl, 4.3 mM $Na_2HPO_4 \cdot 7H_2O$, 1.4 mM KH_2PO_4, pH 7.4) overnight at 4°C. After two washes with water and once with 100 µL buffer A (1× PBS with 0.1 M Tris pH 7.5 and 0.05% vol/vol Tween® 20; Sigma Chemical), we added blocking buffer (buffer A with bovine serum albumin [BSA] at 2.5 mg/mL and sheared denatured herring DNA at 100 µg/mL) for 1 h at room temperature. This was followed by two more washes with buffer A. The PCR DNA in dilution series was captured with 1.5 pmol biotinylated primer in 50 µL 6× standard saline citrate (SSC) by heating for 3 min at 94°C in the Coy thermocycler and quenching on ice. The samples were placed in the wells of the coated microplate and incubated on a shaker for 1 h at room temperature. After 3 washes with buffer A, the developing primer was annealed by adding 0.2 pmol of a nested PGK-1 primer, or the general p(dN)$_6$ primer, both of which labeled at the 3' end with DIG. This step was in 50 µL 6× SSC for 1 h at 37°C, followed by 3 further washes with buffer A. In the case of PCR DNA internally labeled with DIG by adding DIG-11-dUTP analog to the above PCR assay, the developing step with DIG-primer was omitted. Next, 50 µL of anti-DIG-antibody-alkaline phosphatase conjugate, Fab fragments (Boehringer Mannheim, Dorval, Quebec, Canada), at a dilution of 1:1000, were added for 45 min at 37°C, and this was followed by 3 washes with buffer A. Then 50 µL substrate p-nitrophenol phosphate (PNPP) at 1 mg/mL in 10 mM diethanolamine, 5 mM $MgCl_2$, pH 9.0 were added. The yellow color due to phosphatase action after 1–2 h (or overnight) was quantitated by optical density at 405 nm in a Model 3350 plate reader (Bio-Rad,

215

Mississauga, Ontario, Canada). Terminal deoxyribonucleotidyl transferase (TdT) from Promega-Fisher was used for labeling the developing primer-reagent by incubation of 2 pmol primer in 50 µL Na-cacodylate buffer, 1 mM dithiothreitol (DTT), 1 mM $CoCl_2$, pH 7.0 for 30 min at 37°C with 11-dUTP-DIG at 20 µM. The $p(dN)_6$-DIG and other DIG-DNA probes were stored at -70°C.

RESULTS AND DISCUSSION

The strategy for Q-RT-PCR of PGK-1 mRNA is schematically shown in Figure 1; it normally uses "outside" primers, PB3 and PB5. The nested primers, *MM3 and *MM4, which were biotinylated (*) at their 5′ ends, were used as capture primers to attach amplified DNA to streptavidin-coated surfaces. The primers PB1, PB2 and PB4 served as DNA-strand specific probes for detection of captured PGK-1 DNA, and all primers were tested in various combinations with templates from rat and mouse. The primer sequences are listed in Table 1. These primers annealed to target PGK-1 templates with similar efficiencies, as deduced from agarose gel electrophoresis of amplification products at 25 and 30 cycles, which resulted from the use of, for example, the reverse PB5 primer combined with forward primers such as

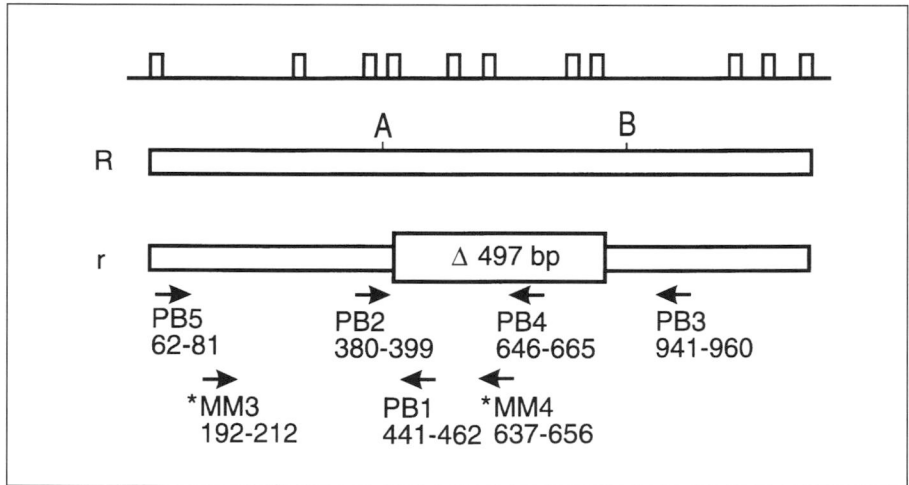

Figure 1. Design of the PGK-1 primers for Q-RT-PCR. Schematic of the PGK-1 gene, its cDNA (R) and internally deleted DNA construct (ΔPGK-1 or r) which is used as an internal competitive standard in the Q-RT-PCR assay. The r_{pgk} resulted from removal of the ApaI (A)-BglII (B) restriction fragment from cloned cDNA (sequence coordinates A: 326 and B: 823). Horizontal arrows indicate polarity of the primers denoted with their data bank coordinates. The use of PB3 and PB5 with R_{pgk} and r_{pgk} PCR templates results in amplification products of about 900 and 400 bp, respectively. These are the post-PCR targets for capturing with the MM primers; asterisk indicates the biotin group attached to their 5′ termini. MM3 and MM4 are located inside and outside of the gap in ΔPGK-1, respectively, and are specific for R_{pgk} and R_{pgk} plus r_{pgk}, respectively. When the captured PGK-1 DNA is not labeled, additional PB primers or $p(dN)_6$ can be end-labeled with DIG for quantitation of captured DNA.

PB1, PB4 or MM4 with its biotin group or combinations of these primers. After Q-RT-PCR for measurement of PGK-1 mRNA levels with PB3 and PB5, the relative amounts of amplified PCR products from cDNA (R_{pgk}) vs. competitive ΔcDNA (r_{pgk}) must be determined (3,5). The cDNA sample is characterized by the amount of internal standard DNA needed to reach the point of equivalence in titration-like experiments. Separation and quantitation of the PCR products, R_{pgk} and r_{pgk}, may be achieved by agarose gel electrophoresis and laser densitometry (3). The tedious gel separation may be bypassed by capturing the specific PCR products (1,8,9). Because the MM4 capture primer is located within the gap of ΔPGK-1, it will react with PCR products from cDNA only and capture R_{pgk}. The MM3 primer will capture PCR products from only cDNA and internal standard, R_{pgk} and r_{pgk}. The PGK-1 amplification products could be internally labeled with DIG by use of the DIG-11-dUTP nucleotide in the PCR assay, or amplified DNA could be labeled in a post-PCR step by annealing with DIG-labeled oligonucleotides. In both cases, the amount of DIG label captured in the wells of microplates could be quantitated with alkaline phosphatase-linked antibodies to DIG. Alternatively, radioactive R_{pgk} and r_{pgk} PCR products could be captured on streptavidin-coated magnetic beads (Promega-Fisher), but these attempts were not pursued. We first studied internal DIG-labeling of PGK-1 PCR DNAs with DIG resulting from use of the PB3/PB5 primers with a cardiac cDNA preparation. Figure 2 shows that the deoxyribonucleotide analog in amounts varying from 14–0 µM was readily incorporated in the PCR products. In Figure 2, lanes 2–4, DIG incorporation was seen to slightly reduce the DNA mobility on agarose gels but not the yield of amplified DNA.

Figure 3 shows DNA quantitation in the form of binding curves. The DNA preparations from Figure 2 and primers defined above were used in the capturing assays with the biotinylated MM3 primer. The amount of DNA bound in the wells of a microplate is reflected by OD_{405} on the y-axis. The x-axis indicates the amount of DNA. The input amount of PCR DNA was estimated to be close to 100 pg (based on OD_{260} readings of the stock DNA and by densitometry readings vs. lambda DNA standard), and it was used in 5-fold serial dilutions. In Figure 3, curve D, an added step at 55°C did not increase the annealing efficiency of MM3 to the DIG-labeled DNA as compared with a

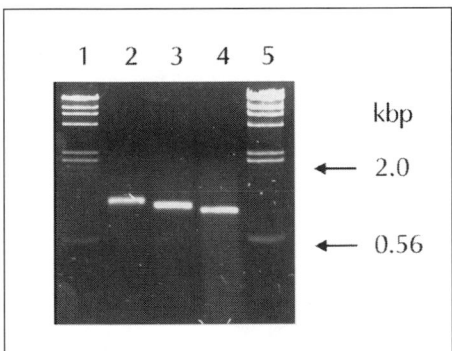

Figure 2. Internal digoxigenin labeling. The PCR products were run on a 1% agarose gel in TBE alongside lambda-DNA size markers in lanes 1 and 5. Results of the use of 14, 7 and 0 µM 11-dUTP-DIG, in lanes 2–4, respectively, in a standard PCR assay with PB3/PB5 primers and a PGK-1 cDNA template prepared from total cardiac RNA.

217

simpler annealing step of MM3 to template in an incubation at 94°C with template and quenching on ice (curve C). Different developing primers were then tested on unlabeled PCR products (Figure 2, lane 4). We saw similar results with MM3-captured DNA and two different but specific development primers, PB2-DIG (curve A) and PB5-DIG. Next, we compared the use of the specific PB2 primer with the general p(dN)$_6$ primer, both of which were labeled with DIG at their 3′ ends. Similar results were seen for two specific capture probes (MM3 and MM4), and Figure 3, curve B, shows a representative binding curve with biotin-MM3. By comparing curves 3A and 3B and 3C and 3D, it can be seen that the developing step afforded more specificity than obtained with PCR DNA internally labeled with DIG (which is also more expensive). This is indicated by the slope of the binding curves. Use of the specific PB2 primer yielded a very steep binding curve over the narrow range of 4 dilutions (Figure 3, curve A). The general p(dN)$_6$ primer resulted in a binding curve that reached the same maximum value, but the range of dilutions covered was greater, covering 8 or more of the 5-fold dilution steps. Thus, relative to the specific PB2 primer, the general p(dN)$_6$ probe covered a much wider range of DNA samples. These results suggest that the general p(dN)$_6$ development step will be a useful modification of current protocols for Q-RT-PCR. Chemical attachment of DIG to the developing primers may

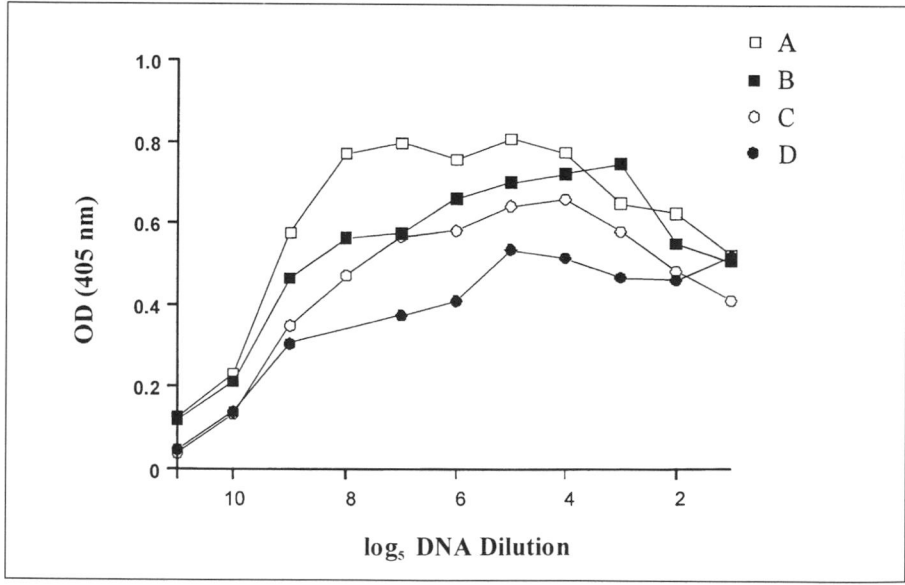

Figure 3. Oligonucleotide capturing. Fivefold serial dilutions were prepared from the DNA samples used in Figure 2, lanes 3 and 4, starting from 0.1 ng DNA. Binding curves of captured PGK-1 DNA are shown as OD$_{405}$ readings plotted against the DNA dilution; PCR DNA was captured with the biotinylated MM3 primer. Curves A and B resulted from unlabeled DNA (Figure 2, lane 4) after developing with PB2 and p(dN)$_6$ primers labeled at their 3′ ends with DIG, respectively. Curves C and D resulted from internally DIG-labeled PCR DNA (Figure 2, lane 3). In curve D, a short 55°C incubation step was added to the annealing step for MM3 binding. The blanks (no DNA) values are plotted on the y-axis itself.

be more efficient than by use of terminal transferase, and this may further increase the detection efficiency of captured DNA.

The results in Figure 3 were obtained after dilutions of the PCR products of up to 5^{10}-fold (or 10 million-fold); hence this is a very sensitive method. Here, an input of 100 pg DNA (900 bp; 9.5×10^7 mol) diluted 5^{10}-fold corresponds to a few mol/assay. The great sensitivity of the method presented here allows for reduced PCR cycle number and may facilitate different mRNA quantitations by a Q-RT-PCR with multiple PCR primer pairs (multiplexing). For the PGK-1 amplification products in this study, different capture primers were equally efficient and were seen to be effectively developed with $p(dN)_6$. Lastly, many of the steps involved can be automated.

In conclusion, tissue-specific gene activity measured as stable mRNA levels in a given cDNA preparation may readily be characterized with a nonradioactive Q-RT-PCR assay with an outside primer set and two nested capture primers (one specific for mRNA, the other for internal standard) in conjunction with a common developing agent such as $p(dN)_6$ primer.

ACKNOWLEDGMENT

Support from the Heart and Stroke Foundation of Ontario (HSFO) and Medical Research Council of Canada is gratefully acknowledged. W.L. and S.K. received HSFO awards, M.M. received intramural support (Hypertension Unit), P.B. is a HSFO Research Scholar. We thank Dr. Z. Rassi and E. Hobden for PCR studies, Dr. R. Milne for use of the plate reader, and Drs. F. Leenen and E. Harmsen for support and discussion.

REFERENCES

1. **Alard, P., O. Lantz, M. Sebagh, C.F. Calvo, D. Weill, G. Chavanel, A. Senik and B. Charpentier.** 1993. A versatile ELISA-PCR assay for mRNA quantitation from a few cells. BioTechniques *15*:730-737.
2. **Boer, P.H.** 1993. Activation of the gene for Atrial Natriuretic Factor during in vitro cardiac myogenesis by P19 embryonal carcinoma cells. Exp. Cell Res. *207*:421-429.
3. **Boer, P.H.** 1994. Activation of the gene for type-B natriuretic factor in mouse stem cell cultures induced for cardiac myogenesis. Biochem. Biophys. Res. Commun. *199*:954-961.
4. **Chomczynski, P.** 1993. A reagent for the single-step simultaneous isolation of RNA, DNA and proteins from cell and tissue samples. BioTechniques *15*:532-537.
5. **Foley, K.P., M.W. Leonard and J.D. Engel.** 1993. Quantitation of RNA using the polymerase chain reaction. Trends Genet. *9*:380-385.
6. **Innis, M.A., D.H. Gelfand, J. Sninsky and T.J. White (Eds.).** 1990. PCR Protocols, A Guide to Methods and Applications. Academic Press, New York.
7. **Mori, N., J. Singer-Sam, C.Y. Lee and A.D. Riggs.** 1986. The nucleotide sequence of a cDNA clone containing the entire coding region for mouse X-chromosome linked phosphoglycerate kinase. Gene *45*:275-280.
8. **Sano, T., C.L. Smith and C.R. Cantor.** 1992. Immuno-PCR: very sensitive antigen detection by means of specific antibody-DNA conjugates. Science *258*:120-122.
9. **Yang, B., R. Viscidi and R. Yolken.** 1993. Quantitative measurement of nonisotopically labeled polymerase chain reaction product. Anal. Biochem. *213*:422-425.

Update to:

Random Primer p(dN)$_6$-Digoxigenin Labeling for Quantitation of mRNA by Q-RT-PCR and ELISA

Jeyanthi Ramamoorthy, Sherissa Microys, Andrew McColgan, Zahra Rassi and Poppo H. Boer[1]

Cardiac Molecular Biology, University of Ottawa Heart Institute; [1]Departments of Pathology and Laboratory Medicine, and Biochemistry, University of Ottawa, Ottawa, Ontario, Canada

ABSTRACT

Q-RT-PCR is an internally controlled quantitative reverse transcriptase-linked polymerase chain reaction to assess gene activity as the accumulated stable transcript level in a sample of total tissue RNA. Internal standardization is necessary because of the exponential nature of PCR; however, when using a homologous DNA or RNA competitor, its PCR product, "r", will inevitably recombine with the mRNA derived PCR product, "R", yielding "R/r" DNA molecules. Co-migration of these PCR products often occurs on neutral agarose gels, and this will cause serious errors in determined mRNA values. Adjustment of the electrophoresis conditions to separate homodimer DNAs, "R/R" and "r/r", from the heterodimer DNA, "R/r", or inclusion of a DNA denaturation step to measure separated "R" and "r" DNA strands, is required to arrive at the correct mRNA value.

INTRODUCTION

In our studies on control of cardiac muscle gene expression, we have extensively made use of quantitative reverse transcriptase-mediated polymerase chain reaction assays, or Q-RT-PCR, to determine amounts of accumulated mRNA in total RNA extracted from small tissue sections. Thus, a variety of cardiac mRNA assays was set up by preparing competitive internal standards that yielded PCR products different in size from that resulting from the mRNA to be measured. For mRNAs encoding phosphoglycerate kinase, PGK-1, (1,3), type-A natriuretic factor (2), renin mRNA (1) and α- and β-isoforms of myosin heavy chain (unpublished), small deletions were created in cloned cDNAs, whereas in assays for mRNAs for type-B natriuretic factor

220

(2) and angiotensinogen (1), genomic DNA constructs were used where the sequences of the oligonucleotide primers were located in different exons separated by a small intron. This setup assured that in each mRNA assay, primers reacted in identical fashion with the templates derived from mRNA and its internal standard, as is required for competitive PCR (6), and that the respective PCR products could be separated by size by electrophoresis in simple agarose gels. To exclude possible variations in reverse transcriptase efficiency, the internal standards are prepared in riboprobe plasmids and synthetic RNA is used as competitor. In a titration type of assay, the amount of mRNA is expressed in terms of added competitor needed to reach the point of equivalence of PCR product DNAs from mRNA and internal standard. As demonstrated here for the case of renin mRNA, and verified for all assays mentioned above, homologous recombination between DNA amplification products from mRNA and internal standard can easily be detected. The comigration of various PCR DNAs on standard agarose gels causes serious error in determined mRNA values.

MATERIALS AND METHODS

The pGEM4 plasmids containing the full-length renin cDNA and an internal deletion construct have been described (1). Large-scale plasmid DNA preparations were purified using Qiagen columns (Chatsworth, CA, USA), and quantitated by OD260/280 measurement. The in vitro transcription reaction with SP6 RNA polymerase was done in a final volume of 20 µL using 10 U enzyme/µg DNA as specified by the supplier (GIBCO-BRL/Life Technologies, Burlington, Ontario, Canada). The renin plasmid DNA was linearized with *Hin*dIII, and 100 ng were transcribed in the presence of 10 mM dithiothreitol (DTT), 20 U RNasin®, 0.5 mM rNTPs, and 0.1 µM random primer pdN6 at 37°C for 1 h. Riboprobes may contain residual DNA template even after repeated DNase I treatment (e.g., 5), and this was the case for renin cRNA. Therefore, endogenous DNA template within this riboprobe transcript preparation was digested with *Rsa*I followed by heat inactivation of the restriction enzyme at 65°C for 10 min. From this reaction, the first strand cDNA was synthesized using avian myeloblastosis virus (AMV) reverse transcriptase (Promega-Fisher, Ottawa, Ontario). DNA amplification occurred in a Perkin-Elmer GeneAmp® 2400 model thermocycler under the following conditions: 1 min at 94°C, 1 min at 60°C and 1 min at 72°C, usually for 30 cycles, prior to which was a hot start at 94°C for 2 min. PCRs were assembled on ice in a laminar flow cabinet. PCR products were analyzed by agarose gel electrophoresis using double-combed minigels in TBE (89 mM Tris, 89 mM boric acid, 2 mM EDTA, pH = 8.0). These gels could contain 0.5 µg/mL ethidium bromide (see below). A portable DS34 Polaroid camera (Cambridge, MA, USA) was used to photograph the fluorescent DNA profiles and band intensity was measured through a Sharp X325 laser scanner (New York, NY, USA). The computer program OLIGO version 4.1 (National BioSciences, Plymouth, MN, USA) was used to design the PB18/22 20-mer oligonucleotides (1) from the renin GenBank® database sequence: RATREN, accession # M16984.

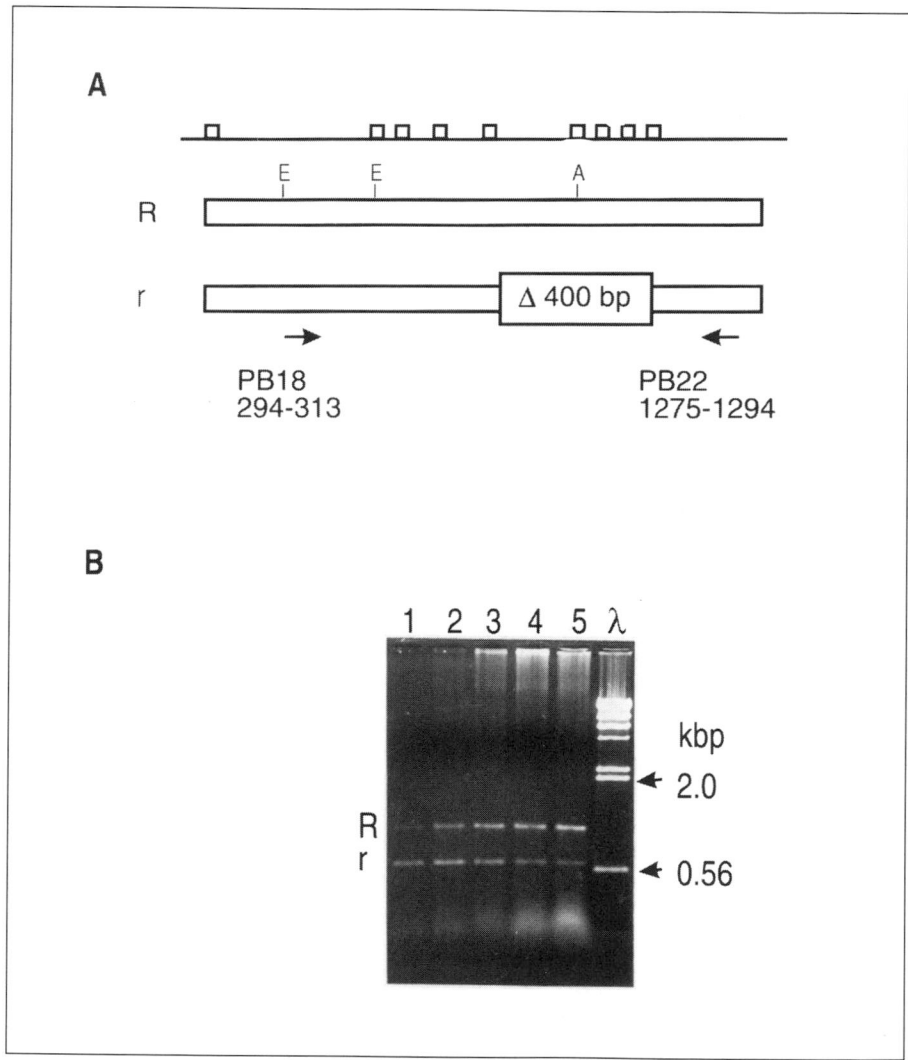

Figure 1. Cardiac renin RT-PCR. (A) schematic of the renin gene (>10 kbp), the renin cDNA ("R"), and the δ renin construct used as internal RT-PCR standard ("r"), and primers PB18 and PB22, denoted in data bank coordinates. Restriction sites are indicated for cDNA: E, *Eco*RI and A, *Apa*I. (B) Representative RT-PCR assay to determine the level of renin mRNA in a control RNA sample from the right ventricle of an adult Wistar rat. In this case, 0.11 fg δ renin DNA was incubated with increasing amounts of cDNA, using 35 cycles of PCR. Lanes 1–5 contained cDNA in twofold serial dilutions, from 25–400 ng RNA, respectively; lane 6, lambda DNA × *Hin*dIII size marker. The crossover point was close to lane 3, and scanning analysis resulted in a value of 90 ng RNA, therefore the relative level was 1.2 fg renin mRNA/μg RNA.

RESULTS

Cardiac Renin and PGK-1 mRNA Levels

The strategy for renin RT-PCR is analogous to that for PGK-1 and is shown in Figure 1A. Indicated are the renin cDNA and Δ deletion construct cloned in the pGEM4 transcription vector with the binding site for SP6 RNA polymerase at left and the *Hin*dIII linearization site at right. Also shown are the data bank coordinates for the PB18/22 primer combination. The amplification products from renin cDNA (R) and internal standard (r) are about 1200 and 800 bp long, as seen in Figure 1B. The cardiac renin mRNA levels are very low, much lower than in kidney, where renin mRNA levels were similar to those for PGK-1. Thus, both the number of PCR cycles and amounts of input cDNA were increased because of these low mRNA levels; the maximum PCR cycle number used was 37 and the maximum cDNA input was about 600 ng RNA; thereafter discrete DNA products were disappearing in the increased background fluorescence. In Figure 1B, the δ renin internal standard, r_{REN}, was held constant and competed with increasing amounts of cDNA, R_{REN}. The fluorescence in the low-molecular-weight region correlates with RNA input and is ascribed to broken RNA fragments after thermocycling. The formation of so-called primer-dimer PCR products, which could take primers away from the competition process, appears to be very low in renin and other mRNA assays (1–3). From Figure 2B, the relative renin mRNA level was determined to be 1.2 fg/µg right ventricular RNA. Among the various chambers of the heart, relative renin mRNAs were highest in the right atria (3.5 fg/µg) and lowest in the left ventricle (0.5 fg/µg); the latter amounts to approximately 0.025% of the PGK-1 mRNA level. The cardiac renin mRNA levels are about 300- to 1000-fold lower than in the kidney that, in our estimates, contains approximately 1 pg renin mRNA/µg RNA. The rat renin mRNA levels determined here are similar to those previously determined for

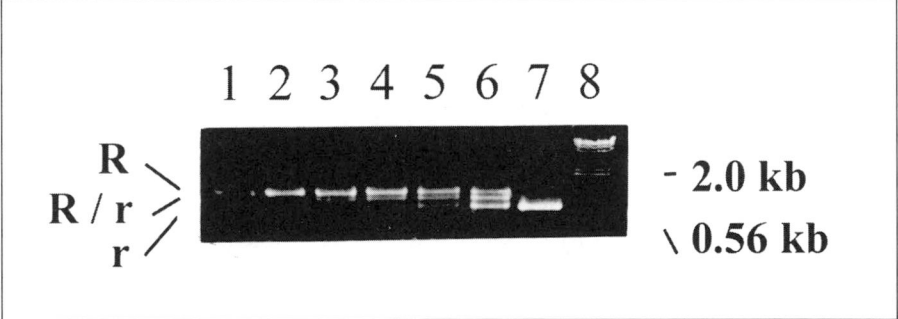

Figure 2. Resolution of homo- and heterodimer DNA. Ethidium bromide was included in the agarose gel. The cDNA was derived from 40 ng left ventricular RNA: lane 1, input; lanes 2–7, titration with twofold increasing amounts of δ RNA (1.1 fg in lane 5); lane 8, lamba DNA × *Hin*dIII with sizes indicated at right. At left, homodimer DNAs are labeled "R" and "r", heterodimer DNA is indicated by "R/r".

cardiac atria and kidney from the rat (4); however, as discussed below, these values are subject to certain correction factors.

As PCR products from target mRNA and competitive internal standards are homologons, we have wondered about the occurence of DNA recombination products. A very different picture occurs when ethidium bromide is added to the agarose gel; now three amplification products can be observed when a similar renin mRNA Q-RT-PCR is conducted (Figure 2). A detailed quantitative analysis supports the notion of co-migration of "R/r" heterodimers with "R/R" homodimers to give rise to the upper DNA band labeled "R" in Figure 1. The presence of ethidium bromide resolves the two DNA products into "R" and "R/r" presumably by the differential amount of intercalating dye taken up by the two different DNA molecules. Relative to "R/R" homodimer, the heteroduplex DNA in the renin Q-RT-PCR assay is an omega-like shaped molecule with the non-homologous DNA portion being looped out. The apparent co-migration of "R/r" heterodimers and "R/R" homodimers occurs in all mRNA assays we devised. Thus the co-migration artifact is independent of experimental design, but more easily detected when size differences between "R" and "r" are bigger. Moreover, when genomic or "g" competitor DNAs are used for quantitation of mRNA, "R", we find that "g/R" heteroduplexes normally also display retarded electrophoretic mobility and often co-migrate with g DNA. This leads to underestimated mRNA values because now part of the "R" DNA will be sequestered in upper "g" DNA band. Again, addition of ethidium bromide was shown to normalize electrophoretic migration and the correct point of equivalence could be determined. The magnitude of the error due to this co-migration may be in excess of 100%; the absolute cardiac renin mRNA values determined in Figure 2 differ by twofold, depending on the presence of ethidium bromide in the experiment. The same is true for other mRNA assays.

DISCUSSION

The Q-RT-PCR mRNA assay is evolving rapidly due to recent improvements in primer design, nonradioactive methodology, and particularly by use of internal standards, so that more quantitative results can be obtained. The simple RT-PCR method described in Figure 1 is a semiquantitative one, and three main correction factors are needed to convert the relative mRNA levels to absolute ones: (*i*) The efficiency of the RT step is not corrected for, and, being less than 100%, this leads to an underestimated mRNA level; (*ii*) size differences, and hence fluorescence intensity differences, between PCR signals from mRNA and shorter internal standards, lead to correction factors of 1.5 for the renin mRNA assay, and cause underestimated mRNA levels; (*iii*) the cDNA templates initially are single-stranded while competitors are double-stranded DNA, causing the value for their point of equivalence to be an overestimated. However, none of these factors is expected to be different for experimental versus control RNA samples; moreover, values for normalized mRNA levels or fold up-regulation will be accurate because the correction factors cancel in the equation for treated/control values.

Heteroduplex formation between target and competitor PCR product strands does occur in Q-RT-PCR assays, and in order to avoid errors in mRNA determinations, co-migration of PCR products must be avoided. Results from Figure 2 yield absolute mRNA values because RNA internal standards were used, and the artifactual co-migration of PCR products was avoided. The alternative is to include a DNA denaturation step and determine amounts of single-stranded DNA molecules, for example by use of strand-specific oligonucleotides (3). The use of non-homologous internal standards (7) also avoids problems due to DNA recombination.

ACKNOWLEDGMENTS

We thank William Lear for discussion and we thank Feras Jallad for Q-RT-PCR tests. Operating grant support from the Medical Research Council of Canada (MRC) and Heart and Stroke Foundation of Ontario (HSFO) is gratefully acknowledged. S.M. received MRC summer studentships, and P.B. is a HSFO Research Scholar.

REFERENCES

1. **Boer, P.H., M. Ruzicka, W. Lear, E. Harmsen, J. Rosenthal and F.H.H. Leenen.** 1994. Stretch-mediated activation of cardiac renin gene. Am. J. Physiol. 267:H1630-H1636.
2. **Lear, W. and P.H. Boer.** 1995. Rapid activation of the type-B versus type-A natriuretic factor gene by aortocaval shunt-induced cardiac volume overload. Cardiovasc. Res. 29:676-681.
3. **Lear, W., M. McDonnell, S. Kashyap and P.H. Boer.** 1995. Random primer p(dN)6-digoxigenin labeling for quantitation of mRNA by Q-RT-PCR and ELISA. BioTechniques 18:78-83.
4. **Lou, Y.-k., D.T. Liu, J.A. Whitworth and B.J. Morris.** 1995. Renin mRNA, quantified by PCR, in renal hypertensive rat tissues. Hypertension 26:656-664.
5. **Rundquist, B.A. and J.M. Gott.** 1995. RNA editing of the coI mRNA throughout the life cycle of Physarum polycephalum. Mol. Gen. Genet. 247:306-311.
6. **Siebert, P.D. and J.W. Larrick.** 1992. Competitive PCR. Nature 359:557-558.
7. **Siebert, P.D. and J.W. Larrick.** 1993. PCR MIMICS: competitive DNA fragments for use as internal standards in quantitative PCR. BioTechniques 14:244-249.

Section III

Examples of Specific Applications of Quantitative PCR

Gene Expression Analysis by a Competitive and Differential PCR with Antisense Competitors

Eric de Kant, Christoph F. Rochlitz and Richard Herrmann

Freie Universität, Berlin, Germany

ABSTRACT

We report a sensitive method for the reproducible and accurate measurement of gene expression from small samples of RNA. This method is based on a combination of two PCR techniques: First, an endogenous reporter gene and the gene of interest are simultaneously amplified in one tube after random-primed reverse transcription (RT) of RNA (differential RT-PCR). Second, exogenous homologous fragments of both genes with artificially introduced mutations are added and co-amplified in the same reaction (competitive PCR). The first-strand cDNA and the mutated antisense homologues of the reporter as well as the target gene compete for their respective primers and are therefore amplified with equal efficiencies. After PCR, restriction enzyme digestion allows visualization of the quantitative differences between the four resulting reaction products by gel electrophoresis and/or HPLC. The ratios of products that competed during PCR provide the quantitative information. The initial amount of a specific cDNA can be calculated from any competitor/cDNA ratio of reliably measurable PCR product amounts. Extensive competitor titration to experimentally approach the equilibrium is therefore unnecessary. The differential counterpart of competitive and differential RT-PCR (CD-RT-PCR) allows expression of the levels in reference to a reporter gene. MDR1 expression was determined in tumor cells by CD-RT-PCR.

INTRODUCTION

The introduction of PCR has enormously facilitated qualitative examination of nucleic acids in every research field involving genetic analysis. The development of polymerase chain reaction (PCR)-based approaches for quantification of DNA and RNA clearly offers an additional advance in

(Reprinted from BioTechniques 17:934-942, 1994)

molecular genetics.

With the increasing knowledge of genetic aspects of human diseases, techniques are needed to identify and quantify presumably important genes as well as their transcripts from patients' tissue samples. However, small amounts of tissue often limit broad examination of gene expression for routine clinical investigation. Sensitivity of conventional methods of mRNA quantification such as Northern blot hybridization, dot blot hybridization and RNase protection mapping is not high enough for detection of low-abundance transcripts or transcripts in samples with limited cell numbers.

PCR opens ways for genetic analysis on even minute amounts of DNA or mRNA from biopsies, needle aspirations or paraffin-embedded histological sections (8). Quantitative PCR analysis is, however, compared with qualitative PCR analysis, more complicated because of two features inherent in in vitro amplification. First, during the exponential phase, minute differences in a number of variables (even tube-to-tube variations) can greatly influence reaction rates, with a substantial effect on the yield of PCR products. Second, as a consequence of the consumption of reaction components and generation of inhibitors, the amplification enters a plateau phase. At this point, the reaction rate declines to an unknown level. Different proposals have been made to solve these problems, like the analysis of the kinetics of the amplification reaction (4,7,16,18) and use of competitive templates (3,9,10,22).

Another major problem in quantitative PCR analysis is the inaccuracy due to variations in the amount of starting RNA. In small samples, measurement of the total amount of RNA or genomic DNA is even impracticable. The sample loading problem is also characteristic of Northern analysis. For the latter, the solution has been found by presenting the level of expression of the gene of interest in reference to a constitutively expressed reporter gene. Similarly, in a PCR this can be done by the simultaneous amplification of two different genes in one reaction vessel. This approach called differential PCR was presented before (4,8,17,18,20). However, the use of differential PCR alone does not solve the additional above-mentioned PCR-specific complications and can therefore only be used as a semiquantitative assay.

In this report, we describe that quantitative measurement of nucleic acids using PCR can be achieved by combining both differential and competitive PCR. This was accomplished by the design of a differential PCR that includes co-amplification of competitive templates for both of the two amplified genes. In this way the PCR assay is internally controlled for errors in comparison between samples as well as for the efficiency of the reaction. The principles of competitive processes were carefully taken into account to serve as a guideline for the production of perfectly functioning competitive templates. The results we obtained prove that we successfully developed a quantitative PCR technique that is not only accurate and yields reproducible results but also simplifies gene expression analysis compared with previously described approaches.

230

MATERIALS AND METHODS

Cell Lines

The drug-sensitive SW1573 squamous lung cancer cell line and its drug-resistant derivative, SW1573/2R160, were kindly provided by Dr. H.J. Broxterman (Free University Hospital of Amsterdam, The Netherlands). These monolayer cell lines have been described (2) and were routinely cultured in Dulbecco's medium (Life Technologies, Gaithersburg, MD, USA).

RNA Purification and cDNA Synthesis

Total cellular RNA was purified using RNAzol™ (Biotecx Laboratories, Houston, TX, USA). RNA (up to 0.5 µg) was randomly reverse-transcribed in 20 µL for 40 min at 42°C using a hexadeoxyoligonucleotide primer mixture (5 µM) (Boehringer Mannheim, Mannheim, FRG) and Moloney murine leukemia virus RNase H⁻ Reverse Transcriptase (10 U/µL) (Life Technologies) in 1× PCR buffer, 10 mM dithiotheitol (DTT), 2 U/µL RNasin® (Promega, Madison, WI, USA), and 1 mM each of dATP, dGTP, dCTP and dTTP.

PCR

cDNA representing 25 ng RNA or less was amplified in 1× PCR buffer (10 mM Tris-HCl pH 8.3, 50 mM KCl, 1.5 mM $MgCl_2$, 0.001% gelatin) by repeated cycling at 94°C for 50 s, 55°C for 50 s and 72°C for 50 s, using a DNA Thermal Cycler (Perkin-Elmer, Überlingen, Germany). Initial denaturation was lengthened to 5 min. The amplification efficiencies of cDNAs from the multidrug resistance gene MDR1 and β2-microglobulin (β2M) were manipulated by titrating the primer concentrations so that both genes yielded quantifiable amounts of PCR products over a large range of different expression levels when co-amplified (17,20). MDR1 and β2M primers were used at concentrations of 0.5 µM and 0.1 µM, respectively, in differential reverse transcription PCR (RT-PCR). The reaction mixture further contained 0.2 mM of each deoxyribonucleoside triphosphate (dNTP) and 25 U/mL *Taq* DNA polymerase (Perkin-Elmer). Competitive and differential (CD-RT-PCR) was performed as a Hot Start PCR (6) in 25 µL, and the reaction mixture was overlaid with mineral oil.

Primers used for CD-RT-PCR were synthesized according to sequences published elsewhere (5,11) and were 20–22 bp in length. The sequence regions (in nucleotide numbers) of the primers and their products were as follows: For the MDR1 gene, a 5′ (352–372) and a 3′ (590–609) primer resulted in a 258-bp cDNA fragment; 5′ (1446–1469) and 3′ (1612–2249) primers of the β2M gene gave rise to a 188-bp cDNA fragment.

CD-RT-PCR was performed by the addition of serially diluted and exactly defined amounts of antisense mutant competitor DNA of MDR1 as well as β2M in up to three separate reaction vessels each containing primers for both genes and an aliquoted reaction mixture with a fixed unknown amount of cDNA. After 35 cycles, the amplification reaction products were heat-denatured and allowed to cool down slowly to promote stabilization of heteroduplexes based on random re-annealing of mutated and non-mutated sequences. The concept of CD-RT-PCR starting at cDNA synthesis up to analysis (see below) is schematically shown in Figure 1.

Preparation of Antisense Competitive Mutant Templates

To create or remove a unique *Eco*RI restriction enzyme site for synthesis of competitive MDR1 and β2M templates, respectively, we used a previously described site-specific mutagenesis technique based on PCR (13) with a

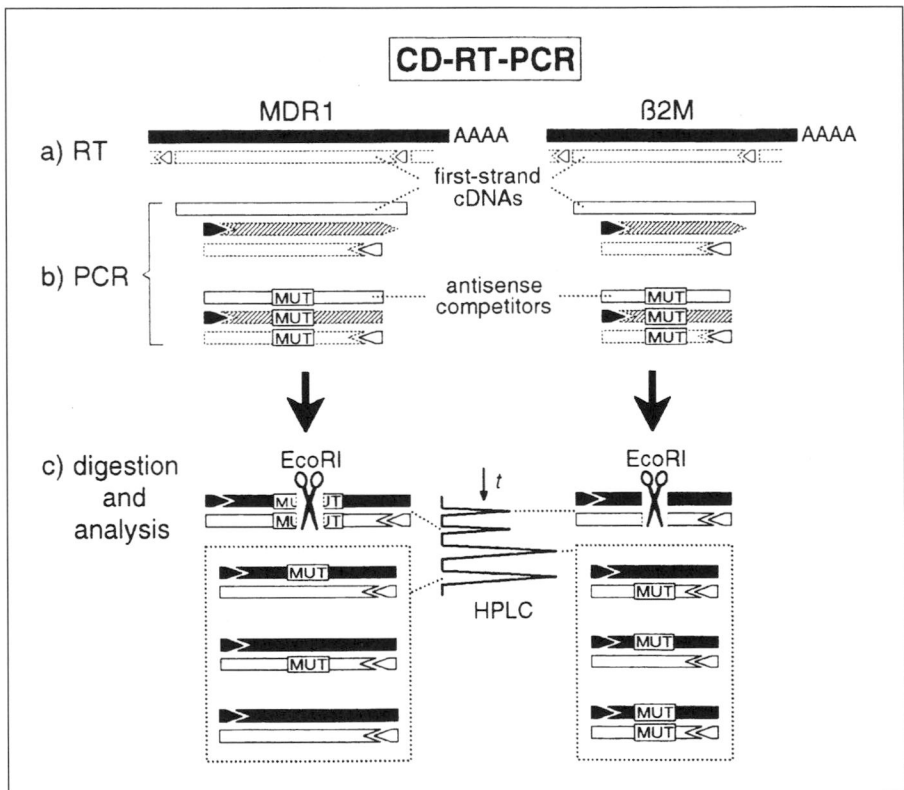

Figure 1. A schematic outline of the concept of CD-RT-PCR. Each reaction step and HPCL analysis are performed simultaneously on MDR1 and β2M within a single tube. The solid and open symbols represent sense and antisense sequences and PCR primers (larger arrow symbols), respectively. a) Transcripts are randomly reverse-transcribed (RT) using hexadeoxynucleotides (short open arrow symbols). b) First-strand cDNAs and mutant (MUT) antisense competitors are amplified with equal efficiencies. c) *Eco*RI digestion renders 1 out of 4 equally efficiently formed duplexes distinguishable from the others. HPLC measurement of the proportions of duplexes that are digested (in this case, 0.25) provides a measure of the competitor/cDNA ratios (in this case, 1/1). The first two peaks each represent coinciding halves after digestion and the last two peaks each represent mixtures of non-digestible homo- and hetroduplexes.

slight modification. In brief, two separate, mutated amplification products, generated by mutagenic primers carrying one to four base pair changes, were excised from a preparative agarose gel. Remelted gel slices with the two mutant fragments, which are homologous around the mutated sequence, were mixed without purification in a new PCR. In the first few cycles, the overlapping mutant fragments were recombined and, subsequently, after the addition of the flanking primers, they were amplified as full-length mutated β2M and MDR1 fragments. The mutagenic and the flanking primers were chosen so that the unique restriction enzyme sites resided exactly in the middle of the amplified full-length fragments (Figure 1).

Biotinylated 5′ (sense) primers were used for the purpose of separating the antisense strands of the recombined PCR products. The PCR-generated, biotinylated DNA was directly bound to streptavidin-coupled polystyrene magnetic beads (Dynal, Oslo, Norway) and alkali-denatured. The immobilized sense strands were reused several times for the solid-phase synthesis of antisense DNA. After at least 5 cycles of annealing and extension of the antisense primers both at 58°C, extensive washing and alkali-denaturation, DNA was obtained as antisense fragments. The quality of antisense competitors was verified after electrophoresis on a 10% non-denaturing polyacrylamide gel. Complementary single-stranded DNA fragments can be easily distinguished, based on the different sequence-dependent conformations that influence the electrophoretic mobility (19). The amounts were estimated by densitometric scanning of a positive image of a silver-stained polyacrylamide gel using a concentration marker. The neutralized competitors were diluted to different concentrations. Mixtures of each concentration of competitor DNA and 10 ng/µL of mechanically sheared salmon sperm DNA as a carrier were heat-denatured and slowly cooled down before they were stored in siliconized vials.

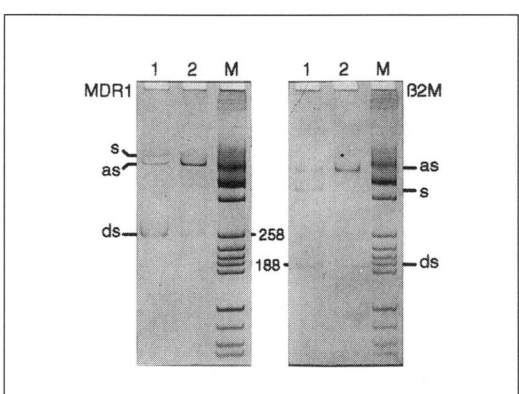

Figure 2. Double-stranded and single-stranded competitors. Denatured PCR samples of both MDR1 (left panel) and β2 (right panel) were loaded on a non-denaturing polyacrylamide gel according to the SSCP technique (19). Lanes 1: double-stranded (ds) PCR competitor fragments. Numbers between the panels refer to their length (in bp). Lanes 2: sense (s) depleted samples of the respective sequences. The remaining mutant antisense (as) fragments were used as PCR competitors for first-strand cDNA. Lanes M: pBR322 DNA-*Hae*III digest as molecular weight marker.

Restriction Enzyme Digestion, Gel Electrophoresis and HPLC

After completion of the PCR process and stabilization of heteroduplexes, the reaction mixture was directly digested without previous DNA precipitation in a potassium glutamate buffer (0.5× KGB) (12) and 0.25 U/μL *Eco*RI for 15 min at 37°C. Of this cleavage reaction, 12 μL, containing 7 μL of the original PCR, were loaded either on gel or on the chromatographic system. KGB influences electrophoresis properties negatively. DNA was therefore separated on 6% agarose gels (NuSieve® 5:1 agarose; FMC BioProducts, Rockland, ME, USA).

Samples were analyzed for quantification by use of high-performance liquid chromatography (HPLC) (14). The chromatographic system from Knauer consisted of a solvent delivery system, a UV detector and an autosampler equipped with a 20-μL loop. Data were collected using a Shimadzu integrator (Kyoto, Japan). Samples were injected automatically every 15 min onto an ion-exchange DEAE-NPR column (Perkin-Elmer), 35 × 4.6 mm i.d., packed with 2.5-μm particles. DNA was separated by employment of a gradient program for the binary mobile phase as follows: reservoir A contained 1 M NaCl and 25 mM Tris-HCl pH 9.0 (Boehringer Mannheim); reservoir B contained 25 mM Tris-HCl pH 9.0; the mobile phase was linearly changed from 70% to 60% B in 10 s, then from 60% to 48% B in 3 min and from 48% to 42% B in 5 min; 100% A was used for a 1-min cleanup before reequilibration (5 min, 70% B). UV absorbance was measured at 260 nm. Under these conditions, all fragments were completely separated and the reproducibilty of retention times and peak areas were as described by Katz and Dong (14).

RESULTS

Mutated Antisense DNA as Competitor for First-Strand cDNA

The MDR1 and β2M competitors for CD-RT-PCR were synthesized as site-specific-mutagenized PCR fragments. In MDR1 competitors a unique *Eco*RI restriction site was created, and in β2M competitors it was removed (see Figure 1). In contrast to the wild-type (wt) β2M cDNA template, the wt MDR1 sequence contains no *Eco*RI site between the sites where

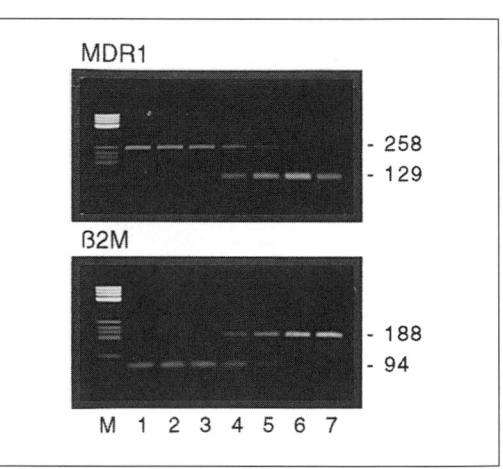

Figure 3. Competitive RT-PCR titrations loaded on agarose gels. In 2 separate experiments, different amounts of antisense MDR1 or β2M competitors were added to a constant amount of SW1573 cDNA. In the upper panel the amount of MDR1 competitor was 1, 3, 10, 100 and 300 fg, respectively, in lanes 1–6, and 300 fg in lane 7 but without cDNA. In the lower panel the amount of β2M competitor was 0, 10, 30, 100, 300, 1000 and 3000 fg in lanes 1–7. After PCR, all samples were digested with *Eco*RI. Samples were electrophoresed and stained with ethidium bromide. Numbers refer to the length (in bp) of the resulting full-length and halved DNA fragments. Lane M: pBR322 DNA-*Hae*III digest as molecular weight marker.

primers anneal. After denaturation, both strands of the competitors were physically separated, and only the antisense strands were used as competitors. Figure 2 shows amplified cDNA of MDR1 and β2M both as denatured double-stranded (ds) fragments and as sense-strand-depleted DNA. Of the two bands representing sense and antisense strands, only the antisense strands were left. They were checked for their quality as shown in Figure 2 and used as competitors for first-strand cDNA as depicted in Figure 1.

Quantification of a Specific cDNA Is Independent of the Relationship between Concentrations of cDNA and Antisense Competitor DNA

To test the competitive nature of the antisense competitors, different known amounts of MDR1 and β2M competitors were spiked into two series of PCR samples with equal amounts of SW1573 cDNA. After digesting the reaction mixture with *Eco*RI, part of it was loaded on the gel (Figure 3). With higher competitor concentrations, the relative amount of the lower digested bands in the MDR1 samples (lanes 1 to 6) increased, and in β2M samples (lanes 2 to 7), the relative amount of the upper undigested bands increased. As controls for digestion, the *Eco*RI digestible templates of both MDR1 (competitor) and β2M (wt cDNA) were amplified separately. *Eco*RI digestion in 0.5× KGB was complete within 15 min as shown in lane 7 of the upper panel and lane 1 of the lower panel. Thus, direct digestion by mixing PCR samples with KGB and the restriction enzyme offers a rapid and effective alternative to digestion in an *Eco*RI-specific restriction buffer after DNA precipitation. The *Eco*RI restriction sites were positioned exactly in the middle of the PCR products (as depicted in Figure 1). The lower bands represent the cleaved fragments of which both halves migrate equally fast. As a consequence of the overlap of these fragments, the intensity of a given amount of either digested or undigested ethidium bromide-stained DNA is the same.

After the last denaturation step of these competitive

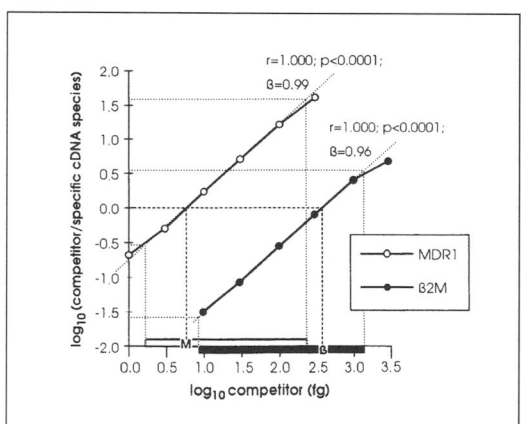

Figure 4. Logarithmic plot of competitive RT-PCR titrations. Samples from the experiments shown in Figure 3 were computed and plotted after HPLC analysis. The two dotted lines of both MDR1 and β2M enclose the area where $0.05 < d < 0.95$ (d = proportion of digested fragments). Regression analysis (r), with calculation of the level of significance (p) and the slope (β) of the regression line, was performed within this area. Horizontal bars show concentration range of the competitors from which specific cDNA amounts can be accurately calculated. The 2 points of equivalence ($d = 0.25$) are indicated by the dashed lines and the one-letter gene symbol within the bars. Open circles and M in open bar: MDR1; closed circles and β in closed bar: β2M.

235

PCRs is completed, there are four different kinds of duplexes in dimerization of the renaturing DNA. This is also shown in Figure 1. However, only the duplexes with the *Eco*RI recognition sequence in both strands can be digested. In the ideal situation where denaturation is complete and homo- and heteroduplexes are formed equally well, only one in four duplexes can be digested in case wt cDNA and competitor DNA are present in a one to one ratio. Thus, the proportion of digested fragments after PCR and restriction digestion will then be 0.25 (Figure 1). We reasoned that, in case there is a perfect competition during PCR amplification and duplex formation, every possible cDNA/competitor ratio could predict an exactly defined fraction of duplexes that can be digested. This proportion of digestible fragments (d) is given by the proportion of single-stranded molecules possessing the restriction enzyme recognition sequence (RE +) that duplicated the complementary sequence also having this sequence. This results in the following equation: $d = (RE +)^2$. Considering this quadratic distribution principle of duplex formation, it is possible to compute ratios of wt cDNA and competitor DNA after performance of a competitive PCR and measurement of the digestible proportion of the PCR products. The proportion of $RE+$ DNA is given by $(RE +) = \sqrt{d}$ and the proportion of DNA fragments without the recognition sequence (RE -) by $(RE - = 1- \sqrt{d}$. An unknown amount of a specific cDNA (Y) prior to PCR can be mathematically processed by combining the formulas given above on the conditions that the initial amount of competitor (C) for this cDNA species is known and that d can be measured experimentally after PCR. The equation for this calculation is:

$$Y = C([1 - \sqrt{d}]/\sqrt{d})^n,$$

where $n = +1$ or -1 depending on whether the competitor possesses ($RE+$) or lacks (RE -) the unique *Eco*RI restriction

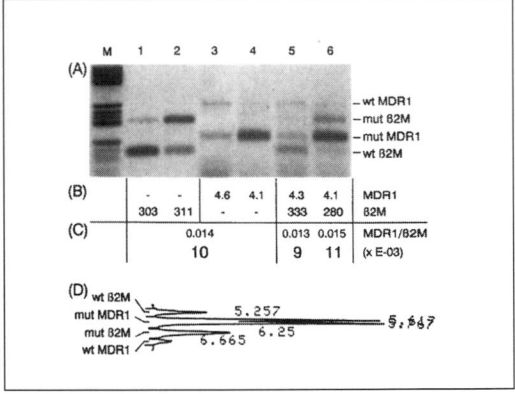

Figure 5. Two different competitive PCRs in one tube. Equal amounts of SW1573 cDNA were aliquoted into 6 separate PCRs. A) A negative image of an ethidium bromide-stained agarose gel of PCR products after *Eco*RI digestion. β2M primers and competitors were added to the reaction mixtures of lanes 1, 2, 5 and 6, and MDR1 primers and competitors to samples in lanes 3–6. The amounts of competitors were 20 fg in lanes 1, 3, and 5, and 200 fg in lanes 2, 4 and 6. Amplified wt endogenous cDNAs (and co-migrating heteroduplexes of β2M) are indicated as wt and the mutated competitor amplification products (and co-migrating heteroduplexes of MDR1) as mut. Lane M: pBR322 DNA-*Hae*III digest as molecular weight marker. B) The computed amounts of cDNA template in fg. C) The computed MDR1 expression ratios as related to β2M; the lower numbers are MDR/β2M ratios (× E-03) corrected with respect to the molecular weight of the 2 fragments. D) HPLC analysis of the digested reaction mixture shown in Panel A, lane 6. All fragments were eluted between 5 and 7 minutes, as indicated.

site. In samples where $0.05 > d > 0.95$, small differences in the level of d have a great influence on the outcome of the computed amount of cDNA. For this reason and because $0.05 < d < 0.95$ can always be accurately measured as long as the total amount of DNA loaded on HPLC is high enough, in most experiments only levels between 0.05 and 0.95 were seriously taken into account.

The PCR samples from the experiment shown in Figure 3 were analyzed with HPLC to measure the proportion of digested DNA. From these experimental values and the known amounts of competitor added prior to PCR, the absolute amounts of MDR1 and β2M cDNA species were computed. In samples where $0.05 < d < 0.95$, the mean values were 5.93 fg with a SD of 0.14 for MDR1 and 360.2 fg with a SD of 28.5 for β2M. The MDR1 and β2M levels varied 3% and 12%, respectively, at most from the mean values.

If it is true that the quantification of a specific cDNA species is basically independent of the amount of antisense competitor used in PCR, then a linear relationship should be found between the amount of competitor and the competitor/cDNA ratio. To demonstrate this relationship, the values from the experiment shown in Figure 3 were plotted in Figure 4. Regression analysis revealed highly statistically significant ($P < 0.0001$) linear relations and slope (β) values close to 1 for both MDR1 and β2M. Thus, within a wide range of competitor concentrations ($>2 \log^{10}$ units), the amount of a specific cDNA can be determined with high accuracy. Competitor titrations can therefore be reduced to a minimum.

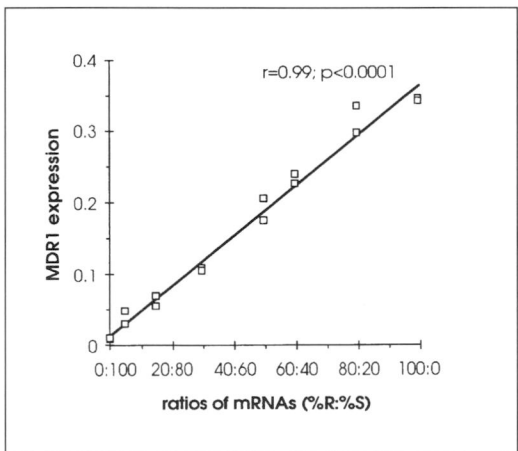

Figure 6. MDR1 expression in dilutions of 2 different mRNAs. Total RNA of SW1573 (S) and SW1573/2R160 (R) was diluted to ratios as indicated. CD-RT-PCR was performed in duplicate for each dilution. MDR1 expression as the computed ratio of molar amounts of MDR1 and β2M cDNA templates is plotted. Linear regression analysis (r) with calculation of the level of significance (p) was performed.

The Competitive Nature of Amplification and the Efficiency of Duplex Formation Are Not Influenced by the Presence of Other DNA Sequences in the Same Tube

The prerequisite for the assessment of MDR1 gene expression related to the internal standard β2M when using competitive PCR is the co-amplification of two sequences for each gene in one reaction vial, as shown in Figure 1. We tested whether the simultaneous determination by competitive PCR of two different cDNA species in a single tube has an influence on the outcome of either one or the other competitive reaction.

Figure 5 shows the experiment in which two separately performed competitive PCRs for MDR1 and β2M were compared to the CD-RT-PCR combining these PCRs. In every PCR a constant amount of cDNA was added. The computed amounts of MDR1 and β2M cDNA, based on the HPLC-measured proportion of digested fragments and the known amounts of competitors, were virtually the same in all reactions. Thus, we can conclude that both the competitive nature of the amplification processes and the efficiency of duplex formations are not impeded in CD-RT-PCR. The combination of two competitive PCRs, from which absolute cDNA amounts of two specific cDNA species within one cDNA pool can be accurately determined, makes expression analysis by CD-RT-PCR independent of absolute amounts of total cDNA. MDR1 gene expression is therefore presented in a molar relation to the internal standard β2M (see MDR1/β2M ratio in Figure 5).

CD-RT-PCR Quantification of cDNA Ratios Is Linear with mRNA Expression

For the assessment of the linearity of the cDNA values as obtained by CD-RT-PCR with mRNA expression values, a dilution series of RNA from low and high MDR1 expressing cell lines was made. Total RNAs from the parental SW1573 (S) cell line with low expression and the drug-resistant derivative SW1573/2R160 (R) with high expression were spectrophotometrically measured and mixed to different ratios. After reverse transcription of these mixtures, CD-RT-PCR was performed and values were plotted in Figure 6. The gene expression levels of MDR1 (as molar cDNA ratios of MDR1 and β2M) were 9.8×10^{-3} (E-03) and $344 \times$ E-03, respectively, for S and R. Thus, the drug-resistant cell line showed a 35× higher MDR1 expression. This is in the same range as the approximate 100× higher level reported by Baas et al. (2). Regression analysis revealed a highly significant linear relationship ($r = 0.99$; $P<0.0001$) between the extent to which MDR1 expression was diluted and the MDR1 expression levels measured with CD-RT-PCR.

Unlike Antisense Competitors, Double-Stranded Competitors Do Not Function like Perfect Competitors for First-Strand cDNA

As mentioned previously, measurement of DNA amounts should be independent of the amounts of competitor DNA in a competitive PCR. In the ideal situation, both the regression coefficient (r) and slope (β) are 1 in a plot as shown in Figure 4.

When double-stranded competitors were used in different concentrations to perform CD-RT-PCR, the calculated amount of a specific cDNA in each sample unpredictably varied depending on how the amounts of competitor DNA and cDNA were related to each other. Though in most experiments with double-stranded competitors there was a linear relationship with $r>0.98$ and $P<0.01$, it was never as strong as with antisense competitors. More importantly, the slope deviated from 15% to 30% from $\beta = 1$. This implies that the calculated amounts of a specific cDNA from different samples with different amounts of double-stranded competitors varied up to threefold within one experiment even though the proportion of digested fragments could be reliably measured ($0.05 < d < 0.95$).

To illustrate the reproducibility of CD-RT-PCR, MDR1 expression was measured four times in separate experiments with different master mixtures, different cDNA preparations and different amounts of antisense competitors. The mean value was 10.6 ($\times$ E-03) with a SD of 0.9. None of the values from these independent experiments varied more than 10% from the mean. When performed with double-stranded competitors, the lowest and highest values varied more than 50% from the mean of four independent experiments.

Clearly, the independence of the amount of competitor and the reproducibility could only be guaranteed when the type of competitors used for CD-RT-PCR was single-stranded antisense.

DISCUSSION

Quantification using PCR is a complicated matter. The efficiency of the amplification process depends on even minute differences in a large number of variables. The initial concentration of a DNA template prior to PCR can therefore only be estimated from the yield of amplified product when either the efficiency is measured or the assay is controlled for the variations in the efficiency using a standard that is equally dependent on these variations.

For the calculation of the efficiency during the exponential phase, the kinetics of amplification need to be analyzed by withdrawing part of the reaction after successive cycles or by serially diluting the amount of template DNA in the PCR (4,7,16,18). Not only is this an extremely laborious task since the exponential range should actually be determined for every single sample (16), but in many cases, it also requires an additional magnification to allow detection and quantification of the yield of PCR product that might be relatively low in the exponential phase (4,7,15,16,18). An alternative to controlling variable amplification rates in developing a quantitative PCR method is to co-amplify exogenous competitive standards. The ratios of target to standard should be preserved independent of the reaction rate when amplification is equally efficient (3,9,10,21,22). This competitive PCR can therefore also be evaluated in the plateau phase, yielding higher PCR product amounts. Among other kinds of competitors, site-specific mutated cDNAs have been used as competitive templates for expression analysis (9).

Though competitive PCR controls for intertube variations in PCR, it does not control for errors in comparison between samples. In analogy to Northern analysis, internal standards have been used in PCR to correct for sample loading variations (4,8,20). The internal standards correct for degradation of RNA and for the spectrophotometric measurement and pipetting errors that cause these variations. Before PCR, an endogenous, constitutively expressed mRNA sequence and the target gene are reverse-transcribed in one tube to correct for variations in reverse-transcription efficiency. Subsequently, both genes are co-amplified in a single reaction tube. The level of target gene expression is reflected in the ratio between the resulting product amounts of the two genes. This type of PCR, named differential PCR (8), is adequate for the semi-quantitative determination of expression in a series of samples (20). It merely ranks the samples performed within a single experiment in order of their expression rates. When the kinetics of amplification of both the target and the reference gene are analyzed (4,16,18), it is possible to learn more about how the expression ratios of different samples are related to each other.

The present report describes a method for quantification of gene expression by combining competitive PCR and differential RT-PCR. Though the basic idea of combining two PCR techniques was also developed independently by others (1), we avoided a very important inadequacy in the design of the competitive counterpart. Considerable attention was paid to the principles of competitive PCR and to how they could be exploited for the development of a simplified accurate quantitative PCR technique.

During a perfectly performed competitive process, the ratio between the processed amounts of a certain component and its competitor is always constant. In competitive PCR the addition of a known amount of competitor should therefore always correctly predict the unknown amount of cDNA by determination of the competitor/cDNA ratio after PCR, independent of the efficiency of the amplification process and of how the concentrations of competitor and cDNA are related. The results show that the choice of antisense DNA fragments as the alleged perfect competitors for first-strand cDNA, which is the starting template in RT-PCR, was of vital importance. The results with double-stranded competitors were clearly less reproducible. More importantly, when double-stranded competitors were used, the quantification of a specific cDNA unpredictably varied depending on the competitor/cDNA ratio. The results lack linearity and/or a slope that does not approach 1 (9,21) in a graph such as shown in Figure 4. It is not correct to assume that only those competitor/cDNA ratios that measure 1 after PCR would have been preserved at this level during the whole amplification reaction. Therefore, even after extensive double-stranded competitor titration, we feel that cDNA levels calculated by extrapolation from the intersection at the point of equivalence to the competitor titration range (21) are highly questionable. The inability of double-stranded DNA fragments to behave as perfect competitors is probably due to different efficiencies of the sense and antisense primers used in PCR. The differences in the relative amounts of sense and antisense templates between first-strand cDNA and double-stranded competitors that appear in the first few cycles could substantially contribute to the initial amplification reaction rate.

We clearly demonstrate that the only constraint for accurate cDNA quantification by CD-RT-PCR is the reliability of the measured molar ratio of amplified cDNA and antisense competitors but not the level of the ratio in itself. It is important to stress that this is not only dependent on the performance of a truly competitive PCR. A trustworthy estimation of the molar ratio after PCR also relies on whether the amplification products of competitors and cDNA can be discriminated efficiently and quantitatively. The sensitivity and linearity of HPLC analysis of PCR products (14) added greatly to the accuracy of the determination of competitor/cDNA ratios. We used competitors that were homologous to the native cDNA sequences with only few sequence modifications to allow discrimination by restriction enzyme analysis. We have chosen to use homologues for two reasons. First, they are expected to be amplified equally efficiently. Second, dimerization of the four possible kinds of duplexes is expected to take place equally efficiently. Taken together, this renders relative amounts of different duplexes exactly predictable given a certain competitor/cDNA ratio. Conversely, competitor/cDNA ratios can be determined from the measured proportion of duplexes that is digested, provided that the above-mentioned conditions are fulfilled and digestion is complete. Indeed, what could be expected theoretically was confirmed by the observations we made. Differential PCR was presented before (17,20). It comprises the measurement of an internal standard, which is the optimal way to correct for sample loading variations. In CD-RT-PCR this is achieved by performing two competitive amplification reactions, one for the target and one for a reference gene, within a single tube. From our results (Figure 5), it can be concluded that both competitive PCRs in combination behave like they do separately.

In CD-RT-PCR, expression values are presented as cDNA ratios of the target and the reference gene. The ratios are linearly related to mRNA expression levels (Figure 6). This is achieved by using randomly annealing hexamer primers in reverse transcription. The use of hexamer primers also reduces the effects of sequence complexity and mRNA secondary structure. Furthermore, in combination with CD-RT-PCR, it overcomes problems because of the variable reverse transcription between different samples.

In conclusion, the accuracy, linearity and reproducibility of CD-RT-PCR guarantees a reliable quantification of gene expression. Within a wide range of competitor amounts, the relative amounts of the four types of PCR products can be determined. This simplifies the procedure because in most cases analysis of only one tube will be necessary. The sensitivity of this PCR-based method and its capacity to accurately measure normalized mRNA expression make it possible to analyze either mRNAs of extremely rare abundance or mRNAs in small samples with unknown amounts of total RNA. It may therefore be extended to the rapid and direct examination of biopsy specimens or aspiration samples of human solid tumors and other tumors for evidence of, e.g., MDR1 expression. The same holds true for the measurement of gene dosage and genomic alterations in samples of DNA. The proposed improvements will make PCR-based nucleic acid quantification more suitable for routine analysis. The combination of differential and competitive PCR as presented in this report takes advantage of both strategies and reduces the shortcomings of each of them separately.

ACKNOWLEDGMENTS

This investigation was supported by a grant for a fellowship under Contract MR4-0148-D (CH) from the Commission of the European Communities (to E.de K.) and by a fund (He 1000/3-2) from the Deutsche Forschungsgemeinschaft (to R.H.).

REFERENCES

1. **Apostolakos, M.J., W.H.T. Schuermann, M.W. Frampton, M.J. Utell and J.C. Willey.** 1993. Measurement of gene expression by multiplex competitive polymerase chain reaction. Anal. Biochem. *213*:277-284.
2. **Baas, F., A.P. Jongsma, H.J. Broxterman, R.J. Arceci, D. Housman, G.L. Scheffer, A. Riethorst, M. van-Groenigen, A.W. Nieuwint and H. Joenje.** 1990. Non-P-glycoprotein mediated mechanism for multidrug resistance precedes P-glycoprotein expression during in vitro selection for doxorubicin resistance in a human lung cancer cell line. Cancer Res. *50*:5392-5398.
3. **Becker-André, M. and K. Hahlbrock.** 1989. Absolute mRNA quantification using the polymerase chain reaction (PCR). A novel approach by a PCR aided transcript titration assay (PATTY). Nucleic Acids Res. *17*:9437-9446.
4. **Chelly, J., D. Montarras, C. Pinset, Y. Berwald-Netter, J.C. Kaplan and A. Kahn.** 1990. Quantitative estimation of minor mRNAs by cDNA-polymerase chain reaction. Application to dystrophin mRNA in cultured myogenic and brain cells. Eur. J. Biochem. *187*:691-698.
5. **Chen, C.J., J.E. Chin, K. Ueda, D.P. Clark, I. Pastan, M.M. Gottesman and I.B. Roninson.** 1986. Internal duplication and homology with bacterial transport proteins in the mdr1 (P-glycoprotein) gene from multidrug resistant human cells. Cell *47*:381-389.
6. **Chou, Q., M. Russell, D.E. Birch, J. Raymond and W. Bloch.** 1992. Prevention of pre-PCR mis-priming and primer dimerization improves low-copy-number amplifications. Nucleic Acids Res. *20*:1713-1723.
7. **Duplàa, C., T. Couffinhal, L. Labat, C. Moreau, J.-M.D. Lamazière and J. Bonnet.** 1993. Quantitative analysis of polymerase chain reaction products using biotinylated dUTP incorporation. Anal. Biochem. *212*:229-236.
8. **Frye, R.A., C.C. Benz and E. Liu.** 1989. Detection of amplified oncogenes by differential polymerase chain reaction. Oncogene *4*:1153-1157.
9. **Gilliland, G., S. Perrin, K. Blanchard and H.F. Bunn.** 1990. Analysis of cytokine mRNA and DNA: detection and quantitation by competitive polymerase chain reaction. Proc. Natl. Acad. Sci. USA *87*:2725-2729.
10. **Guiffre, A., K. Atkinson and P. Kearney.** 1993. A quantitative polymerase chain reaction assay for Interleukin 5 messenger RNA. Anal. Biochem. *212*:50-57.
11. **Gussow, D., R. Rein, I. Ginjaar, F. Hochstenbach, G. Seemann, A. Kottman and H.L. Ploegh.** 1987. The human beta 2-microglobulin gene. Primary structure and definition of the transcriptional unit. J. Immunol. *139*:3132-3138.
12. **Hanish, J. and M. McClelland.** 1988. Activity of DNA modification and restriction enzymes in KGB, a potassium glutamate buffer. Gene. Anal. Tech. *5*:105-107.
13. **Higuchi, R., B. Krummel and R.K. Saiki.** 1988. A general method of in vitro preparation and specific mutagenesis of DNA fragments: study of protein and DNA interactions. Nucleic Acids Res. *16*:7351-7367.
14. **Katz, E.D. and M.W. Dong.** 1990. Rapid analysis and purification of polymerase chain reaction products by high-performance liquid chromatography. BioTechniques *8*:546-555.
15. **Lundeberg, J., J. Wahlberg and M. Uhlén.** 1991. Rapid colorimetric quantification of PCR-amplified DNA. BioTechniques *10*:68-75.
16. **Murphy, E.D., C.E. Herzog, J.B. Rudick, A.T. Fojo and S.E. Bates.** 1990. Use of the polymerase

chain reaction in the quantitation of mdr-1 gene expression. Biochemistry *29*:10351-10356.

17. **Neubauer, A., B. Neubauer and E. Liu.** 1990. Polymerase chain reaction based assay to detect allelic loss in human DNA: loss of beta-interferon gene in chronic myelogenous leukemia. Nucleic Acids Res. *18*:993-998.

18. **Noonan, K.E., C. Beck, T.A. Holzmayer, J.E. Chin, J.S. Wunder, I.L. Andrulis, A.F. Gazdar, C.L. Willman, B. Griffith, D.D. Von-Hoff and I.B. Roninson.** 1990. Quantitative analysis of MDR1 (multidrug resistance) gene expression in human tumors by polymerase chain reaction. Proc. Natl. Acad. Sci. USA *87*:7160-7164.

19. **Orita, M., Y. Suzuki, T. Sekiya and K. Hayashi.** 1989. Rapid and sensitive detection of point mutations and DNA polymorphisms using the polymerase chain reaction. Genomics *5*:874-879.

20. **Rochlitz, C.F., E. de Kant, A. Neubauer, I. Heide, R. Böhmer, J. Oertel, D. Huhn and R. Herrmann.** 1992. PCR-determined expression of the MDR1 gene in chronic lymphocytic leukemia. Ann. Hematol. *65*:241-246.

21. **Siebert, P.D. and J.W. Larrick.** 1992. Competitive PCR. Nature *359*:557-558.

22. **Wang, A.M., M.V. Doyle and D.F. Mark.** 1989. Quantitation of mRNA by the polymerase chain reaction. Proc. Natl. Acad. Sci. USA *86*:9717-9721.

Update to:

Gene Expression Analysis by a Competitive and Differential PCR with Antisense Competitors

Eric de Kant

Academisch Ziekenhuis Utrecht, Utrecht, The Netherlands

Since the publication of our original report, hundreds of papers have appeared that deal with quantification of the amount of specific nucleic acids in a sample. Many others have also described (empirically generated) theoretical considerations resulting in a wide application of competitive PCR which now seems to be accepted as the most reliable and precise basic approach for quantitative measurement (reviewed in Reference 5).

A prerequisite for the reliability of competitive PCR is that the amplification efficiency of target and competitor sequences are the same. PCR product amounts that are plotted in a graph relating log(competitor/target) to log(competitor) should then result in a straight line with a slope of +1. The importance of these requirements cannot be overemphasized. Even slope curves that differ with 0.1 from +1 may lead to unacceptable variation and erroneous results depending on, for example, ratios of initial target and competitor DNA amounts. Deviations from slopes of unity result from differences in length between targets and competitors (2–4), that may affect efficiencies of denaturation and completion of elongation, or from the use of double-stranded (ds) competitors for quantification of first-strand cDNA (1,4) instead of single-stranded (ss) antisense competitive templates. In the first few PCR cycles, amplification efficiency differences can occur between ss cDNA templates and ds competitive homologues due to differences in annealing characteristics of sense and antisense primers.

In a system design with homologous competitors it is important to carefully analyze the formation of heteroduplexes after PCR. This can be done easily either by mathematical analysis based on measurement of the proportion of endonuclease digestible homoduplexes, provided that duplex formation is equally distributed, or by detection of heteroduplexes (3).

The choice of the type of competitor for gene expression analysis by competitive PCR depends on whether absolute copy numbers or normalized RNA levels have to be determined and is also related to the type of samples to be analyzed. The use of RNA competitors to achieve absolute mRNA quantification is fine as long as the accuracy of RNA concentration measurement and the integrity of the RNA can be guaranteed (3). The levels of target RNA can also be presented in reference to an endogenous constitutively expressed

244

internal standard to achieve normalization of the values in analogy to Northern analysis. This is even mandatory when sample loading variations are to be expected. We demonstrated that normalized expression analysis can be accomplished in a competitive and differential RT-PCR (CD-RT-PCR) that combines randomly primed cDNA with antisense competitive DNA templates of both the target and the reference gene. The advantage of the nonspecific random reverse transcription procedure preceding the addition of gene specific competitors for PCR is that aliquots of the cDNA reaction can be used for expression analysis of any possible gene. Furthermore, a PCR on randomly primed cDNA is controlled for RNA sample degradation and for variability in the efficiency of reverse transcription reactions. Consequently, it is feasible to directly perform the complete procedure on RNA preparations without the need for labor-intensive and time-consuming spectrophotometry and gel electrophoresis to measure quantity and to check the quality of RNA. Moreover, undesirable waste of RNA from small samples would then be avoided.

With the introduction of normalized competitive PCR and automated HPLC analysis to separate and detect PCR products, routine diagnostic and clinical applications of gene expression quantification have become accessible. Detailed protocols of our CD-RT-PCR method will be published in 1997 by Humana Press in the *Methods of Molecular Biology* and *Methods in Molecular Medicine* series.

REFERENCES

1. **Apostolakos, M.J., W.H.T. Schuermann, M.W. Frampton, M.J. Utell and J.C. Willey.** 1993. Measurement of gene expression by multiplex competitive polymerase chain reaction. Anal. Biochem. 213:277-284.
2. **Gilliland, G., S. Perrin, K. Blanchard and H.F. Bunn.** 1990. Analysis of cytokine mRNA and DNA: detection and quantitation by competitive polymerase chain reaction. Proc. Natl. Acad. Sci. USA 87:2725-2729.
3. **Hayward-Lester, A., P.J. Oefner and P.A. Doris.** 1996. Rapid quantification of gene expression by competitive RT-PCR and ion-pair reversed-phase HPLC. BioTechniques 20:250-257.
4. **Siebert, P.D. and J.W. Larrick.** 1992. Competitive PCR. Nature 359:557-558.
5. **Zimmermann, K. and J.W. Mannhalter.** 1996. Technical aspects of quantitative competitive PCR. BioTechniques 21:268-270.

CHAPTER 19

Quantitative RT-PCR on CYP1A1 Heterogeneous Nuclear RNA: A Surrogate for the In Vitro Transcription Run-On Assay

Cornelis J. Elferink and John J. Reiners, Jr.

Wayne State University, Detroit, MI, USA

ABSTRACT

A quantitative reverse transcription polymerase chain reaction (RT-PCR) assay was developed to amplify a region of the CYP1A1 heterogeneous nuclear RNA (hnRNA) transcript encompassing the first intron-exon boundary. The RT-PCR protocol uses a CYP1A1 recombinant RNA internal standard identical to the target hnRNA except for an engineered unique internal restriction site. Its inclusion enables normalization between reactions and a measurement of the absolute number of target hnRNA transcripts. Specificity for the hnRNA was achieved by using intron-directed primers in both the RT and the PCR. Nuclear run-on assays and the hnRNA RT-PCR assay detected an equivalent increase in transcription of Cyp1a-1 in cultured murine Hepa 1c1c7 cells following exposure to 2,3,7,8-tetrachlorodibenzo-p-dioxin (TCDD). The RT-PCR assay also revealed TCDD-dependent transcriptional activation of the Cyp1a-1 gene in murine skin, a tissue unsuited to the nuclear run-on assay because of inherent difficulties associated with the isolation of nuclei. These examples demonstrate that the hnRNA RT-PCR assay is a facile surrogate for the nuclear run-on assay. Moreover, the sensitivity and design characteristics of the RT-PCR assay suggest the potential for its broad application in general transcriptional research.

INTRODUCTION

The predominant method used to examine the transcriptional status of a given gene is the in vitro nuclear run-on assay. However, it embodies several significant constraints that limit its usefulness. The assay is laborious, time-

(Reprinted from BioTechniques 20:470-477, 1996)

Table 1. RT and PCR Primer Sequences

	Primer Sequence	Nucleotide* Position
Oligo(dT) primer	5′-GCTCTAGAGCTTTTTTTTTTTTTTTTTT-3′	Not Applicable
RTase primer	5′-TGCTTAATTCAGAGGGCTACAGATCC-3′	809-834
Forward primer	5′-GGTCCTAGAGAACACTCTTCACTTCAGTC-3′	678-700
Reverse primer	5′-TAATTGGTGTCCCTAAGGAACCC-3′	42-70

*Numbering refers to the *Cyp1a-1* structural gene.

consuming, expensive and involves significant amounts of radioactivity. Most notably, it requires large numbers of isolated nuclei, restricting it to systems where adequate numbers of nuclei can be isolated. Furthermore, obtaining reproducible data is difficult particularly in situations where gene transcription is low. The advent of a more amenable method to examine gene transcription would be desirable.

Quantitative reverse transcription polymerase chain reaction (RT-PCR) amplification of cellular messenger RNA (mRNA) is increasingly used to study gene expression, especially when tissue samples are limiting. However, analyses of mRNA provide only an indirect account of gene transcription, since mRNA levels represent the sum of both transcriptional and post-transcriptional events (e.g., processing, degradation). Heterogeneous nuclear RNAs (hnRNA) are nascent, as yet unspliced transcripts. In principle, measuring hnRNA levels may prove useful in determining a gene's transcriptional state. In fact, several reports suggest that RT-PCR on hnRNA might serve as a complementary approach to the nuclear run-on assay, but adequate experimental support for this idea was never provided (3,6,11,19).

This report describes a simple, quantitative RT-PCR-based strategy to amplify hnRNA for use as a transcriptional index of any gene of interest. We used this strategy to investigate the transcriptional activation of *Cyp1a-1* in cultured murine Hepa 1c1c7 cells and in vivo in murine keratinocytes. Accumulation of the cytochrome P450 CYP1A1 protein in response to xenobiotics such as 2,3,7,8-tetrachlorodibenzo-*p*-dioxin (TCDD, dioxin) is a consequence of transcriptional activation of the *Cyp1a-1* gene (8,18). The well-characterized transcriptional response of *Cyp1a-1* to TCDD in the Hepa cells (18) provided an excellent model in which to develop and validate the RT-PCR transcriptional assay. In turn, the keratinocytes provided a test case for the assay since *Cyp1a-1* induction occurs primarily in differentiating keratinocytes, from which it is virtually impossible to isolate nuclei (12). We consider this application of the RT-PCR strategy a facile surrogate for the in vitro transcription nuclear run-on assay offering several significant advantages over the run-on assay.

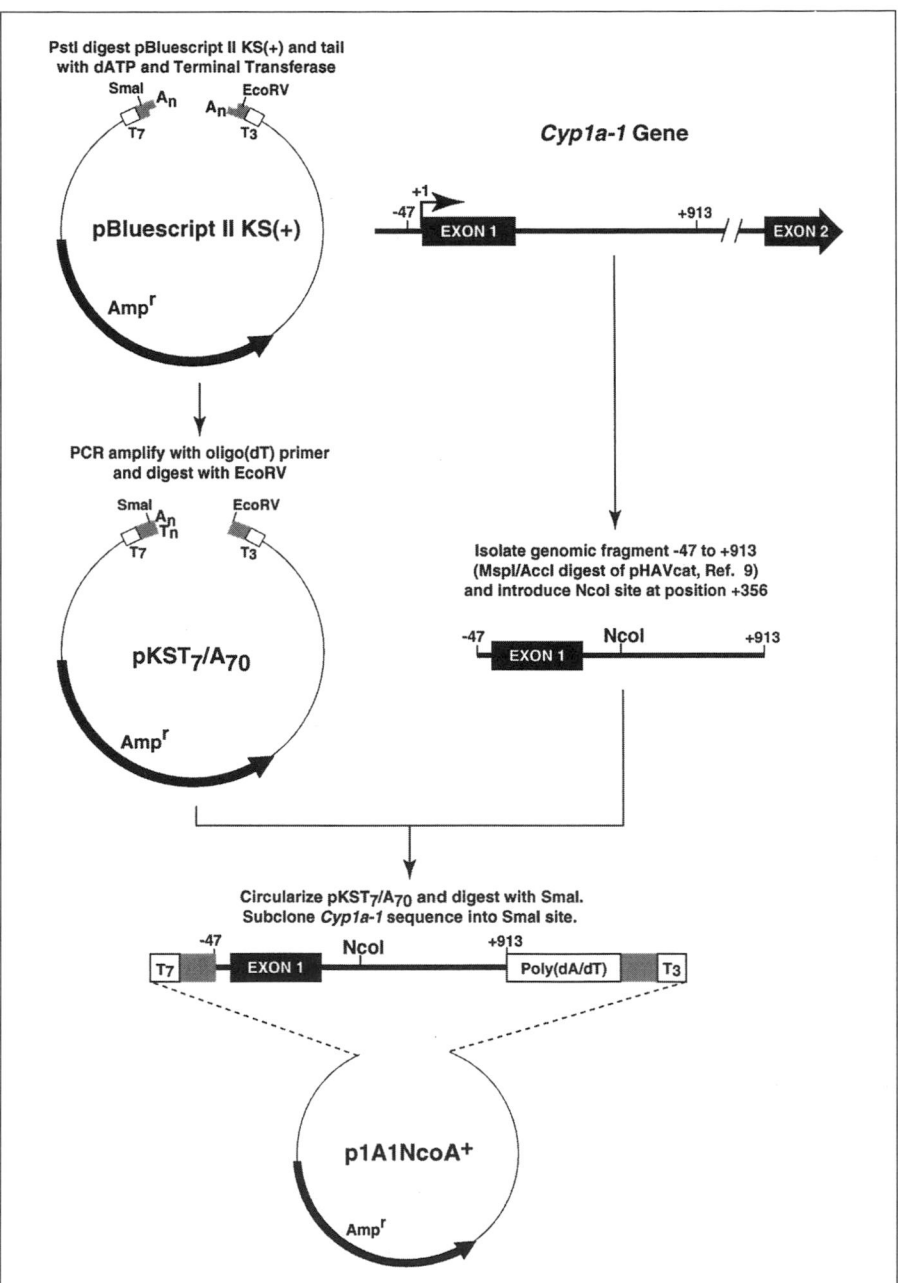

Figure 1. Scheme for the synthesis of the internal standard plasmid. The left half depicts the strategy for adding the polyadenosine tails to pBluescript II KS(+), followed by PCR amplification using an oligo(dT) primer against the polyadenosine tails. Polyadenosines were removed from either vector arm using $EcoRV$ (shown as $pKST_7/A_{70}$) or $SmaI$ (not shown). The $Cyp1a-1$ genomic sequence depicted on the right was subcloned into $pKST_7/A_{70}$ after introducing the $NcoI$ site at position +356 to generate the internal standard rcRNA construct p1A1NcoA+.

MATERIALS AND METHODS

Internal Standard Cloning Vectors

*Pst*I-digested pBluescript® II KS(+) (Stratagene, La Jolla, CA, USA) was tailed with adenosines using terminal deoxynucleotidyl transferase (TdT) and dATP from Boehringer Mannheim (Indianapolis, IN, USA) (Figure 1). Polyadenylated vectors were PCR-amplified using an oligo(dT) primer harboring an *Xba*I restriction site (see Table 1). PCR products were digested with either *Sma*I or *Eco*RV to remove the polyadenosine sequence from either vector arm, and the vector circularized in the presence of T4 DNA ligase (Life Technologies, Gaithersburg, MD, USA). DNA sequence analysis of individual clones using Sequenase® Version 2.0 (United States Biochemical, Cleveland, OH, USA) identified clones with 21–90 nucleotide polyadenosine cassettes in either orientation.

Synthesis of the Cyp1a-1 Internal Standard Sequence

The 960-bp *Acc*I/*Msc*I fragment from pHAVCat (9), representing nucleotides -47 to +913 of the *Cyp1a-1* gene, was end-filled and subcloned into *Sma*I-cut pBluescript II KS(+). A partial *Bsa*HI digestion (*Bsa*HI cuts once within both the insert and vector) of a subclone containing the *Cyp1a-1* sequence in the correct orientation, and subsequent gel isolation of the 3.9-kb product, was followed by dephosphorylation of the DNA ends with calf intestinal phosphatase (Boehringer Mannheim) and ligated in the presence of 1 µg *Nco*I linkers (New England Biolabs, Beverly, MA, USA). Individual clones were screened for integration of an *Nco*I site into the insert (at position 356 in the *Cyp1a-1* gene sequence) by restriction analysis. The *Cyp1a-1* sequence, excised with *Eco*RI and *Bam*HI, was end-filled and introduced into *Sma*I cut $pKST_7/A_{70}$ to generate the RT-PCR internal standard plasmid $p1A1NcoA^+$ (Figure 1).

In Vitro Transcription of the Internal Standard

Ten microliters of *Hin*dIII-linearized $p1A1NcoA^+$ DNA (0.2 µg/µL) were added to a 10-µL volume of reaction mixture containing T7 transcription buffer, 1 mM of each ribonucleoside triphosphate (0.5 mM final concentration) and 2 units T7 RNA polymerase (all from Boehringer Mannheim). Transcription occurred at 30°C for 1 h. Following phenol/chloroform extraction and ethanol precipitation, the recombinant complementary RNA (rcRNA) was purified by oligo(dT)-Sepharose® (Pharmacia Biotech, Piscataway, NJ, USA) chromatography (2), quantified spectrophotometrically and stored at -80°C as single-use aliquots containing 10^8 rcRNA molecules/µL.

Cell Culture and Total RNA Extraction

Hepa 1c1c7 cells were grown in 100-mm plates as previously described

(13) and exposed to TCDD (5 nM, 1 h) or dimethyl sulfoxide (DMSO) (vehicle) for 1 h. For the RT-PCR, α-amanitin (Sigma Chemical, St. Louis, MO, USA) was added to the media at a final concentration of 2 µg/mL, 1 h post-TCDD treatment. The dorsa of outbred female SENCAR mice (National Cancer Institute, Frederick, MD, USA) were shaved 5–7 days prior to use. Mice were sacrificed, and the dorsa were excised at various times following topical treatment of the areas with acetone (200 µL) or TCDD (4 nmol in acetone). After removal of the fat pad, the skin pelts were cut into strips, quick-frozen in liquid nitrogen and ground with a mortar and pestle. RNA extraction from cultured cells and skin used the method of Chirgwin et al. (4), except that the samples were centrifuged at 55 000 rpm for 2.5 h in a TLS-55

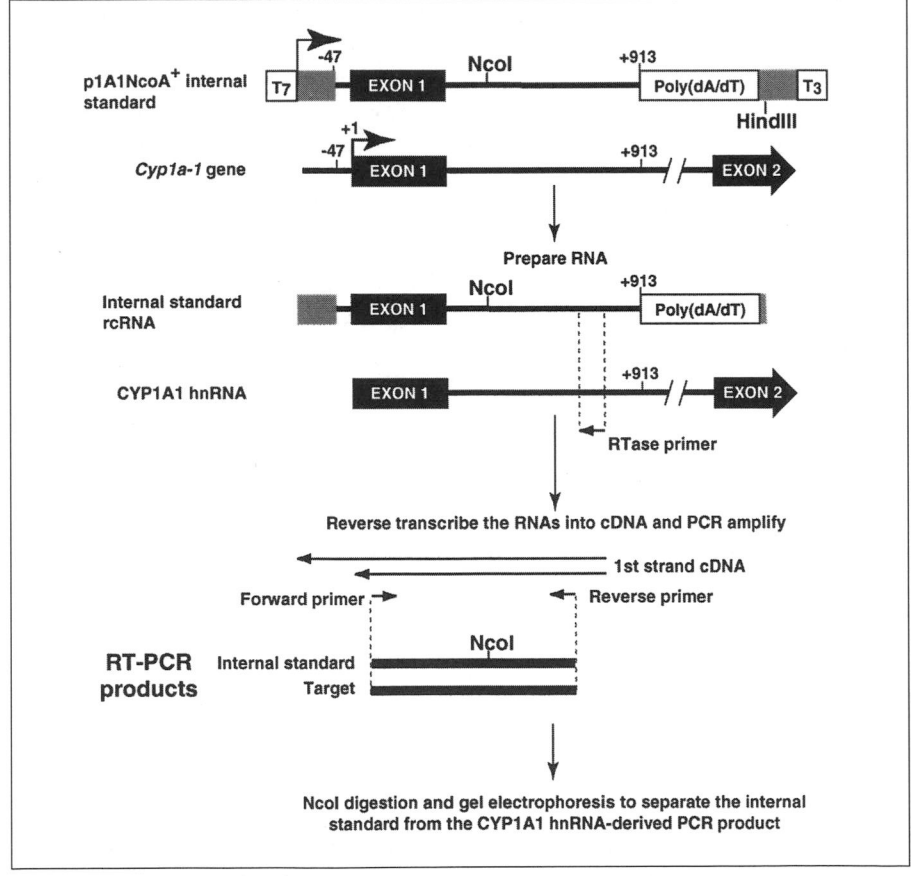

Figure 2. Scheme for the RT-PCR assay. Depicted are the internal standard (p1A1NcoA+) and genomic target DNA sequences for *Cyp1a-1* and corresponding rcRNA and hnRNAs, respectively. A mixture of rcRNA and hnRNA was reverse-transcribed into cDNA using the RTase primer (nucleotides 809–834). PCR used a forward primer (42–70) and reverse primer (678–700) to amplify the cDNA. This was followed by *Nco*I digestion of the internal standard and fractionation of the products on a non-denaturing 5% polyacrylamide gel and visualized by autoradiography.

rotor (Beckman Instruments, Fullerton, CA, USA). RNA was stored at -80°C.

Northern Analysis

Total RNA from mouse keratinocytes (20 μg per lane), or the internal standard rcRNA (1 μg), was electrophoresed in a 1.2% formaldehyde-agarose gel and transferred to a nitrocellulose membrane (Schleicher & Schuell, Keene, NH, USA) by capillary transfer and UV-fixed in a Stratalinker® UV crosslinker (Stratagene). Membranes were prehybridized at 50°C for 3 h and hybridized at 50°C overnight with 10^6 cpm/mL probe in 6× sodium chloride sodium phosphate EDTA (SSPE), 5× Denhardt's, 1% sodium dodecyl sulfate (SDS) and 100 μg/mL tRNA. The CYP1A1 probe was a ^{32}P-random primer labeled (Life Technologies) 700-bp *Cyp1a-1*-encoding *Pst*I fragment (10). A T4 Polynucleotide Kinase (New England Biolabs) ^{32}P-end-labeled RTase primer (see Table 1) served as a probe for the internal standard rcRNA transcript. Filters were washed in 1× standard sodium citrate (SSC) containing 0.1% SDS at RT for 30 min followed by 1× SSC containing 0.1% SDS at 42°C for 30 min and exposed to X-ray film (Kodak XAR-5; Eastman Kodak, Rochester, NY, USA).

Quantitative RT-PCR on hnRNA and Product Analysis

Reverse transcription was carried out using SuperScript™ II (Life Technologies) using conditions described by the manufacturer. The RTase primer was annealed to a mixture containing 1–2 μg Hepa 1c1c7 cells or keratinocyte total RNA and 5×10^6 copies of the internal standard rcRNA. Following a 1 h/37°C incubation, the reaction was terminated at 70°C for 15 min and diluted to 250 μL with water. The cDNA was either used immediately in the PCR or stored at -20°C. PCR was performed using 5 μL of the diluted cDNA in a final reaction volume of 50 μL using a forward primer and a "nested" reverse primer specific for the *Cyp1a-1* gene's first exon and intron, respectively (Table 1). PCR was initiated by adding 25 μL of a 2× master mixture containing 20 mM Tris-HCl, pH 8.3 (at 25°C), 100 mM KCl, 5 mM MgCl$_2$, 0.002% gelatin, 400 μM each of four dNTPs (dATP, dCTP, dGTP, dTTP), 5 μCi dCTP (3000 Ci/mmol; Du Pont NEN, Boston, MA, USA), 400 nM of both the forward and reverse primer and 1 unit AmpliTaq® DNA Polymerase (Perkin-Elmer, Norwalk, CT, USA) to the cDNA (25 μL) at 95°C. PCR was performed using the GeneAmp™ PCR System 9600 from Perkin-Elmer, cycling at 95°C/15 s (strand separation), 60°C/20 s (annealing) and 72°C/45 s (extension) for 30 cycles (except where indicated differently) with a final 10-min extension. Ten units of *Nco*I were added directly to the PCR tubes and incubated at 37°C for at least 1 h. Products were analyzed by fractionating 10 μL of the PCR on a nondenaturing 5% polyacrylamide/TBE (TBE = 45 mM Tris-borate, 1 mM EDTA, pH 8.0) gel followed by autoradiography of the dried gel. The absence of PCR products following the PCR on non-reverse-transcribed total RNA served as a routine check for contaminating genomic DNA.

In Vitro Transcription Nuclear Run-On Assay

Isolation of nuclei from Hepa 1c1c7 cells and the run-on assay were performed essentially as described previously (1,5). Briefly, 1.5×10^7 isolated nuclei (50 µL) were added to 100 µL 75 mM Tris-HCl, 7.5 mM MgCl$_2$, 225 mM KCl, 1.5 mM dithiothreitol, 1.5 mM MnCl$_2$, 3 µM UTP, 1.5 mM ATP, 1.5 mM CTP, 1.5 mM GTP and 100 µCi [α-^{32}P]UTP (3000 Ci/mmol), pH 7.9, and incubated for 30 min at 26°C. The reaction was stopped with 750 µL 0.5% SDS and 900 µL 100 mM Na acetate (pH 5.0). Radiolabeled RNA was prepared as described (16), and radiolabeled ribonucleotide incorporation was determined by TCA precipitation (5) using tRNA as carrier. Denatured, gel-purified insert DNA (2 µg) from p1A1NcoA$^+$ and a β-actin clone (8) was immobilized onto Nytran® membranes (Schleicher & Schuell) by "slot-blotting". Blots were prehybridized overnight in 6× SSPE, 5× Denhardt's, 0.1% SDS, 0.1% Na pyrophosphate, 100 µg/mL tRNA, pH 7.6, at 52°C. Hybridizations were carried out in fresh hybridization solution containing 2×10^6 cpm radiolabeled RNA at 52°C for 72 h. Filters were washed twice in 2× SSC containing 0.1% SDS at room temperature for 30 min and once in 0.2× SSC containing 0.1% SDS at 65°C for 30–60 min. Filters were dried and exposed to X-ray film (Kodak XAR-5).

Data Quantitation

Autoradiographs were scanned using a Silverscanner II flatbed scanner with a transparency adaptor (La Cie, Beaverton, OR, USA), and quantitative analyses were performed on a Macintosh® Quadra 650 computer using the program NIH Image (Version 1.54). This program is written by Wayne Rasband and is available in the public domain. The hnRNA copy number is obtained by acquiring a densitometric value for both the target hnRNA and internal standard. Knowing the internal standard copy number, the signal ratios are used to determine the hnRNA copy number.

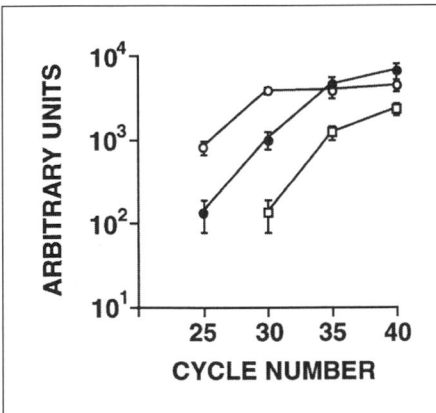

Figure 3. RT-PCR product formation as a function of cycle number and template concentration. In vitro derived internal standard rcRNA transcripts were purified by oligo(dT) chromatography and used in RT-PCR at 6×10^3 (open squares), 6×10^4 (solid circles) and 6×10^5 (open circles) input molecules for 25–40 cycles. Undigested PCR products were gel-fractionated and visualized autoradiographically. Bands were quantified by densitometric analysis on scanned autoradiographs.

RESULTS

Quantitative RT-PCR on CYP1A1 hnRNA

Quantitative RT-PCR most frequently involves inclusion of an RNA internal standard. Plasmid p1A1NcoA$^+$ encodes a portion of the *Cyp1a-1* gene encompassing the first intron-exon boundary, corresponding to the region of interest in the target sequence (Figure 1). The poly(dA/dT) cassette located immediately downstream from the *Cyp1a-1* sequence permits the in vitro synthesis of polyadenylated transcripts from the T7 RNA polymerase promoter. These are readily purified by standard oligo(dT) chromatography (2) and quantitated, and the transcript copy number is calculated from the known sequence length. Chromatography also removes DNA template and any truncated transcripts lacking the poly(A) tail. Northern analysis performed with a [γ-^{32}P]ATP end-labeled oligonucleotide probe (RTase primer) confirmed that the internal standard rcRNA was full-length (results not shown).

The principle of quantitative RT-PCR and rationale behind the design and requirement for internal standards is well documented (15,17). To ensure reliable quantitation with the RT-PCR strategy (Figure 2), the PCR must not exceed the exponential phase of the amplification. Quantitation of RT-PCR products obtained after 25–40 cycles using a 100-fold range of internal standard rcRNA transcripts provided an empirical determination of the exponential phase (Figure 3). Amplification of 6×10^3 to 6×10^5 initial rcRNA copies remained exponential for at least 30 cycles. Based on product formation as a function of cycle number, the PCR efficiency was calculated to be 60% using the formula derived by Trapnell (14). Since the rcRNA internal standard and *Cyp1a-1* hnRNA target sequence are almost identical, RT-PCR amplification of the hnRNA target should occur with the same efficiency. This ensures that the PCR parameters that define exponential amplification of the internal standard also apply to the target sequence.

Cyp1a-1 Transcription Measured by the Run-On and RT-PCR Assays

Nuclei for the transcription run-on assay and total RNA for RT-PCR analyses were isolated from Hepa 1c1c7 cell cultures treated with TCDD under conditions that maximally induce *Cyp1a-1* transcription. The two methods detect quantitatively equivalent increases in TCDD-induced *Cyp1a-1* gene expression (Figure 4). The minor difference in fold induction is not statistically significant according to a two-sample Student's *t* test. Whereas run-on data only provide information on relative transcriptional rates, the RT-PCR internal standard permits determination of the absolute number of hnRNA molecules per μg of total RNA. In fact, based on a total RNA recovery of 50 μg from 5×10^6 cells, we are detecting 1–2 and 8–10 copies per cell of the CYP1A1 hnRNA in untreated and TCDD-treated cells, respectively.

Since hnRNAs are transient intermediates, their concentration at any time

254

reflects a steady-state equilibrium between transcription and processing (splicing). Therefore, changes in the hnRNA level may reflect alterations in transcription or the rate of splicing to mature transcripts. α-Amanitin is a potent inhibitor of RNA polymerase II (3). CYP1A1 hnRNA processing at the first intron-exon boundary was determined to have a $T_{1/2}$ = 34 min in α-amanitin-treated cultures (Figure 5). Since CYP1A1 hnRNA levels remain constant for at least 4 h in the absence of α-amanitin (results not shown), the loss of transcript is not simply due to a spontaneous decrease in *Cyp1a-1* transcription. These results are consistent with a constant turnover of nascent transcripts and demonstrate that transcription and hnRNA processing can be readily dissociated in this assay. Experiments using α-amanitin concentrations ranging from 1–10 μg/mL consistently gave the same $T_{1/2}$ value, suggesting that *Cyp1a-1* transcription was maximally suppressed even at the low α-amanitin concentration (results not shown). A previous report indicated that RNA polymerase II transcription is inhibited by 98% in the presence of 2 μg/mL α-amanitin (3).

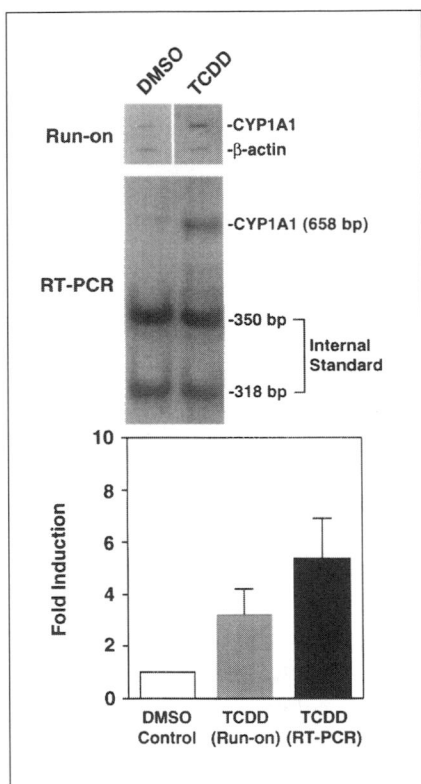

Figure 4. Comparison between the transcription run-on and RT-PCR assay. Hepa 1c1c7 cells treated with DMSO (vehicle) or 5 nM TCDD (in DMSO) for 1 h were used to isolate nuclei or total RNA. The top panel presents the results obtained by the run-on assay with nuclei and RT-PCR assay on hnRNA from the same batch of cells. The bottom panel presents quantitative analyses of the data obtained in three independent experiments. Run-on data and the RT-PCR data were normalized against β-actin and the internal standard PCR product, respectively. Error bars denote the standard error of the mean.

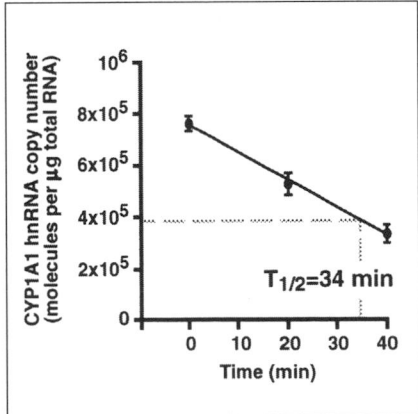

Figure 5. CYP1A1 hnRNA half-life. Hepa 1c1c7 cells were treated with 5 nM TCDD for 1 h prior to the addition of 2 μg/mL α-amanitin. Total RNA was extracted thereafter at the indicated times and subjected to quantitative RT-PCR. Data from three independent experiments were normalized against the internal standard and represent the mean value ± standard error of the mean.

Cyp1a-1 Expression in Mouse Keratinocytes

The original objective behind the development of this RT-PCR assay stems from our desire to study *Cyp1a-1* transcription in skin. Previous investigation in the laboratory has demonstrated that the TCDD-mediated induction of *Cyp1a-1* occurs mostly in differentiating keratinocytes (12). Northern analysis reveals that steady-state CYP1A1 mRNA levels are dramatically elevated in murine skin within 1 h following topical administration of TCDD (Figure 6). The RT-PCR assay similarly detected an increase in CYP1A1 hnRNA content following TCDD application, which was consistent with a persistent increase in gene transcription. Therefore, the RT-PCR assay provides a tool to examine *Cyp1a-1* gene transcription in these cells as a function of their differentiation while simultaneously demonstrating its potential as a surrogate for the transcription run-on assay.

DISCUSSION

This report describes a quantitative RT-PCR assay for the analysis of nascent CYP1A1 hnRNA transcripts. Specificity for CYP1A1 hnRNA is obtained by using an intron-directed primer in the RT reaction and a nested PCR primer (i.e., reverse primer) complementary to an upstream intronic sequence (Figure 2). Using a specific intronic primer, as opposed to oligo(dT) in the RT reaction, avoids amplification of mature mRNAs and the need to reverse-transcribe very long sequences. It also obviates possible complications arising from intron processing and removal prior to hnRNA polyadenylation. Quantitativeness of the assay is achieved by inclusion throughout the RT-PCR protocol of a CYP1A1 rcRNA internal standard that is nearly identical to the target hnRNA transcript. The sequence similarity should ensure that both RNAs are copied with equal efficiency during the RT reaction and during the exponential phase of the PCR. Consequently, absolute quantitation of CYP1A1 hnRNA sequences in the test RNA sample can be calculated simply from the ratio between the target and internal standard PCR products, irrespective of the efficiencies of the RT reaction or PCR. The internal standard also renders inter-sample comparisons more meaningful since it obviates the confounding effects of differences in RT and PCR efficiencies that may occur between samples.

Introduction of a unique *Nco*I restriction site into the CYP1A1 internal standard rcDNA provides a method whereby the PCR products corresponding to the internal standard and target hnRNA can be resolved from one another. However, accurate quantitation of the target transcript assumes that the internal standard is digested to completion following exposure to *Nco*I, since uncut internal standard co-migrates with the target sequence leading to overestimates of the target transcript. In our experience, based upon performing the PCR as replicate samples, incomplete digestion of the internal standard

PCR product has not proven to be a problem. A second potential complication to quantitation is heteroduplex formation between internal standard and target sequence PCR products. Such products cannot be cut. This represents a legitimate concern only if the PCR is allowed to proceed beyond the exponential phase, when DNA templates can reanneal without replication. This issue is easily addressed by limiting the PCR to the exponential phase of the reaction (Figure 3). Others (15,17) report using a truncated internal standard that is distinguishable from the target sequence without the need for DNA digestion. The potential for an altered internal standard sequence to possess

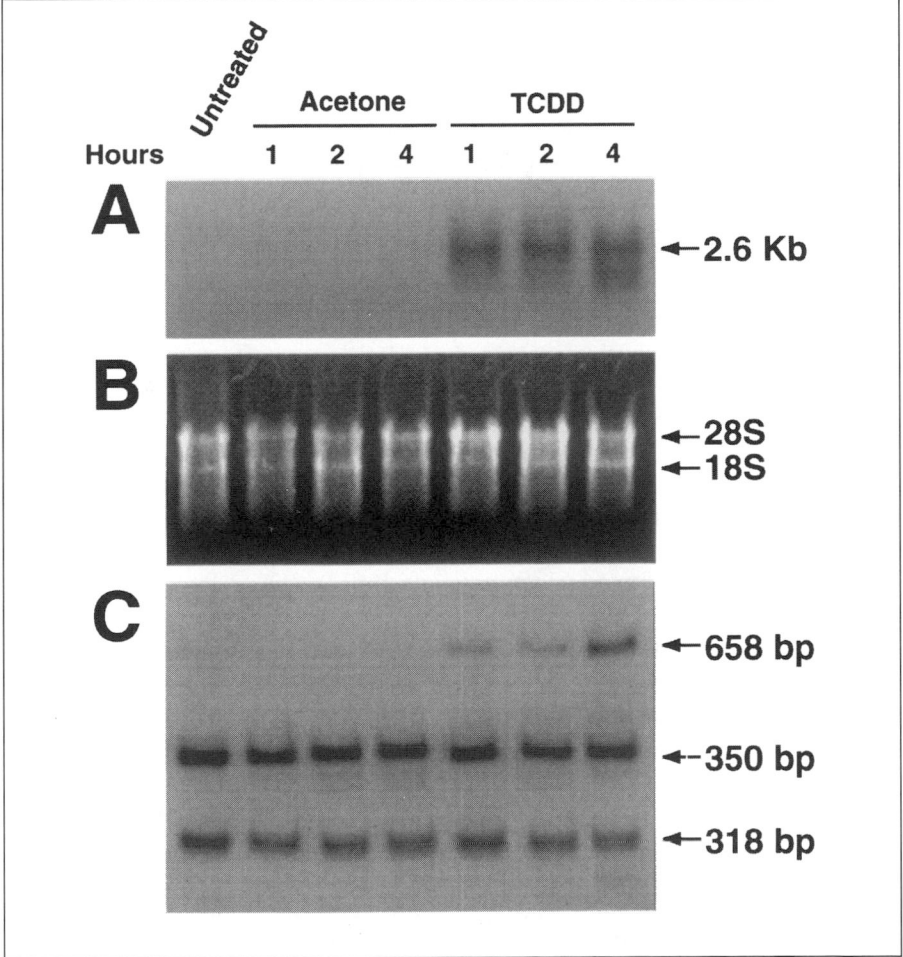

Figure 6. *Cyp1a-1* induction in mouse keratinocytes. Total RNA was extracted from the skins of control mice (lane 1) or mice treated for 1–4 h with either acetone (lanes 2–4) or 4 nmol TCDD (lanes 5–7). 20 µg of total RNA were fractionated on a 1.2% agarose-formaldehyde gel and stained with ethidium bromide (Panel B) prior to transfer to nitrocellulose for detection of CYP1A1 mRNA (Panel A). Quantitative RT-PCR on CYP1A1 hnRNA (Panel C) was performed on 1 µg total RNA.

different RT and PCR efficiencies from the target sequence, thereby compromising assay quantitation, renders this strategy less desirable.

RT-PCR and nuclear run-on assays of *Cyp1a-1* transcription in Hepa cultures gave quantitatively similar results in multiple analyses, each performed on a single batch of cells (Figure 4). This supports our contention that the assay is a potential surrogate for the in vitro transcription nuclear run-on assay. Variability between the RT-PCR experiments was less than 30%, suggesting that the assay should readily detect modest changes in transcription rate (e.g., 2-fold). The inherent difficulty of isolating nuclei that retain the true physiological transcription status renders run-on assays more susceptible to experimental variation.

Studies using the RT-PCR assay assume that changes in the amount of a given hnRNA sequence reflect transcriptional events instead of an altered processing rate. Transcription inhibitors such as α-amanitin can be used to dissociate these processes (Figure 5). If hnRNA processing rates are unaffected, then changes in the level of a given hnRNA can be directly attributed to a transcriptional event. Additionally, RT-PCR could be used to also quantify the corresponding mature mRNA levels. In this event, a proportional change in hnRNA and mRNA levels is consistent with a constant processing rate. Conversely, a disproportionate accumulation of either RNA species is indicative of a change in the rate of RNA processing. These relationships are only meaningful, however, if mRNA degradation is unaffected.

Detection of CYP1A1 hnRNA transcripts in murine keratinocytes by the RT-PCR assay paralleled the TCDD-induced increase in mRNA, consistent with a transcriptional response (Figure 6). This system represents a practical example of where run-on assays are unfeasible due to inherent difficulties with the isolation of nuclei; and RT-PCR of hnRNA functions as a surrogate assay for assessment of the transcriptional activation of a gene. In principle, the assay could be adopted to analyze any hnRNA providing that genomic sequence information and suitable primer sites are available. In addition, the polyadenosine cassette-containing cloning vectors we generated will simplify the preparation of internal standard rcRNA sequences. By using suitable PCR primers, an internal standard can be prepared containing both a unique restriction site and appropriate restriction sites for insertion into the vector MCS.

The exquisite sensitivity of the RT-PCR assay renders it suitable even with small tissue samples comprising only tens of thousands of cells. The RT-PCR studies reported here were performed on 1–2 μg total RNA, representing an estimated $1–2 \times 10^5$ cells. However, only a fraction of the cDNA—equivalent to 20 ng total RNA—was amplified by PCR. In situations where cell number or absolute RNA concentrations cannot be determined, quantitative genomic Southern analysis using a repetitive DNA sequence probe (e.g., Alu sequence) has been used to enumerate human cells with the intent of establishing absolute cellular transcript levels (14). Such an approach might facilitate the use of the hnRNA RT-PCR assay to study the transcriptional activation of genes in tissues of limited sample size, such as embryonic tissue or in the analyses of clinical biopsy samples. The hnRNA RT-PCR assay may also be a useful tool for addressing other issues. For instance, in hnRNA splicing studies, both the rate and the sequential order of intron splicing in a given gene could be determined without reliance on the traditional approaches of Northern analysis, S1 nuclease or RNase protection assays. Alternatively, expression of a reporter construct in transient transfection studies could be used to measure actual transcriptional events, without concern that RNA stability, translation, protein stability or enzyme activity of a reporter is affected by experimental manipulation.

ACKNOWLEDGMENTS

This work was supported by Grant CA34469 awarded by the National Institutes of Health. The technical assistance of Nancy Hong is greatly appreciated. We thank Drs. R. Hines and T. Kocarek for their comments on the manuscript.

REFERENCES

1. **Alterman, R-B.M., S. Ganguly, D.H. Schulze, W.F. Marzluff, C.L. Schildkraut and A.I. Skoultchi.** 1984. Cell cycle regulation of mouse H3 histone mRNA metabolism. Mol. Cell. Biol. *4*:123-132.
2. **Aviv, H. and P. Leder.** 1972. Purification of biologically active globin messenger RNA by chromatography on oligothymidylic acid-cellulose. Proc. Natl. Acad. Sci. USA *69*:1408-1413.
3. **Chang, C-D., P. Phillips, K.E. Lipson, V.J. Cristofalo and R. Baserga.** 1991. Senescent human fibroblasts have a post-transcriptional block in the expression of the proliferating cell nuclear antigen gene. J. Biol. Chem. *266*:8663-8666.
4. **Chirgwin, J.M., A.E. Przybyla, R.J. MacDonald and W.J. Rutter.** 1979. Isolation of biologically active ribonucleic acid from sources enriched in ribonucleases. Biochemistry *18*:5294-5299.
5. **Elferink, C.J., S. Sassa and B.K. May.** 1988. Regulation of 5-aminolevulinate synthase in mouse erythroleukemic cells is different from that in liver. J. Biol. Chem. *263*:13012-13016.
6. **Ferrari, S. and R. Battini.** 1990. Identification of chicken Calbindin D_{28K} pre-messenger RNA sequences by the polymerase chain reaction. Biochem. Biophys. Res. Comm. *168*:430-436.
7. **Green, M.R.** 1991. Biochemical mechanisms of constitutive and regulated pre-mRNA splicing. Annu. Rev. Cell Biol. *7*:559-599.
8. **Israel, D.I. and J.P. Whitlock, Jr.** 1984. Regulation of cytochrome P_1-450 gene transcription by 2,3,7,8-tetrachlorodibenzo-p-dioxin in wild type and variant mouse hepatoma cells. J. Biol. Chem. *259*:5400-5402.
9. **Jones, P.B.C., D.R. Galeazzi, J.M. Fisher and J.P. Whitlock, Jr.** 1985. Control of cytochrome P_1-450 gene expression by dioxin. Science *227*:1499-1502.
10. **Negishi, M., D.C. Swan, L.W. Enquist and D.W. Nebert.** 1981. Isolation and characterization of a cloned DNA sequence associated with the murine *Ah* locus and 3-methylcholanthrene-induced form of cytochrome P-450. Proc. Natl. Acad. Sci. USA *78*:800-804.
11. **Owczarek, C.M., P. Enriquez-Harris and N. Proudfoot.** 1992. The primary transcription unit of the human α2 globin gene defined by quantitative RT/PCR. Nucleic Acids Res. *20*:851-858.
12. **Reiners, J.J. Jr., A.R. Cantu, G. Thai and A. Schöller.** 1992. Differential expression of basal and hydrocarbon-induced cytochrome P450 monooxygenase and quinone reductase activities in subpopulations of murine epidermal cells differing in their stages of differentiation. Drug Metabol. Dispos. *20*:360-366.
13. **Reiners, J.J. Jr., A. Schöller, P. Bischer, A.R. Cantu and A. Pavone.** 1993. Suppression of

cytochrome P450 *Cyp1a-1* induction in murine hepatoma 1c1c7 cells by 12-0-tetradecanoylphorbol-13-acetate and inhibitors of protein kinase C. Arch. Biochem. Biophys. *301*:449-454.

14. **Trapnell, B.C.** 1993. Quantitative evaluation of gene expression in freshly isolated human respiratory epithelial cells. Am. J. Physiol. *264*:L199-L212.

15. **Vanden Heuvel, J.P., F.L. Tyson and D.A. Bell.** 1993. Construction of recombinant RNA templates for use as internal standards in quantitative RT-PCR. BioTechniques *14*:395-398.

16. **Vannice, J.L., J.M. Taylor and G.M. Ringold.** 1984. Glucocorticoid-mediated induction of α_1-acid glycoprotein: evidence for hormone-regulated RNA processing. Biochemistry *81*:4241-4245.

17. **Wang, A.M., M.V. Doyle and D.F. Mark.** 1989. Quantitation of mRNA by the polymerase chain reaction. Proc. Natl. Acad. Sci. USA *86*:9717-9721.

18. **Whitlock, J.P. Jr.** 1987. The regulation of gene expression by 2,3,7,8-tetrachlorodibenzo-*p*-dioxin. Pharmacol. Rev. *39*:147-161.

19. **Yang, M. and M. Kurkinen.** 1994. Different mechanisms of regulation of the human stromelysin and collagenase genes. Eur. J. Biochem. *222*:651-658.

Update to:

Quantitative RT-PCR on CYP1A1 Heterogeneous Nuclear RNA: A Surrogate for the In Vitro Transcription Run-On Assay

Cornelis J. Elferink and John J. Reiners, Jr.

Institute of Chemical Toxicology, Wayne State University, Detroit, MI, USA

The PCR assay developed in our laboratory has been applied to three independent problems. These studies all use the same approach and vector systems described in the original report, to examine CYP1A1 expression in a variety of cell and tissue systems. First, we have used the assay to demonstrate that the transcriptional activation of *Cyp1a-1* in cultured primary murine keratinocytes by TCDD is regulated as a function of the differentiation status of the cultures (C. Jones and J. Reiners, manuscript in preparation). Second, Northern and Western blot analyses demonstrate an induction of CYP1A1 in murine skin surrounding tumors generated in a two-stage carcinogenesis protocol, but no induction in the tumors themselves. This differential response reflects an absence of transcription of the *Cyp1a-1* gene in the tumors, as analyzed by the hnRNA RT-PCR assay (C. Jones et al., manuscript in preparation). Third, previous studies from this laboratory (1) suggest that cells in the G2/M phase of the cell cycle lose their responsiveness to TCDD. Use of the hnRNA RT-PCR assay has shown that it is transcriptional activation of the *Cyp1a-1* gene that is suppressed in this phase of the cell cycle, rather than turnover of CYP1A1 hnRNA, which occurs with a half-life similar to that seen in asynchronous cells. The assay is currently being used to assess *Cyp1a-1* transcriptional activation in other phases of the cycle.

REFERENCES

1.**Schöller, A., N.J. Hong, P. Bischer and J.J. Reiners, Jr.** 1994. Short and long term effects of cytoskeleton-disrupting drugs on cytochrome P450 *Cyp1a-1* induction in murine hepatoma 1c1c7 cells: suppression by the microtubule inhibitor nocodazole. Mol. Pharmacol. *45*:944-954.

CHAPTER 20

Competitor Template RNA for Detection and Quantitation of Hepatitis A Virus by PCR

Biswendu B. Goswami, Walter H. Koch and Thomas A. Cebula

Food and Drug Administration, Washington, DC, USA

ABSTRACT

PCR was used to introduce a 63-bp deletion into the putative RNA replicase coding sequence of hepatitis A virus. RNA was synthesized in vitro from the deletion mutant cloned into a transcription vector. Upon amplification by PCR, cDNA made from the competitor RNA generated an amplified fragment that could be easily distinguished from the product generated from wild-type hepatitis A virus genomic RNA by gel electrophoresis, when the same primers were used, without further manipulation. The competitor RNA was used as a positive control in PCR-based detection of very low copy numbers of hepatitis A virus genomic RNA in the presence of unrelated hard-shell clam RNA. When the competitor RNA was used for competitive PCR to quantitate wild-type RNA, the presence of one template at a 10-fold to 100-fold higher level almost completely inhibited product formation from the underrepresented template. The competitor RNA should be useful as a control for reverse transcription and PCRs to determine hepatitis A virus genome RNA when accidental contamination of test samples by a wild-type positive control template would compromise the results.

INTRODUCTION

Polymerase chain reaction (PCR) is being used increasingly to determine expression and quantitation of mRNAs, particularly where test sample size is limited and extreme sensitivity is required because of a low copy number of target sequence (1,18). PCR is also used to determine the presence and expression of viral genes (1,18). For viruses that do not grow well in cell culture, PCR may provide the only opportunity to identify the genetic material and expression of such viruses. One such virus is the human hepatitis A virus (HAV), which is the major cause of infectious hepatitis worldwide. A significant percentage of the reported cases of infectious hepatitis is attributed to the consumption of raw or inadequately cooked shellfish such as clams and oysters (4,14). Wild-type HAV grows very poorly in cell culture, usually as a

(Reprinted from BioTechniques 16:114-121, 1994)

persistent infection, without any cytopathic effect (7,11). A need therefore exists to develop a direct, sensitive and reliable method for the detection of HAV in environmental samples.

During the course of experiments designed to establish a reverse transcription coupled PCR (RT-PCR) method for the detection of HAV genomic RNA in hard-shell clams (*Mercenaria mercenaria*), we occasionally encountered accidental contamination of test materials by an in vitro synthesized positive control RNA. This problem became more acute when many amplification cycles were required to detect an extremely low number of target sequences present in the test sample. Other researchers have encountered similar problems in this area (1,6,18). The use of positive control templates is necessary to ensure that the RT and the PCRs have functioned as expected and to generate a standard curve, from which the concentration of the target sequence in the test sample can be deduced (2,5,10,12,17). Ideally, such positive control templates should be reverse-transcribed and amplified in the same tube (competitive PCR) to eliminate the so-called tube effect (2,17), which is defined as the variability in the yield of amplified product within the same dilution (5). However, this is difficult to achieve in practice, because a suitable control RNA template has to be synthesized for every test RNA. Three general approaches have been used to obtain positive control templates: (*i*) an unrelated endogenous RNA has been used to generate a standard curve (10,12); (*ii*) a plasmid has been constructed, with primers for several mRNAs symmetrically arranged on both sides of an unrelated sequence, which on RT-PCR generates an amplified product differing in size from the amplification product generated by the test RNAs (17); and (*iii*) a mutant template containing a restriction enzyme site not present in the wild-type template has been used (2,5). In this case, to distinguish between the test and the control template, the PCR-amplified DNA is digested with a restriction enzyme before analysis by gel electrophoresis.

We have constructed a plasmid for the in vitro synthesis of a competitor RNA for use as a positive control during RT-PCR detection of HAV. The competitor RNA harbors a 63-base deletion compared to wild-type HAV genomic RNA. Upon amplification by using the same primer pair, the competitor RNA generates a PCR product, which is easily distinguished from the wild-type PCR product in an agarose gel without further manipulation. A similar approach has recently been reported to quantitate human immunodeficiency virus (HIV) RNA and DNA (9,13,15).

MATERIALS AND METHODS

Materials

A full-length cDNA clone of wild-type HAV (strain HM-175, GenBank® locus HPA) was kindly provided by Dr. S. Emerson. A 926-bp *Bgl*II fragment from this clone was introduced into the *Bam*HI site of plasmid pGem®-4 Vector (Promega, Madison, WI, USA) and named pHAV *Bgl*II. Avian myeloblastosis virus (AMV) reverse transcriptase, RNasin® ribonuclease inhibitor, *Taq* DNA polymerase, deoxyribonucleo side-5′-triphosphates and oligo (dT)$_{16}$ were obtained from Promega. Restriction enzymes, T$_4$ DNA ligase, and competent cells were obtained from GIBCO BRL/Life Technologies, (Gaithersburg, MD, USA). The wild-type HAV (as a cell-free supernatant from infected cells) was kindly provided by Dr. B.H. Robertson. All oligodeoxynucleotide primers were synthesized and purified by Dr. M. Trucksess, US Food and Drug Administration. SP6 RNA polymerase and RNase-free DNase were purchased from United States Biochemicals (Cleveland, OH, USA). QIAGEN® Plasmid Kits and PCR Purification Spin Kits were obtained from Qiagen (Chatsworth, CA, USA).

Oligonucleotides for PCR

Primers 1 and 2 were used to amplify a 330-bp fragment from the plasmid pHAV *Bgl*II. The sequence of the forward primer (primer 1) was 5′ ctagaG GATCCTAGATCTTGATATGG 3′ (nucleotides [nt] 6217–6231), and the reverse primer (primer 2) was 5′caatgAAGCTTTGAGTAATCCAGAGG 3′ (nt 6510–6524). A 202-bp fragment from the same plasmid was amplified by using primers 3 and 4. The sequence of primer 3 was 5′gttaAAGCTTGATCCT-GATAGACAG 3′ (nt 6609–6623), and the reverse primer (primer 4) was 5′ctagaGGATCCTATTGATAAGAGTG 3′ (nt 6775–6789). Each primer has a restriction enzyme site at the 5′ end (italics) and a clamp sequence, 4 or 5 nucleotides long, 5′ to the restriction site (lower case). For primers 1 and 4 a *Bam*HI site was used, whereas for primers 2 and 3, a *Hin*dIII site was used. Primers 5 and 6 were used to amplify a 489-bp product from wild-type HAV genomic RNA and a 426-bp product from the competitor RNA template (Figure 1). The sequence of primer 5 was 5′ ATGCTATCAACATG-GATTCATCTCCTGG 3′ (nt 6256–6283), and the sequence of primer 6 was 5′ CACTCATGATTCTACCTGCTTCTCTAATC 3′ (nt 6716–6744). The positions of all six primers are shown in Figure 1. Their sequences were based on the published sequence of HAV (strain HM-175) cDNA (3) Primers 5 and 6 were chosen because of their high melting temperature (80°C). At the annealing temperature of 63°C used in these experiments, these primers can theoretically accommodate up to three mismatches in their complementary regions, which may arise as a result of mutations in the HAV genome (16). In practice, the effect of mismatches is not entirely predictable; the position of the mismatch, the nature of the sequence being amplified and probably other variables all play a role. We have not encountered any mispriming at this annealing temperature when HAV genomic or in vitro synthesized RNA, tRNA, calf liver RNA or a total RNA preparation isolated from hard-shell clams (*M. mercenaria*) was used as a template.

The scheme used to introduce the deletion in the wild-type HAV genome is depicted in Figure 1. Initially, two fragments of 330 and 202 bp were amplified from the plasmid pHAV *Bgl*II, which contains a *Bgl*II fragment spanning nt 6218 to 7143 of a full-length cDNA of HAV genomic RNA (3). Primers 1 and 2 were used to amplify the 330-bp fragment, and primers 3 and 4 were used to amplify the 202-bp fragment. The conditions for amplification were modified because all four primers have only 15 bases complementary to the sequence being amplified. Thus, PCRs were cycled 10 times at 94°C for 90 s, 35°C for 60 s and 72°C for 90 s. After 10 cycles, the annealing temperature was raised to 63°C. These conditions generated the expected size fragments, with no evidence of mispriming. The two PCR products were purified by using the QIAGEN PCR Purification Spin kit and ligated with T_4 DNA ligase. The ligation reaction was phenol-extracted and precipitated with 1 μg of carrier tRNA. The ligated products were digested with *Bam*HI to eliminate concatamers, and molecules joined in the correct orientation were amplified selectively, using primers 1 and 4. The resultant product was 532 bp in length because of the presence of non-HAV sequences contained in primers; it has two *Hin*dIII sites in tandem. This product was again digested with *Bam*HI to generate cohesive ends and ligated to *Bam*HI-digested psp64 (Poly A) DNA. The ligated product was used to transform *Escherichia coli* DH5α™ (GIBCO BRL/Life Technologies) to ampicillin resistance. One clone named pHAVΔ6 was checked for the presence of the expected deletion by using primers 5 and 6. As expected, this primer pair generated a 426-bp product from the deletion mutant, as opposed to a 489-bp product from the plasmid containing the wild-type sequence.

In Vitro RNA Synthesis

The plasmid pHAVΔ 6 was purified by using a QIAGEN Plasmid Kit. To synthesize RNA, we digested plasmid with *Eco*RI and synthesized RNA by using SP6 polymerase as previously described (8). The template DNA was removed by digestion with RNase-free DNase, and the RNA was recovered by phenol extraction and ethanol precipitation. The RNA was further purified by absorption to oligo (dT) cellulose, eluted with water, precipitated with alcohol, air-dried, dissolved in a small volume of water and quantitated by absorbance at 260 nm.

Isolation of HAV Genomic RNA

Wild-type HAV (as cell-free extract of infected cells) was kindly provided by Dr. B.H. Robertson. Viral RNA was isolated by proteinase K digestion, followed by phenol extraction and alcohol precipitation in the presence of tRNA carrier. RNA was quantitated as above.

RT

RT was performed in 20 μL total volume, which contained 50 mM Tris-HCl, pH 8.3, 75 mM KCl, 10 mM $MgCl_2$, 1 mM each of the four deoxyribonucleoside triphosphates (dNTP): 10 mM dithiothreitol, 1 U/μL RNase inhibitor, 0.5 μg of oligo $(dT)_{16}$ and 15 U AMV reverse transcriptase. Indicated amounts of either pHAVΔ 6 RNA or HAV genomic RNA were added as a template for cDNA synthesis. In addition, each reaction contained 1 or 5 μg of cellular total RNA isolated from the hard-shell clam *M. mercenaria*, which was previously judged to be free of HAV sequence by RT-PCR. Reactions were incubated at 22°C for 10 min followed by 50 min at 42°C, 5 min at 99°C and 5 min at 4°C. The reaction tubes were then centrifuged for 5 min at

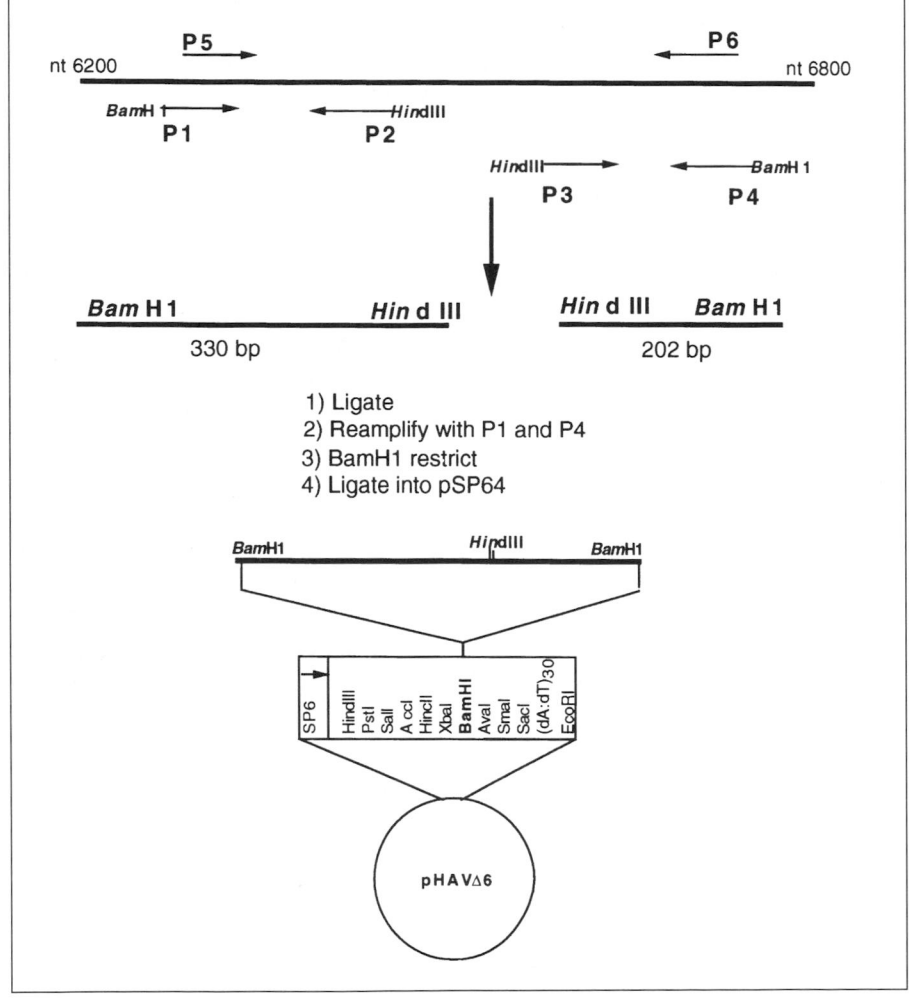

Figure 1. Scheme for the generation of the deletion mutant by PCR. The 330- and 202-bp fragments were joined by blunt-end ligation; hybrid molecules joined in the correct orientation were amplified by using outer primers 1 and 4. The SP6 promoter is upstream of the *Bam*HI site of the 330-bp fragment in the plasmid pHAVΔ 6; two *Hind*III sites are in tandem from the original primers 2 and 3. RNA from this plasmid is synthesized in vitro as the sense strand with SP6 polymerase.

267

full speed in an Eppendorf centrifuge (Brinkmann Instruments, Westbury, NY, USA).

PCR

PCR amplification was performed in 50- to 100-μL volumes, which contained 5–10 μL of cDNA pool from RT reactions or appropriately diluted plasmid DNA or DNA from previous PCR, 3 mM $MgCl_2$ 200 μM of each dNTP, 0.5 μM of each amplification primer, and 1.5 U *Taq* DNA polymerase. The mixture was denatured initially for 3 min at 95°C, followed by 40 cycles, each consisting of 90 s at 94°C, 90 s at 63°C and 120 s at 72°C. These parameters were modified for initial amplification of the 330-bp and 202-bp fragments, as discussed above. A final extension was performed at 72°C for 10 min. From each PCR, 20 μL were analyzed by 1.6% agarose gel electrophoresis in the presence of 1 μg/mL ethidium bromide.

RESULTS AND DISCUSSION

PCR was used to introduce a 63-bp deletion into the 3′ end of the wild-type HAV genome. A difference of 84 bp between the 5′ ends of primers 2 and 3 formed the basis of the deletion (Figure 1). Because of the presence of a total of 21 extra bases at the 5′ ends of primers 2 and 3, which were not removed during the blunt end ligation of the 330- and 202-bp fragments, the final extent of deletion was 63 bp. When RNA was synthesized in vitro from the deletion mutant and used as a template for RT-PCR, the amplified product reflected this size difference and could be easily distinguished from the wild-type product (Figure 2). No product was obtained when reverse transcriptase was omitted from the reaction mixture. The sensitivity of the RT-PCR procedure was such that as little as 0.025 fg (ca. 72 molecules) of the competitor transcript in the RT reaction was easily observed in the presence of 1 μg of unrelated RNA without Southern blotting (Figure 2). In this paper, the quantity of RNA used for PCR indicates that cDNA corresponding to

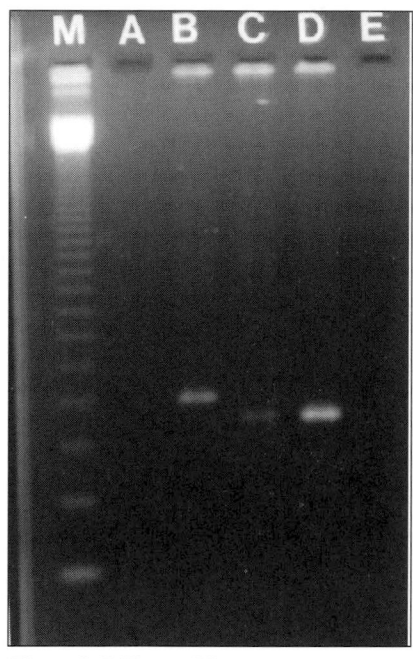

Figure 2. PCR amplification cDNA synthesized from wild-type HAV genomic RNA or a competitor transcript synthesized in vitro. 7.5 pg of HAV genomic RNA (lanes A and B) or 72 molecules (lane C), or 720 molecules (lanes D and E) of the in vitro transcript were reverse-transcribed in the presence of 1 μg of *M. mercenaria* RNA as described in Methods. 5 μL of each cDNA pool were then amplified using primers 5 and 6. Reverse transcriptase was omitted from lanes A and E. Lane M is 123-bp DNA ladder (GIBCO BRL).

that amount of input RNA was actually amplified.

To investigate the utility of the competitor transcript to quantitate HAV genomic RNA in a test sample, 0.96 fg (or 2774 molecules) of the competitor transcript or 30 pg of HAV genomic RNA preparation (including carrier) were reverse-transcribed in the presence of 5 μg of *M. mercenaria* RNA. The reaction mixtures were appropriately diluted and cDNA pools representing 0.3 to 6 pg of the genomic RNA and 3 to 48 molecules of the competitor transcript were amplified (Figure 3). The number of HAV genomic RNA molecules in the test sample was estimated to be 3 molecules per 0.3 pg of total RNA, based on PCR product yields. However, exact quantitation will require use of labeled primers or a labeled dNTP during PCR.

We also investigated whether the in vitro transcript and the genomic RNA could be reverse-transcribed and amplified in the same tube to control for the so-called tube effect observed by other investigators (5,12). Either 0.75 pg or 7.5 pg of the genomic RNA was mixed with 9, 90 or 900 molecules of the competitor transcript and 1 μL of *M. mercenaria* RNA and reverse-transcribed. One-fourth of each reaction (5 μL) was then amplified with primers 5 and 6 and analyzed by gel electrophoresis (Figure 4). PCR products from both templates were observed when both templates were present during RT (Figure 4, lanes C–H). The relative abundance of the two products was roughly equal when a comparable amount of the two templates were present during RT (lanes D and H). However, when one template was present at a 10-fold to 100-fold higher level, formation of product from the less abundant

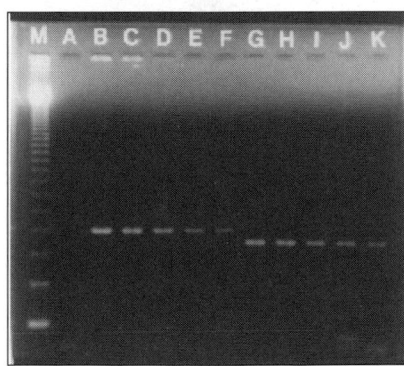

template was inhibited. Similar competitive suppression of product formation from one template by a related or unrelated template in the same tube has been observed by others (9,10,13,15). This phenomenon restricts the usefulness of the single tube approach as a diagnostic tool at very low template concentrations. At high template concentrations, such suppression of product formation from the less abundant template has not been reported (17). No products were seen when either *M. mercenaria* RNA or tRNA was used as a template or when templates were omitted from the PCR.

Figure 3. Quantitation of HAV genomic RNA by PCR. 30 pg of HAV genomic RNA or 2774 molecules of the in vitro synthesized transcript were reverse-transcribed in the presence of 5 μg of *M. mercenaria* RNA. 5 μL of serial 1:2 dilutions of cDNA pools were amplified by using primers 5 and 6. Lanes B to F, HAV genomic RNA; lanes G to K competitor in vitro RNA. Lane A contained 40 fg of competitor in vitro template and 5 μg of *M merce-naria* RNA without reverse transcriptase. Lane M is 123-bp DNA ladder (GIBCO BRL).

Quantitation of specific mRNAs by the PCR technique is becoming increasingly popular because of its extreme sensitivity and because it requires a very limited quantity of test samples (9,13, s15,18). However, quantitation by PCR requires the generation of standard curves with each set of experiments because the final signal obtained by PCR is a function of initial template concentration and the efficiency of the PCR, which varies from experiment to experiment, or even from tube to tube in the same experiment (5,17). In the case of quantitation of RNA by PCR, the efficiency of the initial RT step to convert RNA to cDNA must also be taken into consideration (1,18). Several studies have used unrelated endogenous RNAs such as β-actin or β$_2$-microglobulin to account for the efficiencies of the reactions and to generate standard curves (10,12). However, results obtained by this method can be evaluated only during the exponential phase of the reaction. Because the efficiency of utilization and elongation by *Taq* DNA polymerase may vary significantly from one primer sequence to another, this method requires extensive initial experimentation to determine the suitable range of template concentration for the control and the test RNAs. The two templates cannot be amplified in the same tube to account for the tube effect, because templates and especially primers have different melting temperatures, and one of the reactions reaches immature plateau (10,12). To overcome these difficulties, control templates have been constructed that use the same primers but form a product that can be distinguished from the product of the test template after restriction enzyme digestion or by the difference in size of the two products (2,5,9,13,15,17). Errors can arise in the former case as a result of incomplete digestion or heteroduplex formation (2,5) and, in the later case, because of differences in the efficiency of elongation of the two templates (17).

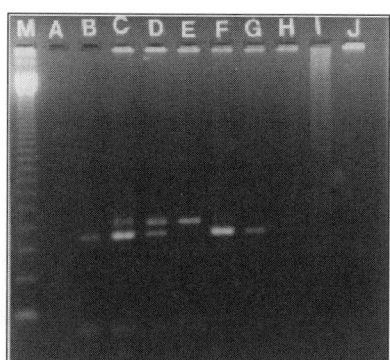

In designing a positive control template for the identification of HAV by RT-PCR, we have tried to eliminate most of these variables. The competitor template RNA is only 63 bases shorter than the wild-type genomic RNA. This difference is enough to allow the product of the control RNA to be readily separated from that of the test RNA, but is not large enough to significantly affect elongation rate. The need for digestion with restriction enzymes is eliminated. In addition, primers were chosen from the 3′ end of the HAV genome to take advantage of the poly (A) tail of the HAV genomic and in vitro RNAs. Use of oligo (dT) during the RT step significantly increased the efficiency of identifying HAV-specific RNA (either genomic or in vitro) in the presence of a high concentra-

Figure 4. Co-amplification of HAV genomic RNA and the competitor in vitro RNA. 7.5 pg (lanes C–E) or 0.75 pg (lanes F–H) of the genomic RNA was mixed with either 9 (lanes E and H), 90 (lanes D and G) or 900 lanes C and F) molecules of the competitor RNA and 1 μg of *M. mercenaria* RMA and reverse-transcribed. 5 μL of each reaction mixture were then amplified using primers 5 and 6. Lane A, PCR reagents mix without any template. Lane B, pHAV Δ 6 plasmid DNA. Lane I, 10 μg tRNA; lane J, 5 μg *M. mercenaria* RNA reverse-transcribed and amplified as for HAV genomic or in vitro RNAs.

tion of unrelated RNAs compared to the use of random primers (data not shown). The competitor RNA can be used for the same tube RT-PCR identification and quantitation of HAV, but care must be taken not to overwhelm the wild-type template present in the test sample, which may otherwise significantly affect product formation. Recently, competitive PCR has been used to quantitate HIV-specific RNA in human blood (9,13,15). Suppression of product formation from the underrepresented template was observed in all instances (9,13,15). To alleviate this problem, several different concentrations of the control template were reverse-transcribed and amplified with the same amount of the test RNA template (9,13,15). We believe this makes competitive PCR a labor-intensive and time-consuming, albeit extremely sensitive and reliable method for RNA quantitation. The competitor HAV RNA we have described enables the researcher to avoid accidental contamination of test samples while using a positive control RNA in routine diagnostic assays and quantitation of clinical or environmental samples suspected to contain HAV. More demanding quantitation, however, is equally feasible with the same competitor RNA.

REFERENCES

1. **Arnheim, N. and H. Ehrlich.** 1992. Polymerase chain reaction strategy. Annu. Rev. Biochem. *61*:131-156.
2. **Becker-Andre, M. and K. Hahlbrock.** 1989. Absolute mRNA quantification using the polymerase chain reaction (PCR). A novel approach by a PCR aided transcript titration assay (PATTY). Nucleic Acids Res. *17*:9437-9446.
3. **Cohen, J.I., J.R. Ticehurst, R.H. Purcell, A. Buckler-White and B.M. Baroudy.** 1987. Complete nucleotide sequence of wild-type hepatitis A virus: comparison with different hepatitis A virus and other picornaviruses. J. Virol. *61*:50-59.
4. **Desenclos, J-C.A., K.C. Klontz, M.H. Wilder, O.V. Nainan, H.S. Margolis and R.A. Gunn.** 1991. A multistate outbreak of hepatitis A caused by the consumption of raw oysters. Am. J. Public Health *81*:1268-1272.
5. **Gilliland, G., S. Perrin, K. Blanchard and H.F. Bunn.** 1990. Analysis of cytokine mRNA and DNA: detection and quantitation by competitive polymerase chain reaction. Proc. Natl. Acad. Sci. USA *87*:2725-2729.
6. **Kwok, S. and R. Higuchi.** 1989. Avoiding false positives with PCR. Nature *339*:237-238.
7. **Lemon, S.M., L.N. Binn and R.H. Marchwicki.** 1983. Radioimmunofocus assay for quantitation of hepatitis A virus in cell cultures. J. Clin. Microbiol. *17*:834-839.
8. **Melton, D.A.** 1987. Translation of messenger RNA in injected frog oocytes. Methods Enzymol. *152*:288-296.
9. **Menzo, S., P. Bagnarelli, M. Giacca, A. Manzin, P.E. Veraldo and M. Clementi.** 1992. Absolute quantitation of viremia in human immunodeficiency virus infection by competitive reverse transcription and poly-merase chain reaction. J. Clin. Microbiol. *30*:1752-1757.
10. **Murphy, L.D., C.E. Herzog, J.B. Rudick, A.T. Fojo and S.E. Bates.** 1990. Use of the polymerase chain reaction in the quantitation of *mdr*-1 gene expression. Biochemistry *29*:10351-10356.
11. **Nasser, A.M. and T.G. Metcalf.** 1987. Production of cytopathology in FRhK-4 cells by BS-C-1 passaged hepatitis A virus. Appl. Environ. Microbiol. *53*:2967-2971.
12. **Noonan, K.E., C. Beck, T.A. Holzmayer, J.E. Chin, J.S. Wunder, I.L. Andrulis, A.F. Gazdar, C.L. Willman, B. Griffith, D.D. Von Hoff and I.B. Robinson.** 1990. Quantitative analysis of *MDR*1 (multidrug resistance) gene expression in human tumors by polymerase chain reaction. Proc. Natl. Acad. Sci. USA *87*:7160-7164.
13. **Piatak, M. Jr., K-C. Luk, B. Williams and J.D. Lifson.** 1993. Quantitative competitive polymerase chain reaction for accurate quantitation of HIV DNA and RNA species. BioTechniques *14*:70-81.

14.**Richards, G.P.** 1987. Shellfish-associated enteric virus illness in the United States, 1934-1984. Estuaries *10*:84-85.

15.**Scadden, D.T., Z. Wang and J.E. Groopman.** 1992. Quantitation of plasma human immunodeficiency virus type 1 RNA by competitive polymerase chain reaction. J. Infect. Dis. *165*:1119-1123.

16.**Wahl, G.M., S.L. Berger and A.R. Kimmel.** 1987. Molecular hybridization of immobilized nucleic acids: theoretical concepts and practical considerations. Methods Enzymol. *152*:399-407.

17.**Wang, A.M., M.V. Doyle and D.F. Mark.** 1989. Quantitation of mRNA by the polymerase chain reaction. Proc. Natl. Acad. Sci. USA *86*:9717-9721.

18.**Wright, P.A. and D. Wynford-Thomas.** 1990. The polymerase chain reaction: miracle or mirage? A critical review of its uses and limitations in diagnosis and research. J. Pathol. *162*:99-117.

Update to:

Competitor Template RNA for Detection and Quantitation of Hepatitis A Virus by PCR

Biswendu B. Goswami and Thomas A. Cebula

Food and Drug Administration, Washington, DC, USA

Recently we described a method (3,4) for the detection and quantitation of hepatitis A virus (HAV) in environmental samples such as shellfish by competitive polymerase chain reaction (PCR). Subsequently, other laboratories have described PCR-based methods for the detection of a variety of human enteric viruses (1,2). An important application of PCR-based detection of viruses is the ability to detect viruses during epidemic outbreaks. Since most viruses have antigenic and/or genetic variants that predominate in particular geographical regions, a method to identify such variants is extremely useful in determining a strategy to combat such outbreaks. Currently, genotyping is carried out by sequencing PCR-amplified viral genetic material. Sequencing can be conveniently carried out on small (typically 150- to 200-bp) amplified regions of the viral genome. Because the targets for sequencing represent only about 2% of the total genome, genetic variants arising in an epidemic may not be detected unless the virus has a high mutation rate and mutations are distributed evenly throughout the genome. We have recently described an alternative to sequencing for the identification of genetic variants (5). This procedure is based on single-stranded conformation polymorphism (SSCP) analysis of restriction fragments of PCR-amplified DNA. This procedure, called restriction endonuclease fingerprinting (REF; 6) was able to distinguish 13 of 17 HAV strains analyzed. It is important to note that the 4 strains that were indistinguishable by REF analysis were all tissue culture-adapted clonal isolates of the strain HM175. REF analysis can be conveniently carried out on larger segments of amplified DNA, requires very little material and provides data more quickly than conventional sequencing.

REFERENCES

1. **Ando, T., S.S. Monroe, J.R. Gentsch, Q. Jin, D.C. Lewis and R.I. Glass.** 1995. Detection and differentiation of antigenically distinct small round-structured viruses (Norwalk-like viruses) by reverse-transcription-PCR and Southern hybridization. J. Clin. Microbiol. *33*:64-71.
2. **Atmar, R.L., F.H. Neill, J.L. Romalde, F. Le Guyader, C.M. Woodley, T.G. Metcalf and M.K. Estes.** 1995. Detection of Norwalk virus and hepatitis A virus in shellfish tissues with the PCR. Appl. Environ. Microbiol. *61*:3014-3018.
3. **Goswami, B.B., W.H. Koch and T.A. Cebula.** 1993. Detection of hepatitis A virus in Mercenaria

mercenaria by coupled reverse transcription and polymerase chain reaction. Appl. Environ. Microbiol. *59*:2765-2770.

4.**Goswami, B.B., W.H. Koch and T.A. Cebula.** 1994. Competitor template RNA for detection and quantitation of hepatitis A virus by PCR. BioTechniques *16*:114-121.

5.**Goswami, B.B., W. Burkhardt and T.A. Cebula.** Identification of genetic variants of hepatitis A virus. J. Virol. Methods (In press).

6.**Iwahana, I., K. Yoshimoto and M. Itakara.** 1992. Detection of point mutation by SSCP of PCR-amplified DNA after endonuclease digestion. BioTechniques *12*:64-66.

CHAPTER 21

Quantification of Hepatitis B Virus DNA by Competitive Amplification and Hybridization on Microplates

T. Jalava, P. Lehtovaara, A. Kallio, M. Ranki and H. Söderlund[1]

Orion Corporation, Espoo and [1]VTT, Technical Research Centre of Finland, Espoo, Finland

ABSTRACT

Present methods for quantification of hepatitis B virus (HBV) particles from serum samples are not sensitive enough for some recent clinical applications. We describe a test that allows quantification of HBV DNA in a broad dynamic range from less than 40 to 10^6 molecules based on competitive PCR. The specimen DNA and a known amount of an internal standard (IS) are co-amplified in the same tube with the same primers, one of which is biotinylated. The two biotinylated products can be quantified by hybridization on microplates coated with streptavidin, because their internal sequences are nonhomologous. An adequate standard curve is obtained by amplifying HBV DNA from a plasmid clone together with an IS. The ratio of amplified HBV DNA to IS DNA enables quantification of the original amount of HBV without tedious titrations of each sample with competitor. The lower limit for quantitative analysis with radioactive probes was between 4 and 40 virus particles in a 10-μL serum sample.

INTRODUCTION

The amplification of DNA sequences using the polymerase chain reaction (PCR) is a widespread routine that is being used while methods of applying this technique for quantification are still under development (4). A quantitative PCR methodology would be valuable in several fields of application, for example, when assessing the clinical significance of low levels of virions. Immunochemical determination of hepatitis B viral (HBV) antigens from serum does not necessarily correlate with the presence of complete viral particles, which can only be quantified reliably by measuring viral DNA. However, the quantitative tests available for HBV DNA are not sensitive enough for such applications as follow-up of interferon therapy or liver

(Reprinted from BioTechniques 15:134-139, 1993)

transplantation. The PCR methodology applied so far for HBV (6,10–12) has not been focused on the quantitative aspect.

PCR obeys the function $N = (1+\text{eff})^n N_0$, where n is the number of PCR cycles, N_0 and N the number of molecules before and after amplification, respectively, and eff is the efficiency of the reaction (0 to 1). It is clear that even minor differences in efficiency have dramatic effects on the final yields after several exponential cycles. Quantification is difficult because the efficiency factor is variable both from sample to sample and also within one sample from cycle to cycle. The intersample variation is created by quality variations of the samples to be amplified. Failure to completely remove various PCR inhibitors by specimen pretreatment procedures may lead to underestimation of the quantity, or even to false negatives (7). The intercycle variation depends most significantly on the reannealing rate of the PCR product and the limiting availability of reagents in later cycles (18).

Two major approaches have been used to solve the problems inherent in quantitative PCR. In the first approach, one tries to exploit the logarithmic phase of amplification when the efficiency is not yet affected by the number of cycles (8,13). This assumes that there is no quality variation between the samples. However, for many types of samples it is very difficult to reproducibly prepare DNA free of inhibitors. This approach is therefore unsatisfactory for quantification from clinical samples, such as HBV from serum. In the second approach applied here, a competitive PCR with an internal standard (IS) is used to eliminate the variation of the efficiency factor (1,5,15,19,20). There is no need to restrict the number of cycles and the sensitivity is increased. In the test described here, the HBV DNA and IS are co-amplified in the same tube with the same primers, and thus factors like depletion of reaction components or the presence of inhibitors should decrease the efficiency of amplification of both DNAs by the same factor.

MATERIALS AND METHODS

DNAs

The HBV plasmid pHB320 contained the entire 3182-nucleotide-long genome of HBV ayw subtype (3) cloned in pBR322, and the hybridization probe for HBV was obtained by subcloning a 0.42-kb-long fragment from the core gene region to pBR322. The IS for amplification was a pBR322 clone, containing a 527-nucleotide-long sequence with no homology to HBV DNA, flanked by 25-mer sequences complementary to the HBV amplification primers (14). The hybridization probe for IS was a 0.35-kb-long subclone of the IS fragment in pBR322. The biotinylated oligonucleotide was synthesized using sulfo-NHS biotin ester and purified by HPLC (2).

The probes were labeled to a specific activity of 10^7 cpm/µg by nick translation using ^{35}S-dCTP (Amersham, Little Chalfont, Bucks, England; >1000

Ci/mmol). The unincorporated label was removed by two ethanol precipitations from 2.5 M ammonium acetate.

Amplification

The reaction mixture of 100 μL contained 0.2 μM of each of the two primers, 200 μM of each of the four deoxyribonucleoside triphosphates (dNTPs), 50 mM Tris-HCl pH 8.8, 1.5 mM $MgCl_2$, 15 mM $NH_4(SO_4)_2$, 0.1% Tween® 20, 1 U of *Taq* DNA Polymerase (Promega, Madison, WI, USA) and the DNAs to be amplified. The mixture was overlayed with paraffin oil and amplified for 30 cycles (DNA Thermal Cycler; Perkin-Elmer, Norwalk, CT, USA) using the following cycle parameters: 96°C for 30 s, 55°C for 60 s and 72°C for 60 s. The last extension step was at 72°C for 10 min.

Hybridization

An aliquot of the amplified sample (usually 1/500) and the nick-translated probe DNAs (2×10^{10} molecules per well, 10^7 cpm/μg) were denatured by boiling in 0.2% sodium dodecyl sulfate (SDS) for 5 min. The hybridization and binding were carried out simultaneously on microplates coated with streptavidin (Labsystems, Helsinki, Finland) at 55°C for 2 h in a microplate shaker. Each well contained the target and probe DNAs in 80 μL of 0.6 M NaCl, 1 mM EDTA, 20 mM sodium phosphate pH 7.5, 0.02% bovine serum albumin (BSA), 0.02% Ficoll® 400 (Pharmacia Biotech, Espoo, Finland), 0.02% polyvinyl pyrrolidone, 3.2% dextran sulfate, 0.1% SDS, 14 μg/mL phenol red. One set of samples was prepared for hybridization with HBV and the other with IS-specific probe. After hybridization, the plates were washed 6 times with a 50°C solution of 0.1× standard saline citrate (SSC), 0.2% SDS. The hybridized probes were eluted with 0.2 M NaOH and counted in a scintillation counter. The results are usually given as a mean of three parallel determinations.

RESULTS

Quantification and Standard Curve

Each HBV DNA sample was co-amplified with the same primers in the same tube with a known constant amount of IS. One primer was biotinylated to enable immobilization of the products to streptavidin-coated microplates. The amplified IS fragment was 567 bp long and contained a 527-bp long HBV-unrelated DNA sequence flanked by amplification primer sequences from HBV. The amplified HBV DNA fragment was 560 bp long. The PCR products were denatured by boiling, and each sample was hybridized with the denatured radioactive HBV probe and the IS probe, respectively. The binding and hybridization occurred simultaneously on a microplate, and incubation at 55°C for 2 h with shaking was found optimal for quantitative results.

277

Figure 1 shows that the two amplified, biotinylated DNAs were hybridized and bound to microplates equally efficiently as required for a quantitative test.

The radioactive hybridization probes used in this study provided a direct means for the quantification of HBV and IS DNAs. The plasmid probes were used at as high a concentration as practicably possible, 2×10^{10} molecules (100 ng) per well; and to ensure that the probe was in excess, only a small aliquot (about 1/500) of the amplified samples was taken to hybridization. The sensitivity, quantitativeness and reproducibility of the hybridization assay for the two biotinylated PCR products were studied by binding known amounts of purified PCR product to a microplate. The DNA concentrations of these reference samples had been determined spectrophotometrically after the removal of free primers and nucleotides. The nick-translated ^{35}S-labeled probes with an identical, low specific activity of 10^7 cpm/µg easily detected 10^8 molecules of HBV and IS DNA (signal-to-noise in the range 4–15), and the log-log plot was essentially linear up to 10^{10} molecules (Figure 1). Labeling to an identical specific activity was reproducible, but the HBV probe always gave about a 2-fold higher signal than the IS probe. Presumably this is the difference in the specific activities of the hybridizing molecular species in the heterogeneous HBV and IS probe populations, respectively. The

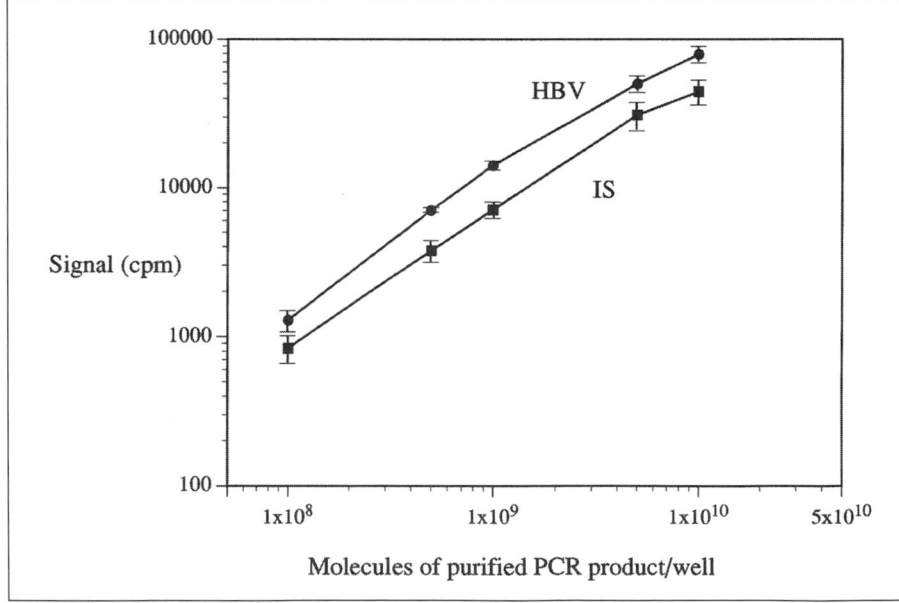

Figure 1. Correlation between the amount of target DNA applied to microplates and the signal obtained after detection by hybridization. The nick-translated probes had a constant specific activity of 10^7 cpm/µg, and 2×10^{10} molecules of probe were used per well. The number of biotinylated, purified DNA target molecules added per well was determined spectrophotometrically. The mean cpm values and standard deviations are from 5 parallel samples.

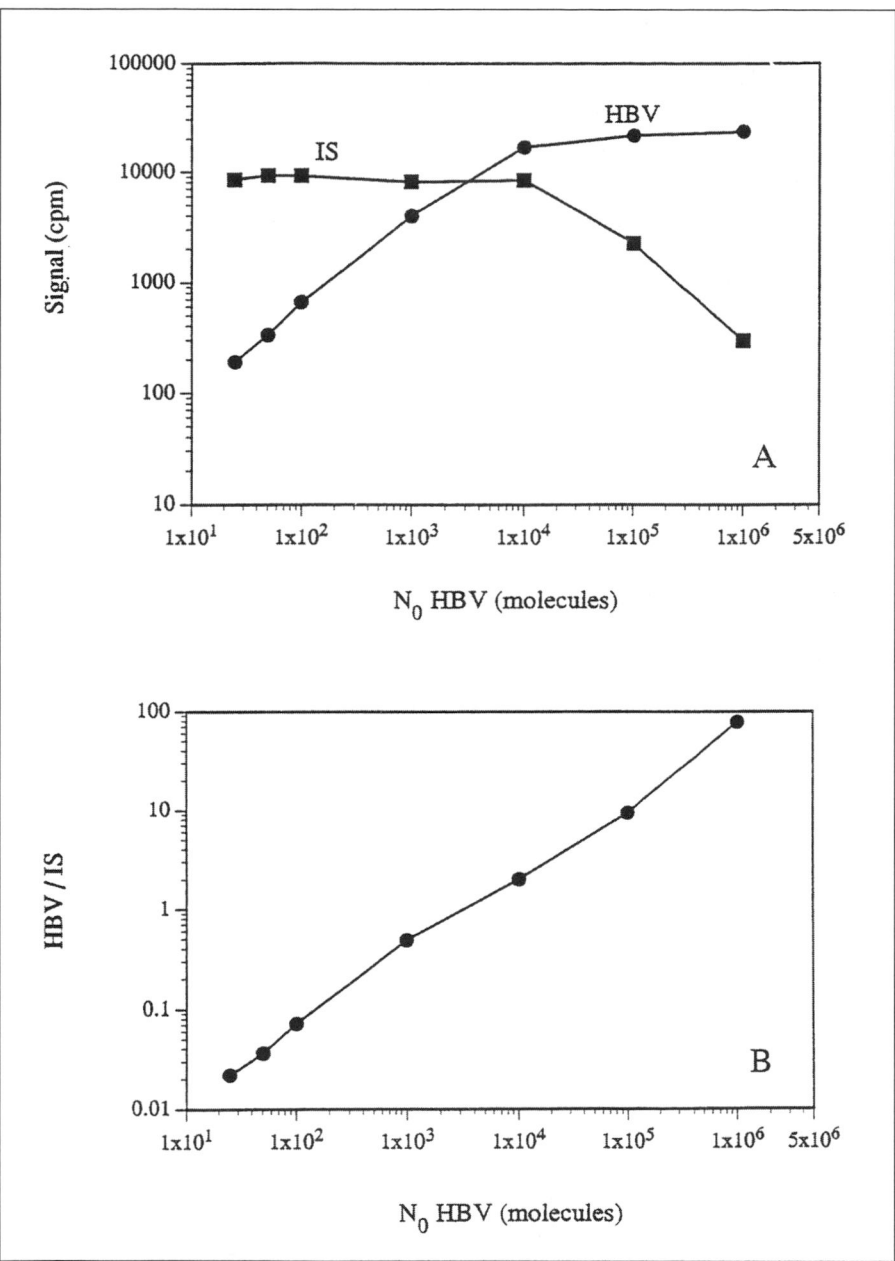

Figure 2. A) Co-amplification of HBV DNA and IS, both from plasmid clones. The initial number of HBV DNA plasmid molecules (N_0) was varied and IS was present in a constant amount of 10^4 molecules. PCR was run for 30 cycles and aliquots of the amplified product were analyzed by hybridization on microplates with HBV- and IS-specific probes, respectively. B) Standard curve for amplification of HBV DNA clone in the presence of 10^4 molecules of IS. The ratio of HBV- and IS-specific hybridization signals after co-amplification is shown as a function of the number of HBV plasmid molecules in the sample before amplification (N_0).

average CV (coefficient of variation) of the cpm signals obtained in Figure 1 was 14%.

Co-amplification of HBV and IS DNAs from plasmid clones is depicted in Figure 2, A and B, where increasing amounts, 25 to 10^6 molecules, of HBV DNA were amplified for 30 cycles together with 10^4 molecules of IS. There was a linear dependence on a log-log scale between the known amount of cloned HBV DNA before amplification and the HBV-specific signal after amplification in the range where the starting quantity N_0 was 25 to about 10^4 molecules. Competition with increasing amounts of cloned HBV DNA decreased IS amplification as expected. When the original amounts of HBV- and IS DNA-containing plasmids were equimolar, the signals obtained after amplification also showed equimolarity, i.e., the 2-fold difference in radioactivity mentioned above was obtained. The actual standard curve for the quantitative test is shown in Figure 2B. It shows the post-amplification ratio HBV/IS, which was determined from the hybridization signals and plotted against the initial number N_0 of HBV DNA plasmid molecules introduced into PCR.

Application to Clinical Samples

To validate the test for clinical samples, we serially diluted a patient serum of known HBV titer with normal human serum. The titer had been determined by the AffiProbe test developed in our laboratory, with the lowest detection limit of 5×10^5 molecules of HBV DNA (9). These serial dilutions

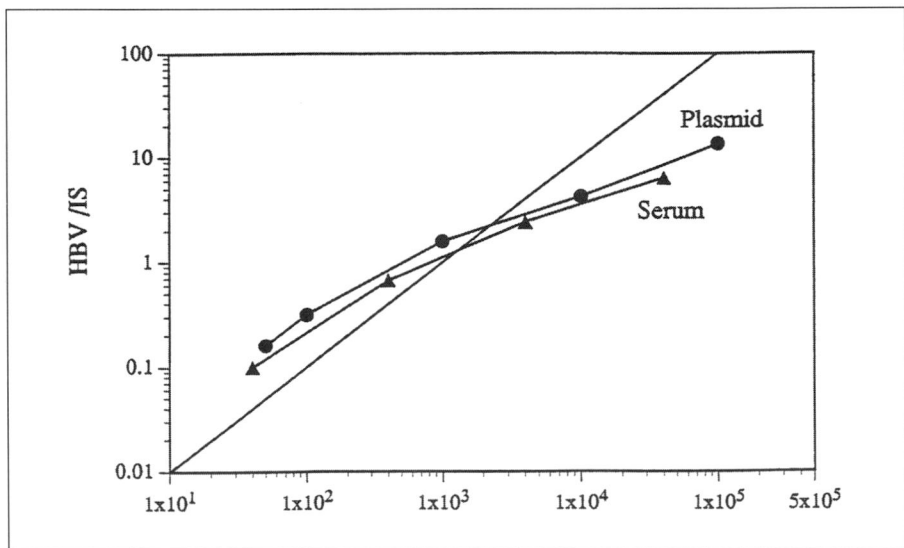

Figure 3. Amplification of DNA from a serial dilution of HB virus containing patient serum with normal human serum. The standard curve from amplification of HBV DNA plasmid clone is shown in comparison. The theoretical curve (straight line) assumes an identical amplification efficiency for HBV and IS DNAs. All samples were in this experiment co-amplified with 10^3 molecules of IS.

with known virus concentrations were boiled for 20 min to release the DNA, then centrifuged, and 10 µL of the supernatants were co-amplified with 10^3 molecules of IS. The amount of IS was lowered from that used in the previous experiment (Figure 2) to increase the sensitivity of detection for HBV. A standard curve representing co-amplification of varying known amounts of HBV DNA from plasmid clone and 10^3 molecules of IS was run in parallel. Figure 3 shows that the experimental standard curve obtained using the control HBV plasmid and the corresponding curve obtained from the dilution series of patient serum were nearly overlapping. This demonstrates that the present quantitative PCR test can be used for serum samples. Competitive amplification from an HBV plasmid clone provides an adequate standard for competitive amplification from the authentic HBV genome, even if it is clear that the two competing DNAs, HBV and IS, are not amplified with an identical efficiency (theoretical curve in Figure 3). Using this method, HBV concentrations lower than 40 viruses can be quantitatively determined from 10 µL of serum. An unknown concentration of HBV DNA in clinical samples can thus be determined by co-amplifying each pretreated sample with IS; this determines the post-amplification ratio HBV/IS for each sample. The final results are interpreted from the experimental standard curve (denoted as "plasmid" in Figure 3), which is run in parallel with the clinical samples. For example, the highest concentration of the clinical sample in Figure 3, containing 4×10^4 viruses per 10 µL serum (4×10^6/mL) on the basis of the reference AffiProbe test, had the HBV/IS ratio of about 5, which corresponds to the result of 2×10^4 viruses per 10 µL when interpreted from the plasmid standard curve. It should be mentioned that the sample pretreatment method used is critical for this quantitative test. The simple boiling method used here is not ideal for all sera, and we have since developed better methods for pretreating clinical samples for the quantitative PCR test (Uusi-Oukari et al., unpublished).

DISCUSSION

The competitive PCR test described here provides a sensitive method with a broad dynamic range for quantification of HBV particles from serum samples. This assay can tolerate samples that are not amplified perfectly due to partial inhibition. False negatives due to complete inhibition of PCR will not occur, since a lack of signal from IS as well as from HBV indicates the failure of the amplification. These properties are clear advantages of the present method. However, the possibility remains that some sequence variants of HBV DNA might exist that are not amplified properly with the primers used, although the primers were selected from a well-conserved region. In our recent version of the test, we have been using amplification primers containing inosin at some sequence positions in order to decrease this risk.

In principle the difference in sequence between target and IS DNA should be minimized to ensure a similar reannealing rate. However, here the entire amplified sequences between the primers were nonhomologous to ensure perfect discrimination in the hybridization step while still being capable of detecting different subtypes of HBV with equal efficiency. The differences in reannealing probably explain the deviation of the experimental standard curve from the theoretical curve, which assumes an identical amplification efficiency for HBV DNA and IS (Figure 3). Since both the standard curve derived from cloned HBV DNA and the curve obtained for the dilution series of virus-containing serum showed similar deviation from the theoretical curve, this was hardly significant for the quantification of HB virus from patient serum. Thus, although self-annealing remained a discriminating factor between the two amplified sequences, it did not appear to seriously disturb the quantitative assay in the conditions used.

The present test is different from the competitive PCR assays described previously, which typically include co-amplification of the sample with a range of different concentrations of competitor, with the drawback that multiple assays are required for titration of each sample (1,5,15–17). Another disadvantage of the titration methods is that the dynamic range and accuracy are mutually exclusive, unless an extremely large number of assays are performed with each sample. A large number of samples can be analyzed simultaneously in the assay described here. The broad dynamic range obtained is partly due to the use of radioactive tracers, but in our present developments, we obtained almost the same sensitivity and range with nonradioactive detection.

ACKNOWLEDGMENTS

The expert technical assistance of Ms. Auli Kähärä-Uppgård is gratefully acknowledged. We also wish to thank the Department of Virology, University of Helsinki, for providing the clinical samples and Ms. Raija Lahdensivu for the reference tests of HBV-positive samples. This work is a part of Eureka project EU 547.

REFERENCES

1. **Becker-André, M. and K. Hahlbrock.** 1989. Absolute mRNA quantification using the polymerase chain reaction (PCR). A novel approach by a PCR aided transcript titration assay (PATTY). Nucleic Acids Res. *17*:9437-9446.
2. **Bengtström, M., A. Jungell-Nortamo and A.-C. Syvänen.** 1990. Biotinylation of oligonucleotides using a water soluble biotin ester. Nucleosides Nucleotides *9*:123-127.
3. **Bichko, V., P. Pushko, D. Dreilina, P. Pumpen and E. Gren.** 1985. Subtype ayw variant of hepatitis B virus. DNA primary structure analysis. FEBS Lett. *185*:208-212.
4. **Chou, S.** 1991. Optimizing polymerase chain reaction technology for clinical diagnosis. Clin. Chem. *37*:1893-1894.
5. **Gilliland, G., S. Perrin, K. Blanchard and H.F. Bunn.** 1990. Analysis of cytokine mRNA and DNA:

detection and quantitation by competitive polymerase chain reaction. Proc. Natl. Acad. Sci. USA 87:2725-2729.

6.**Gupta, B.P., N. Jayasuryan and S. Jameel.** 1992. Direct detection of hepatitis B virus from dried blood spots by polymerase chain reaction amplification. J. Clin. Microbiol. 30:1913-1916.

7.**Harju, L., P. Jänne, A. Kallio, M.-L. Laukkanen, I. Lautenschlager, S. Mattinen, A. Ranki, M. Ranki, V.R.X. Soares, H. Söderlund and A.-C. Syvänen.** 1990. Affinity-based collection of amplified viral DNA: application to the detection of human immunodeficiency virus type 1, human cytomegalovirus and human papillomavirus type 16. Mol. Cell. Probes 4:223-235.

8.**Holodniy, M., M.A. Winters and T.C. Merigan.** 1992. Detection and quantification of gene amplification products by a nonisotopic automated system. BioTechniques 12:36-39.

9.**Jalava, T., M. Ranki, M. Bengtström, P. Pohjanpelto and A. Kallio.** 1992. A rapid and quantitative solution hybridization method for detection of HBV DNA in serum. J. Virol. Methods 36:171-180.

10.**Kaneko, S., S.M. Feinstone and R.H. Miller.** 1989. Rapid and sensitive method for the detection of serum hepatitis B virus DNA using the polymerase chain reaction technique. J. Clin. Microbiol. 27:1930-1933.

11.**Kaneko, S., R.H. Miller, S.M. Feinstone, M. Unoura, K. Kobayashi, N. Hattori and R.H. Purcell.** 1989. Detection of serum hepatitis B virus DNA in patients with chronic hepatitis using the polymerase chain reaction assay. Proc. Natl. Acad. Sci. USA 86:312-316.

12.**Keller, G.H., D.-P. Huang, W.-K. Shih and M.M. Manak.** 1990. Detection of hepatitis B virus DNA in serum by polymerase chain reaction amplification and microtiter sandwich hybridization. J. Clin. Microbiol. 28:1411-1416.

13.**Landgraf, A., B. Reckmann and A. Pingoud.** 1991. Quantitative analysis of polymerase chain reaction (PCR) products using primers labeled with biotin and a fluorescent dye. Anal. Biochem. 193:231-235.

14.**Liang, T.J., K.J. Isselbacher and J.R. Wands.** 1989. Rapid identification of low level hepatitis B-related viral genome in serum. J. Clin. Invest. 84:1367-1371.

15.**Lundeberg, J., J. Wahlberg and M. Uhlén.** 1991. Rapid colorimetric quantification of PCR- amplified DNA. BioTechniques 10:68-75.

16.**Menzo, S., P. Bagnarelli, M. Giacca, A. Manzin, P.E. Varaldo and M. Clementi.** 1992. Absolute quantitation of viremia in human immunodeficiency virus infection by competitive reverse transcription and polymerase chain reaction. J. Clin. Microbiol. 30:1752-1757.

17.**Porcher, C., M.-C. Malinge, C. Picat and B. Grandchamp.** 1992. A simplified method for determination of specific DNA or RNA copy number using quantitative PCR and an automatic DNA sequencer. BioTechniques 13:106-114.

18.**Syvänen, A.-C., M. Bengtström, J. Tenhunen and H. Söderlund.** 1988. Quantification of polymerase chain reaction products by affinity-based hybrid collection. Nucleic Acids Res. 16:11327-11338.

19.**Syvänen, A.-C. and H. Söderlund.** 1993. Quantification of polymerase chain reaction products by affinity-based collection. Methods Enzymol. 218:474-490.

20.**Wang, A.M., M.V. Doyle and D.F. Mark.** 1989. Quantitation of mRNA by the polymerase chain reaction. Proc. Natl. Acad. Sci. USA 86:9717-9721.

21.**Zeldis, J.B., J.H. Lee, D. Mamish, D.J. Finegold, R. Sircar, Q. Ling, P.J. Knudsen, I.K. Kuramoto and L.T. Mimms.** 1989. Direct method for detecting small quantities of hepatitis B virus DNA in serum and plasma using the polymerase chain reaction. J. Clin. Invest. 84:1503-1508.

Update to:

Quantification of Hepatitis B Virus DNA by Competetive Amplification and Hybridization on Microplates

Hans Söderlund

Technical Research Centre, VTT Biotechnology and Food Research, Espoo, Finland

The key element of the method described in the original paper is the use of an internal standard with flanking sequences identical to that of the target and an internal sequence deviating from that of the target for identification. Thus PCR amplification is carried out using the same primer pair for the two sequences. The two sequences should have essentially the same length for equal efficiency in amplification. Optimally the sequences should be similar with only enough differences for detection. When this principle is employed, the test is a robust and easy-to-use quantitative PCR assay. A number of improvements over the published method are easily adopted. Simplified sample treatments are available; improvements on PCR, including novel enzymes have been developed and a number of nonradioactive probes can be used. Repetitive PCR amplification of the same sequence, as is the case in diagnostic applications, increases the risk for false-positive results due to contamination. This problem has been addressed in a test kit adopting the above principle: the kit, Amplicor® HBV Monitor, distributed by Roche Diagnostics Systems (Branchburg, NJ, USA) includes deoxyuridine triphosphate (dUTP) instead of dTTP in the PCR mixture. Thus PCR products can be selectively destroyed by uracil-*N*-glycosylase without affecting sample DNA (1). This kit further utilizes dinitrophenyl (DNP)-labeled oligonucleotide probes that are detected colorimetrically with the aid of an anti-DNP-alkaline phosphatase conjugate. The linear range for quantitative measurements of HBV DNA using this kit is from 10 to 1 000 000 DNA copies per sample.

REFERENCE

1. **Longo, M.C., M.L.S. Beringer and J.L. Hartley.** 1990. Use of uracil DNA glycosylase to control carry-over contamination in polymerase chain reactions. Gene *93*:125-128.

CHAPTER 22

Quantification of HIV-1 Using Multiple Competitors in a Single-Tube Assay

T. Vener, M. Axelsson[1], J. Albert[1], M. Uhlén and J. Lundeberg

KTH, Royal Institute of Technology, and [1]Swedish Institute for Infectious Disease Control, Stockholm, Sweden

ABSTRACT

Methods for quantification of human immunodeficiency virus type 1 (HIV-1) based on competitive PCR and fragment analysis have been developed. Samples containing HIV-1 DNA and known amounts of three cloned competitors were co-amplified by PCR with semi-nested primers. The competitor DNAs contained the same long terminal repeat primer binding sequences as the wild-type DNA, but they are different in internal sequences and length. One of the inner primers was fluorescent-labeled to allow discrimination between the wild-type DNA and the three competitors by fragment analysis using a standard automated sequencer. A calibration curve using the peak area of the three competitors enabled accurate determination of target amount with minimal variations. The method presented here can be used for quantification of HIV-1 in clinical samples and will be useful for monitoring disease progression and treatment effects.

INTRODUCTION

The polymerase chain reaction (PCR) has rapidly become a widely used technique in molecular biology. An important application is to use the PCR for quantitative analysis of pathogens. However, large-scale routine quantification of clinical samples has been hampered by the amount of labor needed for accurate quantification (1–3,5,8). Here, we describe an alternative single-tube assay for viral load analysis of human immunodeficiency virus type 1 (HIV-1) in samples using multiple competitors in a semi-nested amplification.

MATERIALS AND METHODS

Sample Material

DNA from HIV-1$_{MN}$-infected peripheral blood mononuclear cells (PBMC) were diluted in crude cell lysates of uninfected PBMC to contain a various number of viral HIV-1 copies. PBMC were isolated by Ficoll-Paque®

(Reprinted from BioTechniques 21:248-255, 1996)

density centrifugation and lysed as previously described (10). Crude cell lysates were used directly for PCR amplification.

Construction of Competitor DNAs

The assembly of the three competitors was performed by an initial isolation of a 30-bp poly(A) stretch from vector pGEM4pA (kindly provided by Dr. L.G. Lundberg; Reference 9), which was inserted into the *Eco*RI site of pGEM®-4Z vector (Promega, Madison, WI, USA). The subsequently used linkers were assembled and ligated into vectors according to standard methods. The first competitor (No. 1, 89-bp fragment) was constructed by insertion of an upstream linker (28-mer duplex, 5′-AGCTTCAGCTGCT-TTTTGCCTGTACTGGGTCTCGCATG-3′ and 3′-AGTCGACGAAAAAC-GGACATGACCCAGAGC-5′) between *Hin*dIII/*Sph*I restriction sites and a downstream linker (25-mer duplex, 5′-TCAATAAAGCTTGCCT TGAG-TGCTTGAGCT-3′ and 3′-CATGGAGTTATTTCGA -ACGGAACTCAC-GAAC-5′) was inserted between *Kpn*I/*Sac*I restriction sites. The construction of competitor No. 2 (100-bp fragment) was performed by insertion of *Eco*RV linker (17-mer duplex, 5′-CTCCATGATGGATATCGTG-3′ and 3′-GTAC-GAGGTACTACCTATAGCACAGCT-5′) between *Sph*I/*Sal*I restriction sites of construct No. 1. The construction of competitor No. 3 (125-bp fragment) was generated by insertion of a *lac* operator linker (36-mer duplex, 5′-CTAGAAATTGTGAGCGGATAACAATTACGTAACTAGCAT CTG-3′ and 3′-TTTAACACTCGCCTATTGTTAATGCATTGATCGTGACCTAG-5′) between *Xba*I/*Bam*HI restriction sites of construct No. 1. The concentration and purity of generated competitor plasmid DNAs were determined according to Maniatis et al. (6) and by limiting dilution experiments using PCR with HIV-1 specific primers (see Limiting Dilution). The competitor plasmids were subsequently serially diluted in 10 mM Tris-HCl, pH 8.3 and 10 ng/µL yeast tRNA (Boehringer Mannheim, Mannheim, Germany).

Limiting Dilution

Determination of competitor and HIV-1MN cell lysates (both of stock and diluted lysates) concentrations were performed using limiting dilution and nested PCR as previously described (2). In short, the materials were diluted in fivefold steps, and at least ten PCR determinations were performed on each dilution. The DNA copy numbers were calculated by the Poisson distribution formula (2) (i.e., one starting DNA copy corresponds to a dilution step in which 63% of the samples are positive). The basis for the analysis is the ability of the nested PCR to reproducibly detect single HIV-1 molecules.

PCR

The 3′ long terminal repeat (LTR) PCR primer sequences were JA159, 5′-CAGCTGCTTTT TGCCTGTAC-3′ (outer, 432–452); JA160F, 5′-FITC-CTGCTTTTTGCCTGTACTGGGTCTC-3′ (inner, 435–460); JA161B 5′-

biotin-AAGCACTCAAGGCAAGCTTTATTGA-3′ (inner, 524–499); and JA162, 5′-AGCACTCAAGGCAAGCTTTA-3′ (outer, 528–508). Positions are given relative to MN (7) strain of HIV-1. Inner and outer PCRs were carried out in 50 μL containing 5 μL of 10-fold PCR buffer (100 mM Tris-HCl, pH 8.3 at 25°C, 500 mM KCl, 25 mM MgCl$_2$) and 200 μM of each dNTP with 5 pmol of each primer and 1 U *Taq* DNA Polymerase (Perkin-Elmer, Norwalk, CT, USA). The temperature profile consisted of denaturation, 94°C for 5 min linked with the cycle program: 95°C for 30 s; 50°C (outer) or 60°C (inner) for 30 s; 72°C for 30 s. Both outer and inner PCR used 32 cycles (GeneAmp® PCR System 9600; Perkin-Elmer).

Competitive PCR

For quantification by competitive PCR, the three competitors were premixed in different fixed ratios before amplification. Ten microliters of competitor mixtures were added together to the reaction tube with 10 μL of sample.

Immobilization of PCR Products and Fragment Analysis

The resulting PCR products were immobilized onto 40-mL (10 mg/mL) streptavidin-coated paramagnetic beads (Dynabeads®-M280; Dynal, Oslo, Norway) according to Hultman et al. (4). Eluted dye-labeled strands were neutralized and separated on an A.L.F.™ DNA Sequencer (Pharmacia Biotech, Uppsala, Sweden). Quantification and interpretations of the raw data output were facilitated by using Fragment Manager™ software (Pharmacia Biotech).

RESULTS

A method for the quantification of HIV-1 DNA using multiple competitors was developed (outlined in Figure 1). The method is based on the co-amplification of the sample DNA with multiple competitor DNAs that can be discriminated by length. The PCR primers anneal to the 3′ LTR region of HIV-1 and to equivalent LTR linker sequences in the competitor DNA. The resulting PCR products are analyzed by electrophoresis using the internal competitors to create a standard curve used to quantify the amount of target.

Quantification of Increasing Amounts HIV-1

In an initial experiment, various amounts of HIV-1$_{MN}$ DNA determined by limiting dilution (see Materials and Methods) were added to a premix containing different amounts of the three competitors (competitors Nos. 1–3; 3000, 50 and 400 copies, respectively). The PCR was performed in a seminested approach using non-labeled outer PCR primers and labeled inner primers (biotin and fluorescein, respectively). The corresponding length of inner PCR product varied over a short interval: competitors Nos. 1–3; 89, 100 and 125 bp, respectively. Amplification of the target, HIV-1 DNA, resulted in a product of 114 bp, thus within the range covered by the competitors. The rather narrow interval in length between the different products was deliberately chosen to minimize possible amplification differences while still enabling baseline separation. After amplification, the resulting products were purified by immobilization to streptavidin beads. After magnetic separation and a washing step, fluorescent-labeled strands were eluted by alkali, neutralized, denatured and loaded on the A.L.F. DNA sequencer. The quality of the raw data after electrophoresis is shown in Figure 2. Distinct peaks corresponding to the competitors and target were achieved with low background. Figure 2, Panels a–e, depict increasing amounts of added HIV-1 DNA (100–500 copies) to the competitor premix. The chromatograms clearly show the predicted increase in peak areas of the HIV-1 target (114 bp) throughout

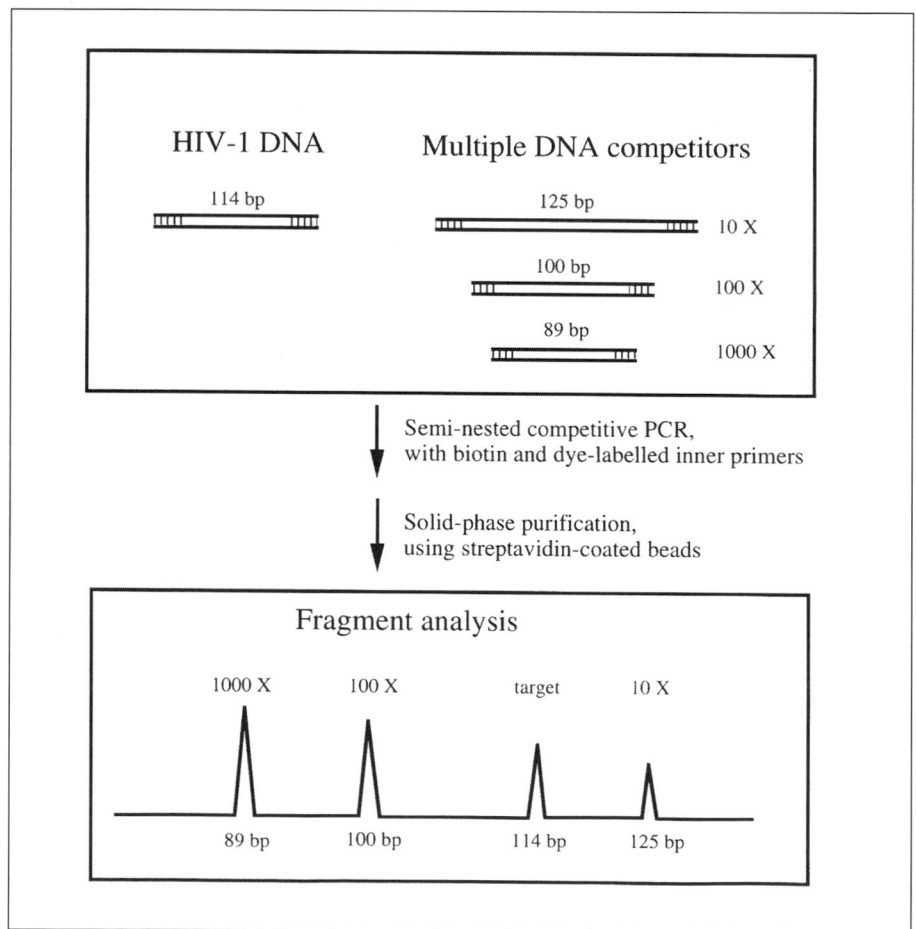

Figure 1. A schematic illustration of the competitive approach for HIV-1 quantification.

the series. Note that a similar peak area was observed in Figure 2, Panel d for HIV-1 target (400 copies added, according to limiting dilution) and peak area of competitor 3 (400 copies added to the competitor premix). Importantly, each of the individual experiments with multiple competitors (corresponding to Figure 2, Panels a–e) could be used to accurately estimate the amount of HIV-1$_{MN}$ using the peak areas of the competitors as an internal standard curve.

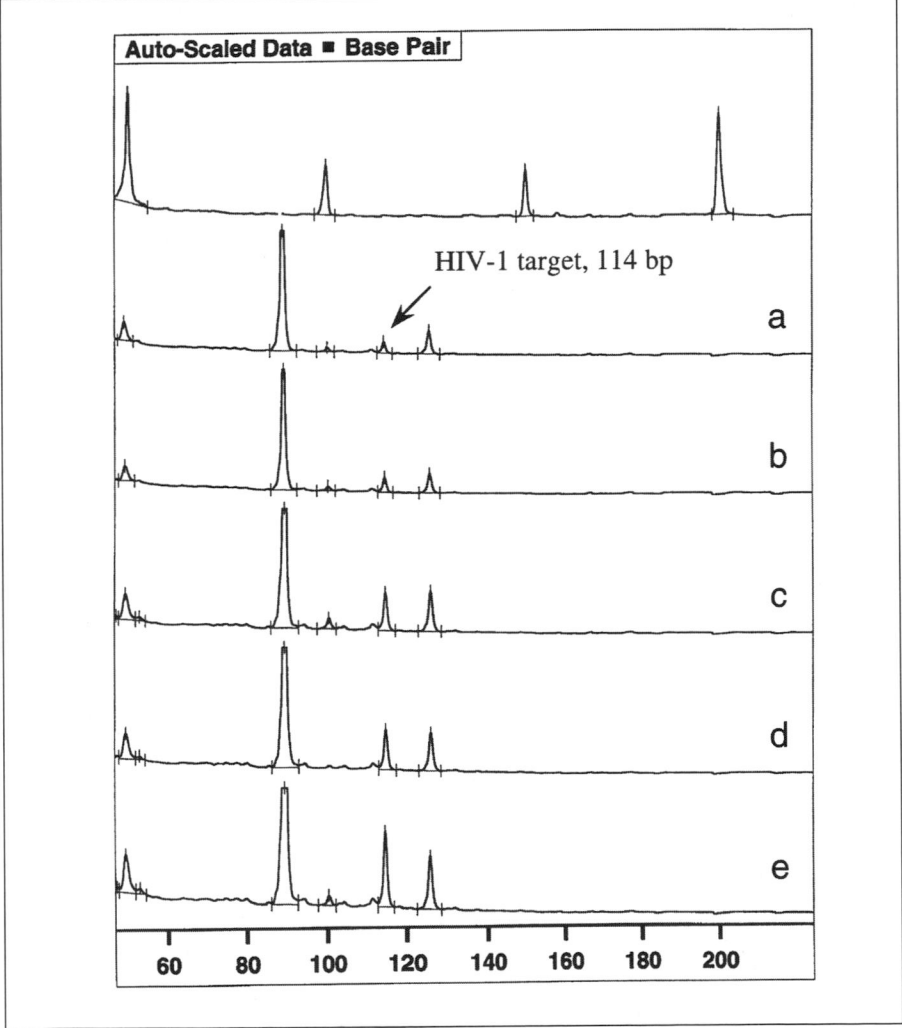

Figure 2. The fragment analysis results with increasing amount of HIV-1$_{MN}$ DNA target. Panels a–e corresponds to 100–500 copies of target (target size 114 bp). The configuration of the competitors: competitor No. 1: 89 bp, 3000 copies; competitor No. 2: 100 bp, 50 copies; competitor No. 3: 125 bp, 400 copies. Panel at the top shows the dye-marked ladder: 50, 100, 150 and 200 nucleotide (nt), and the corresponding length estimates are given below.

Quantification of HIV-1, Competitor Ratios and Sample Ratios

Different configurations of the amount of competitors were analyzed to investigate any amplification bias. Theoretically, the smallest fragment may amplify slightly better than the longest; therefore, a correction factor may be needed. Interestingly, no or little differences could be found by fragment analysis of various competitor configurations. One explanation could be the narrow size interval and relatively short amplicons. Figure 3, A–C, depicts the raw data for some different competitor configurations. Figure 3, Panels a and b show the same sample (500 HIV-1 copies) analyzed on separate occasions, and Figure 3, Panel c corresponds to a 2-fold higher concentration of HIV-1$_{MN}$ (1000 copies). The quality of the resulting chromatograms shows that a rough estimation of the number of targets can be attained by a simple analysis of the raw data, irrespective of competitor configuration. However, to achieve a more accurate determination of the number of HIV-1 targets, the peak areas of the internal standards are used to create a calibration curve. Figure 4 presents the data from Figure 3A, Panels a–c. A good linearity of the calibration curve was observed, and the independent experiments determined the number of targets to: 560 copies (Panel a), 550 copies (Panel b) and 950 copies (Panel c), which is in good agreement with the actual number of target copies added (determined by limiting dilution). Furthermore, the reproducibility was analyzed by repeated determination (13 times) of a known

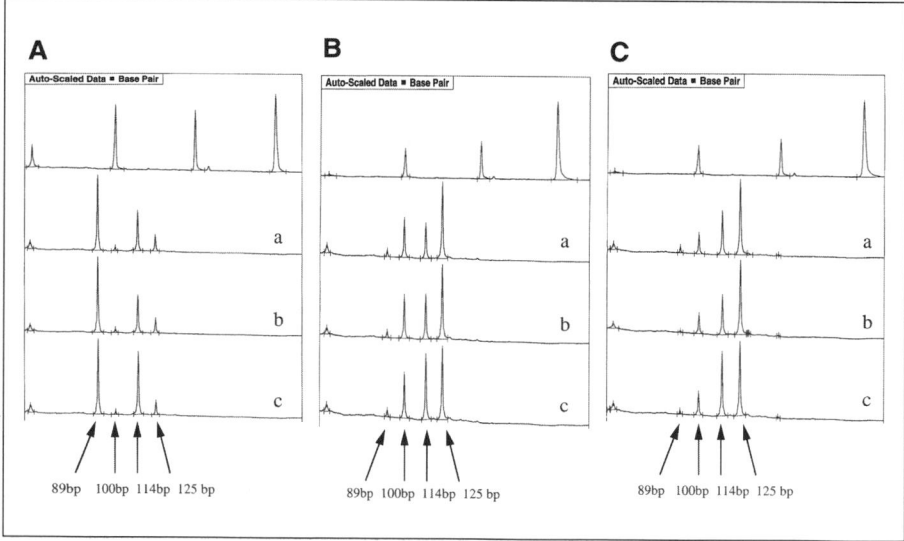

Figure 3. Raw data from fragment analysis with various competitor configurations and duplicated targets. Panel a corresponds to analysis of 500 copies of target (114 bp), Panel b corresponds to analysis of the same sample (500 copies of target) at another occasion, Panel c corresponds to analysis of 1000 copies of target. In the top panel, dye-labeled ladder: 50, 100, 150 and 200 nt. (A) configuration - competitors Nos. 1–3: 1000, 50 and 200 copies, respectively; (B) configuration - competitors Nos. 1–3: 80, 400 and 1000 copies, respectively; (C) configuration - competitors Nos. 1–3: 50, 200 and 1000 copies, respectively.

sample containing 100 HIV-1 copies resulting in an average value of 97 copies ($s = 9.4$).

DISCUSSION

In this study, we have presented a simple and reliable method for quantification of HIV-1 DNA with multiple internal competitors. The need to be in a narrow exponential phase of amplification for proper evaluation is avoided since the PCR was allowed to continue until the reaction is saturated, i.e., until the plateau is reached. Importantly, this ensures high sensitivity as "single-molecule" detection can be achieved, which is of great importance for precise quantification of HIV-1 DNA and RNA.

An important benefit of using this scheme for quantification is that it eliminates the need to use multiple analysis to assess the inflection point of the competitive titration curve that previously has been the dominating approach,

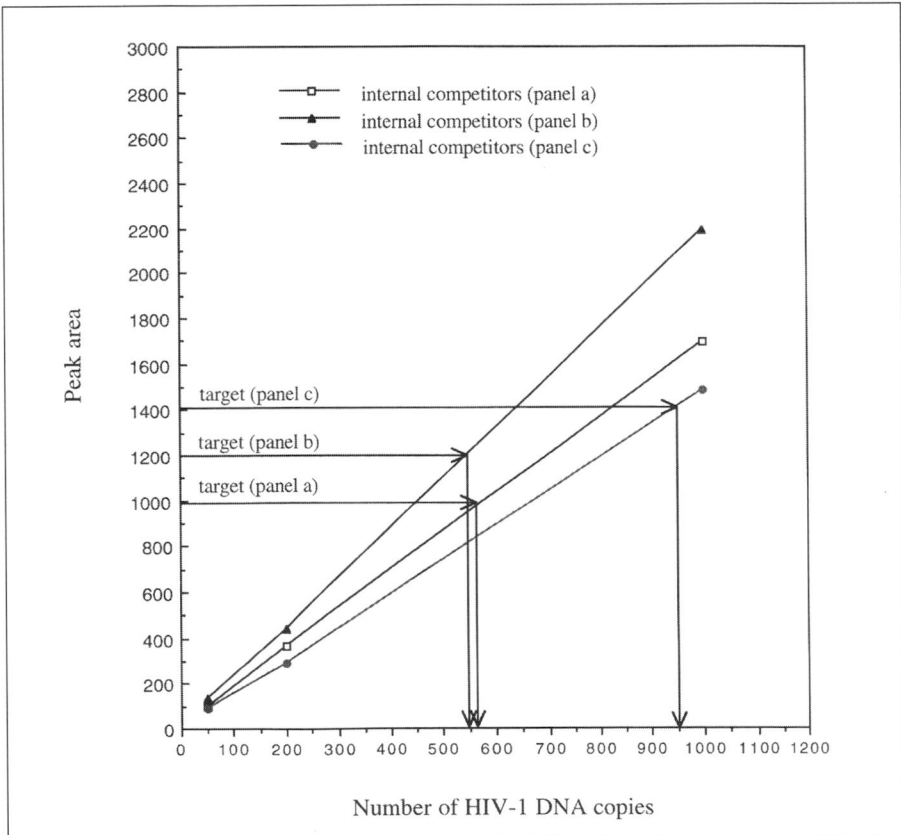

Figure 4. Estimation of amount of HIV-1 target. Experiments, corresponding to Panels a–c (Figure 3A), are interpreted using the calculated peak areas.

while still enabling the use of the same primers for amplification of the sample and the internal controls. Thus, such a single-tube analysis increases throughput drastically. Furthermore, a more accurate analysis or a wider dynamic range can be achieved by increasing the number of competitors. From our results, it appears that competitors consisting of a heterologous DNA fragment flanked by cloned linkers containing the primer annealing sites could be used, provided that the competitor lengths do not differ considerably from the native target sequence length.

For evaluation, we have used fragment analysis on automated laser fluorescent electrophoresis apparatus. The simple PCR product purification using streptavidin-coated solid support resulted in "clean" chromatograms with little noise from other spurious PCR fragments. In conclusion, we have developed an integrated method for quantification of HIV-1 nucleic acids in vivo. The one-tube assay makes the assay rapid and simple as well as suitable for automation. The methods described here will be useful for monitoring disease progression and the effect of treatments in HIV-1-infected individuals and can easily be extended to include RNA quantification.

ACKNOWLEDGMENTS

This work was supported by grants from the Göran Gustafsson Foundation and the Swedish Medical Research Council. We thank Deirdre O'Meara and Jacob Odeberg for critical comments.

REFERENCES

1. **Becker-André, M. and K. Hahlbrock.** 1989. Absolute mRNA quantification using the polymerase chain reaction (PCR). A novel approach by a PCR aided transcription titration assay (PATTY). Nucleic Acids Res. *17*:9437-9446.
2. **Brinchman, J.E., J. Albert and F. Vartdal.** 1991. Few infected CD4+ T cells but a high proportion of replication-competent provirus copies in asympotomatic human immunodeficiency virus type 1 infection. J. Virol. *65*:2019-2023.
3. **Gilliland, G., S. Perrin, K. Blanchard and H.F. Bunn.** 1990. Analysis of cytokine mRNA and DNA: detection and quantification by competitive polymerase chain reaction. Proc. Natl. Acad. Sci. USA *87*:2725-2729.
4. **Hultman, T., S. Ståhl, E. Hornes and M. Uhlén.** 1989. Direct solid phase sequencing of genomic and plasmid DNA using magnetic beads as solid support. Nucleic Acids Res. *17*:4937-4946.
5. **Lundeberg, J., J. Wahlberg and M. Uhlén.** 1991. Rapid colorimetric quantification of PCR-amplified DNA. BioTechniques *10*:68-75.
6. **Maniatis, T., E.F. Fritsch and J. Sambrook.** 1982. Molecular Cloning: A Laboratory Manual, Cold Spring Harbor Laboratory, Cold Spring Harbor, NY.
7. **Myers, G., B. Korber, J. Berkofsky, R.F. Smith and G.N. Pavlakis.** 1991. Human Retroviruses and AIDS 1991. Los Alamos National Laboratory, Los Alamos, New Mexico.
8. **Piatak, M. Jr., K-C. Luk, B. Williams and J.D. Lifson.** 1993. Quantitative competitive polymerase chain reaction for accurate quantitation of HIV DNA and RNA species. BioTechniques *14*:70-81.
9. **Stålbom, B.M., A. Torvén and L.G. Lundberg.** 1994. Application of capillary electrophoresis to the post-polymerase chain reaction analysis of rat mRNA for gastric H+K+-ATPase. Anal. Biochem. *217*:91-97.
10. **Wahlberg, J., J. Albert, J. Lundeberg, A. von Gegerfelt, K. Broliden, G. Utter, E-M. Fenyö and M. Uhlén.** 1991. Analysis of the V3 loop in neutralization-resistant human immunodeficiency virus type 1 variants by direct solid-phase DNA sequencing. AIDS Res. Hum. Retroviruses *7*:983-990.

Update to:

Quantification of HIV-1 Using Multiple Competitors in a Single-Tube Assay

Tatiana Vener, Mathias Uhlén and Joakim Lundeberg
Royal Institute of Technology, Stockholm, Sweden

NEW DEVELOPMENTS

The single-tube quantification procedure has been further refined by the introduction of a new competitor. Additional competitors will allow for a more reliable estimation of the amount of HIV-1 DNA by increasing the number of standards in the internal calibration curve. The fourth competitor was constructed simply by extending competitor No. 2 by ligation of a 36-mer linker. The resulting amplicon, using the primers described in the original paper, yields a fragment size of 136 bp. Hereby, the HIV-1 target (amplicon size of 114 bp) will be centered in the middle between the four competitors. Theoretically the new competitor can also be used to expand the dynamic range of the assay. However, in order to improve the dynamic range, one has to consider the instrument detection range. The available fluorescent-based instruments had been already used at their maximum linear range in the original report. The chromatographic truncated peaks typically depict values out of range due to saturation of the detector and cannot be used reliably in the construction of the internal standard curve. Thus the additional competitor will not add extra value in a single lane analysis. This problem can be partially circumvented by a dilution of the sample after PCR for parallel loading of two separate dilutions.

In Figure 1 we present the raw data for PCR quantification of HIV-1 DNA using a fourth competitor. Here we show the use of an extra competitor for more exact quantification as well as the use of a dilution step of the PCR products to achieve an improved dynamic range. Figure 1A, Panel I, depicts a 2-fold configuration of competitors Nos. 1–4 corresponding to 50, 100, 200 and 400 copies, respectively, and an HIV-1 DNA target corresponding to 150 starting copies. Here competitor No. 4 cannot be used in the construction of a standard curve due to the truncated peak. However, by a 5-fold dilution of the same PCR product (panel II), it can be reliably applied for estimation of the target. Nevertheless, both panels can be used for a direct determination of HIV-1 DNA quantities by comparing with the competitor peak areas. Figure

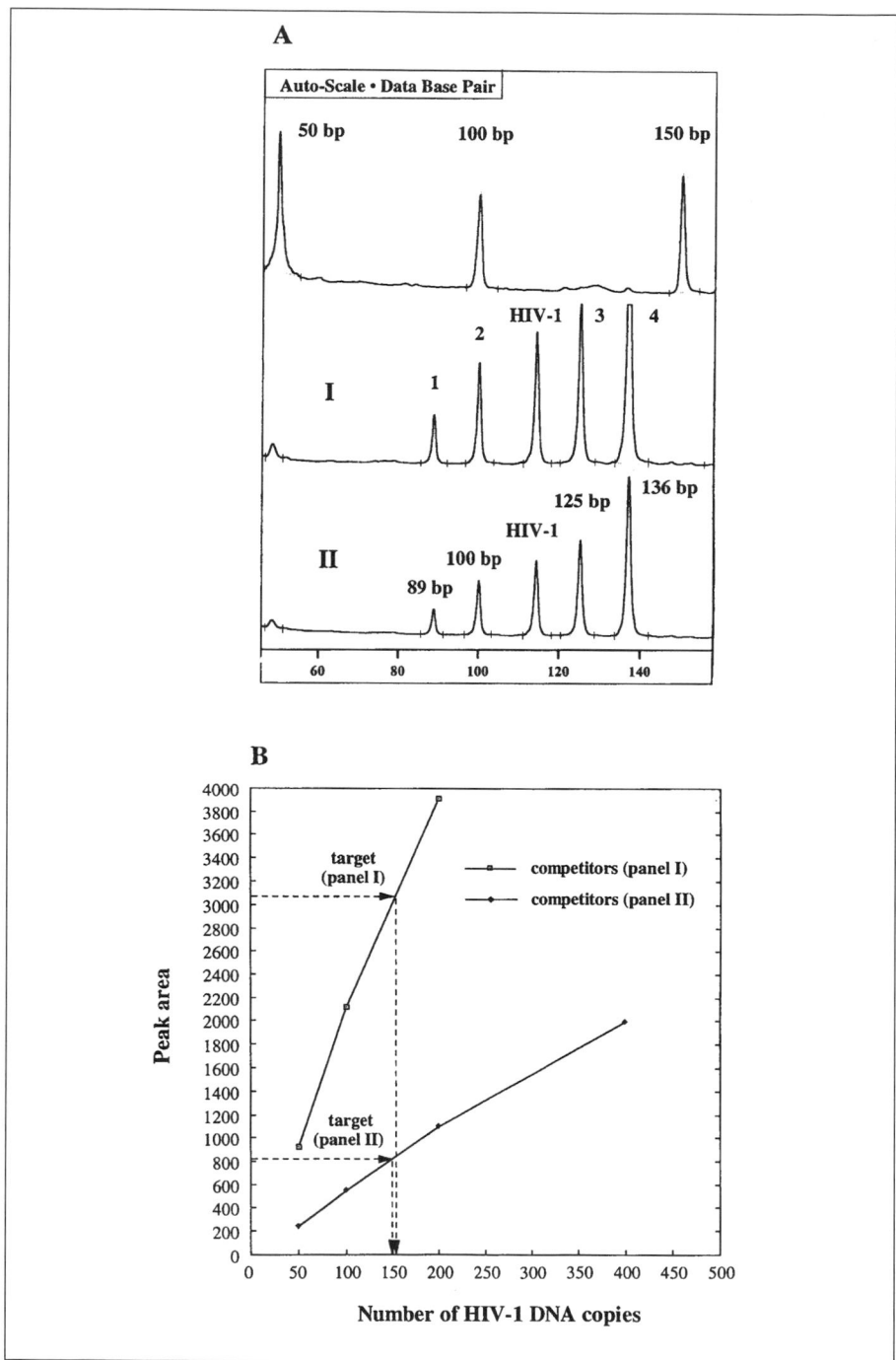

Figure 1. (A) Fragment analysis of competitive PCR. (B) Calibration curves for target quantification.

1B presents the corresponding calibration curves derived from the calculated peak areas of the four competitors facilitating the quantification of HIV-1 target DNA. A good linearity is observed and the independent experiments determined the number of HIV-1 DNA copies to approximately 150, which is a good agreement with the actual 150 target copies added (determined by limiting dilution).

Furthermore, the DNA-based quantification approach has now been successfully converted into a RNA quantification strategy for HIV-1 RNA analysis. For these purposes competitive RT-PCR with four in vitro transcribed RNA competitors has been established, which will improve the monitoring of disease progression and antiviral treatment effects in infected individuals.

CHAPTER 23

Detection and Quantitation of Low Numbers of Chromosomes Containing *bcl*-2 Oncogene Translocations Using Semi-Nested PCR

X-Y. Zhang and M. Ehrlich

Tulane Medical School, New Orleans, LA, USA

ABSTRACT

A PCR assay has been optimized for the detection of one or a few molecules of a translocation-containing human DNA sequence in the presence of a vast excess (7 μg) of the normal human genome. This procedure avoids blot hybridization by the use of two rounds of PCR with 20–22 cycles of amplification per round and by the replacement of one of the two primers from the first round of PCR with a different primer in the second round (semi-nested PCR). We demonstrate that very low numbers of the target DNA molecules can be quantitated by this semi-nested PCR. This method can be used to detect a single DNA molecule from one mutant cell displaying a translocation between the bcl-2 *proto-oncogene region and a J$_H$ immunoglobulin gene sequence [t(14;18)] in a background of normal human DNA from 10^{6}cells.*

INTRODUCTION

As little as a single DNA molecule can be detected by the polymerase chain reaction (PCR) followed by blot hybridization (15). Generally, when such exquisite sensitivity is achieved, the amount of total DNA present, including nonspecific DNA sequences, is not very high. A great excess of such nonspecific DNA can interfere with the amplification of the targeted DNA sequence and the detection of the specific amplification product. If a very large amount of extraneous DNA is present during such single-molecule detection by PCR, blot hybridization is usually employed to visualize only the desired PCR product (7,19,21). By a procedure more amenable to routine analysis than one involving blot hybridization, we wanted to obtain this same level of sensitivity in the presence of approximately 10^6 genome equivalents of human DNA not containing the exponentially amplifiable sequence of

(Reprinted from BioTechniques 16:502-507, 1994)

interest. In this report we describe such a method involving the use of 22 cycles of amplification with one set of primers for the target sequence (first round of PCR) followed by another 20 cycles (second round of PCR) in which one of the original primers is replaced by a radiolabeled primer complementary to part of the amplified sequence (semi-nested PCR). With this procedure, which should be of use in many types of amplification reactions, we were able to quantitate the number of target molecules in the range of 1 to 15 amplifiable molecules in the original sample.

The targeted DNA in this study is the junction fragment containing the translocated *bcl-2* oncogene region from human chromosome 18 and J_H DNA sequences from the immunoglobulin DNA portion of chromosome 14. This translocation, t(14;18), is observed in a majority of follicular lymphomas (12,23). There are two highly preferred sites (hot spots) for the *bcl-2*/J_H translocation (17,22) in the 3' region of the *bcl-2* proto-oncogene and hot spots in the six J_H segments of the immunoglobulin heavy-chain locus (14). Because of these hot spots, it has been possible to establish PCR assays that can detect a large percentage of these translocations (7,9,11,14,17,20). In this study we have substituted semi-nested PCR for PCR followed by blot hybridization to amplify and visualize this translocation, and we have optimized such amplification reactions and used them for quantitation.

MATERIALS AND METHODS

Materials

PCR was carried out in a temperature cycler (Ericomp, San Diego, CA, USA) with primers synthesized on an Applied Biosystems synthesizer (Foster City, CA, USA) and purified by electrophoresis on a 20% polyacrylamide gel before use. For the first round of PCR, we synthesized primer 1, 5'-AGAAGTGACATCTTCAGCAAATAAAC-3', from the sense strand upstream of the *bcl-2* gene's major breakpoint region (MBR; a translocation hot spot of ca. 150 bp within the 3' untranslated portion of the gene) and primer J_H, 5'-ACCTGAGGAGACGGTGACC-3', a consensus human J_H DNA sequence from the antisense strand of the J_H region of t(14;18) chromosomes (5,17,22). For the second round of PCR, we used primer J_H and radiolabeled primer 3, 5'-ACATTGATGGAATAACTCTGTGG-3', from the sense strand upstream of the *bcl-2* gene's mbr; this sequence is located 67 bp downstream of primer 1. Primer 3 (5 pmol) was labeled with [γ-^{32}P]ATP (80 µCi; 3000 Ci/mmol) in a reaction catalyzed by T4 polynucleotide kinase (20 units) and then mixed with approximately 300-fold excess of unlabeled primer 3 to a final specific activity of 1000–3000 cpm/pmol. The reactions were catalyzed by *Taq* DNA polymerase (Promega, Madison, WI, USA) or the AmpliTaq® DNA Polymerase Stoffel fragment (Perkin-Elmer, Norwalk, CT, USA). The intact *Taq* DNA polymerase, from batch to batch, gave consistent single-molecule detection in the presence of a vast excess of nonspecific DNA, unlike the latter enzyme, despite the intact polymerase having some 5' to 3' exonuclease activity (13). The targeted template for PCR was a t(14;18) *bcl-2*/J_H translocation at the mbr locus in DNA from a human B-lymphoma cell line, SU-DHL-4 (Oncogene Sciences, Uniondale, NY, USA). The concentration of the commercial SU-DHL-4 DNA stock solution was confirmed by agarose gel electrophoresis of multiple samples against sized-matched DNA standards upon visualization by fluorescence induced by EtdBr. The background DNA that was added to the PCR mixtures to mimic a human DNA sample containing only a few copies of the translocation product in a high background of normal human DNA was normal cerebellum DNA prepared as described (8). Contamination of the PCR mixtures was prevented by handling PCR products and reaction ingredients in separate rooms with dedicated pipetting devices and reagents.

Reaction Conditions for the First Round (22–25 Cycles) of PCR

To minimize mis-primed DNA amplification, the Hot Start PCR method (2) employing a paraffin wax (AmpliWax™ PCR Gem; Perkin-Elmer) was used in the first round of PCR. The 100 µL reaction mixtures consisted of different numbers of copies of SU-DHL-4 DNA containing the target translocation; 2.5 units of *Taq* DNA polymerase or 10 units of AmpliTaq DNA Polymerase, Stoffel Fragment; dATP, dGTP, dCTP, dTTP (200 µM each); 2.5 mM $MgCl_2$; 20 pmol each of primers 1 and J_H; 7 µg of normal human brain DNA; and 10 mM Tris-HCl, pH 9.0, 0.1% Triton® X-100, 50 mM KCl (for the Promega enzyme) or 12.5 mM Tris-HCl, pH 8.3, 12.5 mM KCl (for the Stoffel fragment). The initial denaturation step was at 95°C for 5 min followed by the indicated number of cycles of 1 min of annealing at 60°C, 1 min of primer extension at 72°C, and 1 min of denaturation at 94°C. In the last cycle, the primer extension time was 10 min.

Reaction Conditions for the Second Round (20 Cycles) of PCR

For the second round of PCR, 5 µL of the undiluted product obtained from the first round of PCR were used in a reaction mixture that was the same as that for the first round except that radiolabeled primer 3 (ca. 2000 cpm/pmol) replaced unlabeled primer 1 and the total reaction volume was 75 µL. Given the much lower sequence complexity of this reaction mixture, to economize, Ampliwax was not used in the second round of PCR, but rather, the mixture was overlaid with two drops of mineral oil and kept on ice until after the 5-µL sample from the first-round PCR product was added through the oil layer to the aqueous layer. The DNA was then subjected to an initial denaturation for 5 min at 95°C followed by 20 cycles of 1 min at 60°C, 1 min at 72°C, and 1 min at 94°C except that the final primer extension was for 10 min at 72°C. The products from the second round of PCR were analyzed by electrophoresis on

299

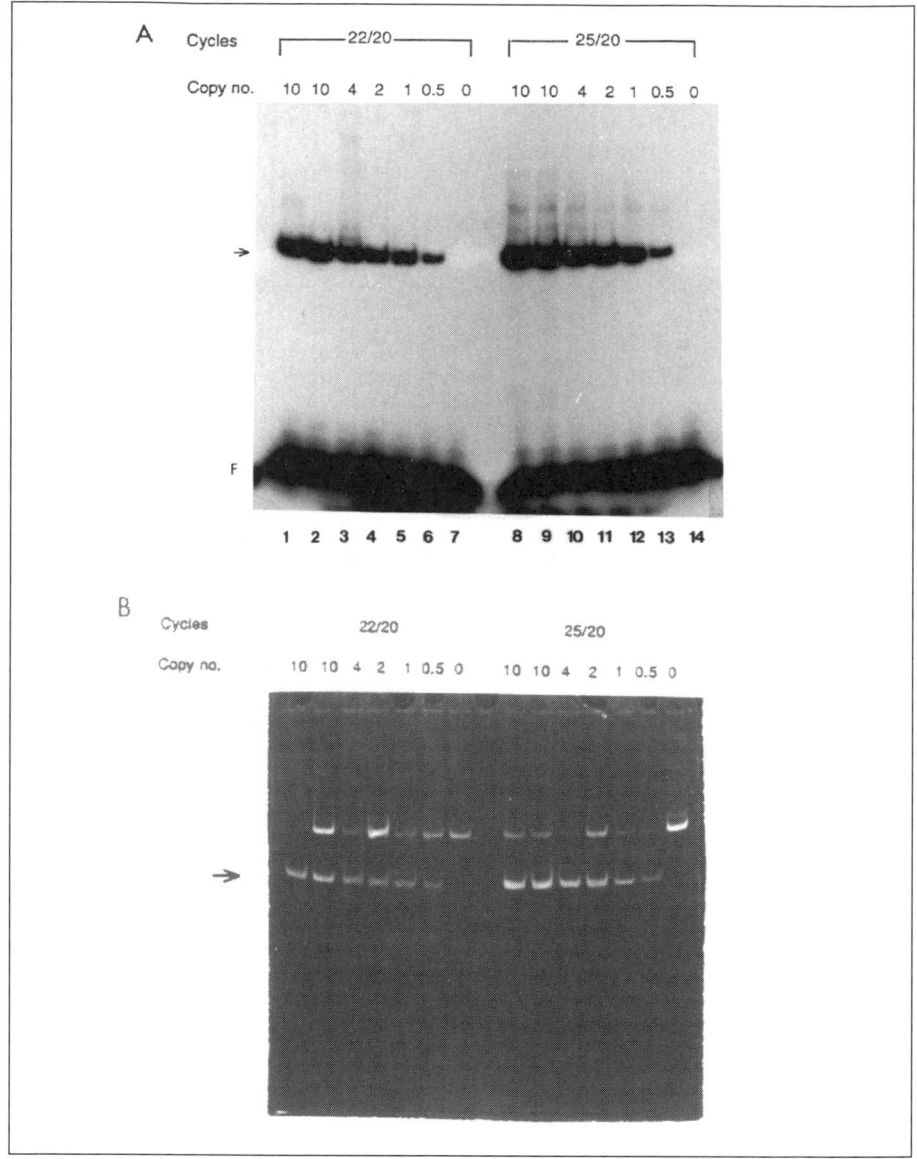

Figure 1. Optimized detection of the amplification of very low numbers of target molecules in the presence of a vast excess of background DNA by semi-nested PCR. The copy number refers to the expected average number of copies of the bcl-2/J_H translocation target molecule that would be present in replicate samples before PCR. Samples containing the indicated copy number of target molecules mixed with 7 μg of normal human brain DNA were subjected to 22 (lanes 1–7) or 25 (lanes 8–14) cycles of amplification with primers 1 and J_H, and then a 5-μL aliquot of this first-round PCR product was amplified for 20 cycles in a second round of PCR with primer J_H and ^{32}P-labeled primer 3. Samples with a copy number of 0.5 will probably have 1 or no target molecules. A) Gel electrophoresis of the untreated products from the second round of PCR followed by autoradiography upon 1 day of exposure of the electrophoresis gel to x-ray film. Free ^{32}P-labeled primer 3 ("F") is indicated. B) EtdBr-induced fluorescence of the gel in A. Arrow indicates the position of the specific amplification product of about 330 bp.

Table 1. Analysis of the Probability of Obtaining Specific Amplification of Very Low Numbers of the *bcl-2* Translocation-Containing Chromosomes by Semi-Nested PCR

	No. of Positive Samples/No. of Tested Samples				
	Average Copy No. of the *bcl-2* Translocation in Tested Samples				
	8	4	2	1	0
Data obtained	10/10 1.0	28/29 0.97	27/31 0.98	26/37 0.70	0/30 0
Data predicted	0.9997	0.982	0.865	0.632	0

"Data obtained" refers to the number of positive samples after semi-nested PCR/the number of tested samples with the given average number of copies of the *bcl-2* translocation. All samples contained a background of 7 μg of normal human DNA (from approximately 10^6 cells) and were subjected to two rounds of PCR as described in Materials and Methods. The products were analyzed by auto-radiography. "Data predicted" is from the Poisson distribution used to predict the number of successful amplifications starting with different average low numbers of *bcl-2* translocation-containing molecules per PCR mixture: $1-e^{-n}$, where n is the average number of the *bcl-2* translocation-containing molecules per reaction mixture. According to the Chi Square test of significance, the observed outcomes were not significantly different from those predicted by the Poisson model (e.g., for the samples with an average copy number of 1 *bcl-2* translocation, $\chi^2 = 0.29$ and $P = 0.59$).

a 5% polyacrylamide gel followed by autoradiography and, where indicated, EtdBr-induced fluorescence. For quantitative PCR, the specific DNA fragments (ca. 330-bp and, when observed, 830-bp products) were excised from the gel and their Cerenkov radiation (i.e., light emission in the absence of added scintillants) was measured in a liquid scintillation spectrometer.

RESULTS AND DISCUSSION

We optimized conditions for PCR detection of a few copies of a DNA containing a chromosome 14/18 translocation at the *bcl-2* proto-oncogene in a background of DNA from approximately 10^6 normal human cells (7 μg). The source of the translocation-containing DNA was SU-DHL-4 cells in which this t(14;18) rearrangement occurred between the translocation-hot spot mbr locus in the 3′-untranslated region of the *bcl-2* proto-oncogene and the fourth of six related J_H regions at the heavy-chain immunoglobulin locus (4,10). Normal human DNA should give no detectable specific amplification products because the primer-complementary sites in the genomic DNA are on 60°–64°C were optimal for the maximum sensitivity and specificity, we tried to detect only a few copies of the specific translocation product after 40 cycles of PCR with a single pair of primers (primers 1 from chromosome 18 and J_H from chromosome 14). Such primers can amplify a high percentage of naturally occurring, lymphoma-associated translocations (1,3,6,22). We could detect, by EtdBr-induced fluorescence, a specific amplification product of the expected size (4), approximately 400 bp, from samples with only two copies of a t(14;18) chromosome in a background of 7 μg of normal human DNA (data not shown). After 30 cycles of PCR with the same J_H primer and a different mbr-specific primer using Southern blotting to visualize the specific products, Negrin et al. (16) obtained a similar specific DNA fragment from as little as one copy of SU-DHL-4 DNA in a background of DNA from 10^5 normal cells. However, when we visualized the PCR products by EtdBr-induced fluorescence, the large amount of background normal human DNA in the samples gave rise to about six bands of nonspecific amplification products (data not shown).

It was desirable to eliminate the background of nonspecific amplification bands so that the procedure could be used to detect newly arising t(14;18) amplification products whose exact size could not be predicted. We first tried

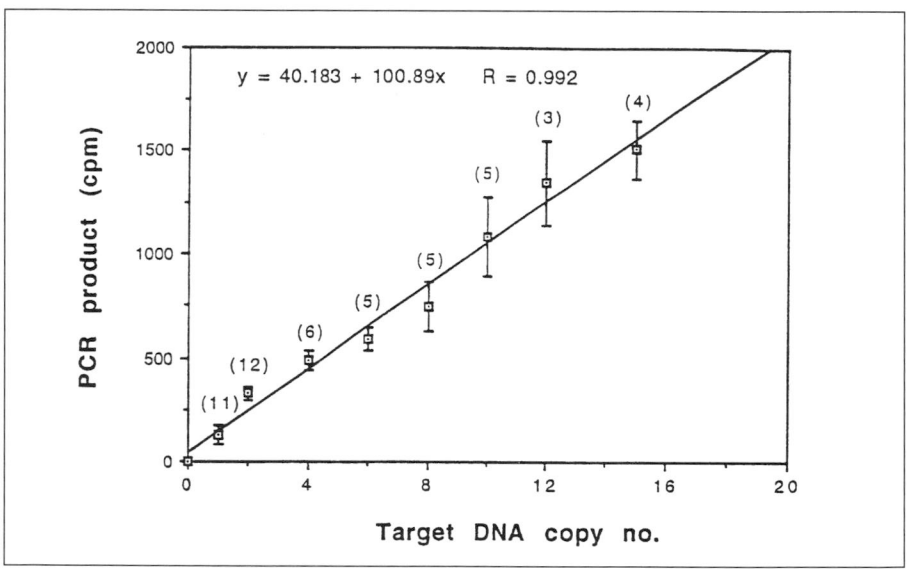

Figure 2. Linear relationship between the amount of ^{32}P-labeled PCR products and the average number of *bcl-2*/J_H translocation target molecules. Two rounds of PCR (22 and 20 cycles) were performed on the indicated average number of *bcl-2*/J_H translocation target molecules mixed with 7 μg of normal human DNA using semi-nested primers as in Figure 1. For each tested average copy number of the target molecules in the initial reaction mixture (including samples with no copies of the translocation product, the "0" data point), 3–12 independent PCR amplifications were performed and the average radioactivity in the specific PCR products (in the ca. 330-bp band and in the ca. 830-bp band, when the latter translocation product was also present) is shown. In parentheses above the linear regression curve are the numbers of independent PCR amplifications used for each of the data points. The bar presents the standard error. The correlation coefficient, R, is 0.992.

Table 2. Effect of Varying the Extent of Dilution of the Products from the First Round of PCR on the Yield of Specific Amplification Products from the Second Round of PCR

Copy in 1st PCR	Dilution Factor in 2nd PCR	Relative cpm Obtained (%)	Predicted Relative cpm (%)
10	1:10	100	100
10	1:100	12.3 12.6	10
10	1:1000	1.2	1
2	1:10	20.9 25.4	2
1	1:10	5.8 8.3	10
1	1:100	0.7	1

Samples that should have an average of 10, 2 or 1 *bcl-2*/J_H translocation target molecules were mixed with 7 µg of normal human brain DNA and amplified for 22 cycles in the first round of PCR. Five microliters of the indicated serial dilutions of this mixture were amplified for 20 cycles of semi-nested PCR as described in Materials and Methods and the second-round PCR products analyzed by gel electrophoresis, autoradiography and quantitation of radioactivity in specific bands. Radioactivity (cpm) in the specific approximately 330-bp product (and, where observed, also in the specific secondary 830-bp product) in a given reaction is expressed as a percent relative to that in the reaction shown in the first line of the table. Where indicated, data from duplicate experiments are given. Predicted relative cpm is expected ratio (expressed as a percent) of cpm for a given sample relative to that from the reaction shown in the first line of the table.

to do this by visualizing amplified DNA by the "oligo extension" assay (18) in which an aliquot of the reaction products after 30–45 cycles of PCR is diluted and subjected to only one cycle of extension of a single primer, namely, the nested, radiolabeled primer 3. The detection limit was approximately 1 molecule of the t(14;18) chromosome; however, the number and electrophoretic mobility of the specific products were unpredictable (data not shown). This could be due to the presence of many single-stranded structures in the products after only a single cycle of extension of a nested primer.

The most reproducible, specific and sensitive method for the detection of only a few molecules of the target DNA in a background of 7 µg of nontarget DNA was semi-nested PCR for two rounds using 22 cycles of amplification with unlabeled primers 1 and J_H in the first round and then 20 cycles with unlabeled primer J_H and labeled primer 3 in the second round (semi-nested PCR; Figure 1A). Fully nested PCRs, in which both primers in the second round are changed to match different positions within the amplified region, should have been at least as good, but it was not possible for the general detection of *bcl-2*/J_H translocations because there was only a very short region of high homology between all six J_H gene segments and, therefore, only one ideal consensus J_H primer was available (7). The main specific radiolabeled product obtained from these semi-nested reactions containing as little as one target DNA molecule was approximately a 330-bp fragment clearly visible after only a single day of autoradiographic exposure (Figure 1A). Sometimes a very low-intensity secondary product of approximately 830 bp, apparently from priming at the J5 rather than the nearer J4 sequence on SU-DHL-4 DNA, was seen (data not shown) as reported by Negrin et al. (16). No radiolabeled products were ever observed in the samples that contained only the background normal human DNA (Figure 1A and the last column of Table 1).

There was a very small amount of product electrophoresing as a smear above the major product band when 20– 25 cycles were used per round of semi-nested PCR (Figure 1A) and much more when 30 or more cycles were used for the first round of PCR (data not shown). This smear was largely eliminated when the PCR products were treated with 10 U/mL of S1 nuclease for 8 min at pH 5.0 (data not shown). When the PCR products were visualized by EtdBr-induced fluorescence, in addition to the specific approximately 330-bp translocation junction fragment, a major nonspecific product of about 580 bp was generally observed, which was also found in samples that contained only normal human DNA (Figure 1B). This amplification product apparently derives from mispriming by just the J_H primer. We observed it in EtdBr-stained gels even when the J_H primer was used with primers for the secondary *bcl-2* translocation hot spot, the mcr locus (6).

Using the above optimal conditions for semi-nested PCR, we determined the relative amounts of radioactivity in the specific radiolabeled PCR products when an average of 1 to 15 target DNA molecules were present in the samples. Each data point shown in Figure 2 was the average of the results from 3 to 12 independent PCR amplifications. Such replicate determinations were especially important in view of stochastic fluctuation at these extremely low copy numbers. The average amount of specific PCR product was nearly linear in the range of about 1 to 15 target molecules in the starting reaction. The proportionality of the average initial copy number of target molecules to the amount of specific product was further examined by a limiting dilution analysis. We compared the intensity of the approximate 330-bp band after the second round of PCR when the products from the first round of PCR were subjected to serial dilutions before the second round of PCR amplifications. The results were close to those predicted (Table 2).

By pooling all of our data for semi-nested PCR of samples containing an average of 1 to 15 target molecules of the t(14;18) translocation, we determined whether the probability of detecting the specific radiolabeled product followed the Poisson distribution (Table 1). The close correspondence between the predicted and empirical results further validate this method for detection of a single copy of the target DNA molecule in the presence of a vast excess (7 μg) of background human DNA not containing the specific translocation.

CONCLUSION

Semi-nested PCR consisting of 22 cycles with two unlabeled primers followed by 20 cycles using an aliquot of the reaction mixture and one unlabeled and one nested labeled primer allowed routine detection of a single translocation-containing molecule of human DNA in the presence of DNA from approximately 10^6 normal human cells. This methodology, which required neither purification of the DNA between the two rounds of PCR nor blot hybridization, also allowed quantitation of very low numbers of the translocation-containing DNA molecules in the presence of a vast excess of background DNA. This type of procedure could be used for routine analyses of induced chromosomal mutations, detection of very low backround levels of chromosomal mutations in disease-free individuals, studies of the persistence of translocation-containing cancer cells in a population of mostly normal cells or the detection of extremely low numbers of viral or other pathogenic DNA molecules in a high background of normal human cells.

ACKNOWLEDGMENTS

We thank Dr. John J. Lefante for his generous help with the statistical analysis. This work was supported in part by grants from the U.S. Army Medical Research and Development Command (91016019), National Institutes of Health (ESO5946) and a grant from the Department of Defense through the Center for Bioenvironmental Studies at Tulane University.

REFERENCES

1. Bakhshi, A., J.J. Wright, W. Graninger, M. Seto, J. Owens, J. Cossman, J.P. Jensen, P. Goldman and S.J. Korsmeyer. 1987. Mechanism of the t(14;18) chromosomal translocation: structural analysis of both derivative 14 and 18 reciprocal partners. Proc. Natl. Acad. Sci. USA. 84:2396-2400.
2. Chou, Q., M. Russell, D.E. Birch, J. Raymond and W. Bloch. 1992. Prevention of pre-PCR mispriming and primer dimerization improves low-copy-number amplifications. Nucleic Acids Res. 20:1717-1723.
3. Cleary, M.L. and J. Sklar. 1985. Nucleotide sequence of a t(14;18) chromosomal breakpoint in follicular lymphoma and demonstration of a breakpoint-cluster region near a transcriptionally active locus on chromosome 18. Proc. Natl. Acad. Sci. USA 82:7439-7443.
4. Cleary, M.L., S.D. Smith and J. Sklar. 1986. Cloning and structural analysis of cDNAs for bcl-2 and a hybrid bcl-2/immunoglobulin transcript resulting from the t(14;18) translocation. Cell 4:19-28.
5. Cotter, F., C. Price, E. Zucca and B.D. Young. 1990. Direct sequence analysis of the 14q+ and 18_q- chromosome junctions in follicular lymphoma. Blood. 76:131-135.
6. Cotter, F., C. Price, E. Zucca and B.D. Young. 1991. Direct sequence analysis of the 14_{q+} and 18_q- chromosome junctions at the MBR and MCR revealing clustering within the MBR in follicular lymphoma. Ann. Oncol. 2:93-97.
7. Crescenzi, M., M. Seto, G.P. Herzig, P.D. Weiss, R.C. Griffith and S.J. Korsmeyer. 1988. Ther-

mostable DNA polymerase chain amplification of t(14;18) chromosome breakpoint and detection of minimal residual disease. Proc. Natl. Acad. Sci. USA *85*:4869-4873.

8. **Ehrlich, M., M.A. Gama-Sosa, L.-H. Huang, R.M. Midgett, K.C. Kuo, R.A. McCune and C. Gehrke.** 1982. Amount and distribution of 5-methylcytosine in human DNA from different types of tissues or cells. Nucleic Acids Res. *10*:2709-2721.

9. **Eick, S., G. Krieger, I. Bolz and M. Kneba.** 1990. Sequence analysis of amplified t(14;18) chromosomal breakpoints in B-cell lymphomas. J. Pathol. *162*:127-133.

10. **Epstein, A. and H.S. Kaplan.** 1974. Biology of the human malignant lymphomas. Cancer *34*:1851-1872.

11. **Gribben, J.G., A.S. Freedman, S.D. Woo, K. Blake, R.S. Shu, G. Freeman, J.A. Longtine, G.S. Pinkus and L.M. Nadler.** 1991. All advanced stage non-Hodgkin's lymphomas with a polymerase chain reaction amplifiable breakpoint of *bcl-2* have residual cells containing the *bcl-2* rearrangement at evaluation and after treatment. Blood *76*:3275-3280.

12. **Hecht, B.K., A.L. Epstein, C.S. Berger, H.S. Kaplan and F. Hecht.** 1985. Histiocytic lymphoma cell lines: Immunologic and cytogenetic studies. Cancer Genet. Cytogenet. *14*:205-218.

13. **Holland, P.M., R.D. Abramson, R. Watson and D.H. Gelfand.** 1991. Detection of specific polymerase chain reaction product by utilizing the $5' \rightarrow 3'$ exonuclease activity of *Thermus aquaticus* DNA polymerase. Proc. Natl. Acad. Sci. USA *88*:7276-7280.

14. **Lee, M.-S., K.-S. Chang, F. Cabanillas, E.J. Freireich, J.M. Trujillo and S.A. Stass.** 1987. Detection of minimal residual cells carrying the t(14;18) by DNA sequence amplification. Science *237*:175-178.

15. **Li, H., U.B. Gyllensten, X. Cui, R.K. Saiki, H.A. Ehrlich and N. Arnheim.** 1988. Amplification and analysis of DNA sequences in single human sperm and diploid cells. Nature *335*:414-417.

16. **Negrin, R.S., H.-P. Kiem, I.G.H. Schmidt-Wolf, K.G. Blume and M.L. Cleary.** 1991. Use of the polymerase chain reaction to monitor the effectiveness of ex vivo tumor cell purging. Blood *77*: 654-660.

17. **Ngan, B.-Y., J. Nourse and M.L. Cleary.** 1989. Detection of chromosomal translocation t(14;18) within the minor cluster region of *bcl-2* by polymerase chain reaction and direct genomic sequencing of the enzymatically amplified DNA in follicular lymphomas. Blood *73*:1759-1762.

18. **Parker, J.D. and G.C. Burmer.** 1991. The oligomer extension "Hot Blot": A rapid alternative to Southern blots for analyzing polymerase chain reaction products. BioTechniques *10*:94-101.

19. **Pisa, E.K., P. Pisa, H.-I. Kang and R.I. Fox.** 1991. High frequency of t(14;18) translocation in salivary gland lymphomas from Sjögren's syndrome patients. J. Exp. Med. *174*:1245-1250.

20. **Said, J.W., A.F. Sassoon, I.P. Shintaku, P. Corcoran and S.W. Nichols.** 1990. Polymerase chain reaction for *bcl-2* in diagnostic lymph node biopsies. Modern Pathol. *3*:659-663.

21. **Stetler-Stevenson, M., M. Raffeld, P. Cohen and J. Cossman.** 1988. Detection of occult follicular lymphoma by specific DNA amplification. Blood *72*:1822-1825.

22. **Tsujimoto, Y. and C.M. Croce.** 1986. Analysis of the structure, transcripts, and protein products of *bcl-2*, the gene involved in human follicular lymphoma. Proc. Natl. Acad. Sci. USA *83*:5214-5218.

23. **Yunis, J.J.** 1982. The chromosomal basis of human neoplasia. Science *221*:227-236.

Update to:

Detection and Quantitation of Low Numbers of Chromosomes Containing *bcl*-2 Oncogene Translocations Using Semi-Nested PCR

Melanie Ehrlich

Tulane Cancer Center, Tulane Medical School, New Orleans, LA, USA

We used the assay described in our original article to determine whether *bcl*-2/J_H translocations are present in cancer-free individuals at the extreme level of sensitivity that our assay offered (one translocation-containing cell per 5×10^6 normal cells). Therefore, we applied this PCR analysis to peripheral blood (PB) samples from 132 individuals ranging in age from 6 to 87 years, most of whom were disease-free volunteer blood donors. The only change that we incorporated into the amplification procedure was to amplify for 24 cycles in the first round and 22 cycles in the second.

Surprisingly, we found that about half of healthy adult blood donors had this translocation at low, but very varied, levels in PB mononuclear cells, namely, 1–900 translocations per 5×10^6 normal PB mononuclear cells (1). We verified that these PCR products were derived from the translocation by replacing one of the *bcl*-2 major breakpoint region (MBR) primers in the second round of PCR with a primer that was 80 bp closer to or 30 bp further from the translocation breakpoints. The amplified DNA samples from *bcl*-2/J_H+ mononuclear cells derived from healthy blood donors had the predicted sizes indicating that they were specific translocation products. DNA sequencing of several of these PCR products revealed that they had the immunoglobulin gene breakpoint junction sequences that are characteristic of V(D)J recombination occurring in pre-B cells. These sequences were joined to MBR sequences from the *bcl*-2 gene. Therefore, we conclude that a large percentage of normal individuals have low levels of *bcl*-2/J_H translocations in circulating cells because of abnormalities in immunoglobulin gene rearrangements during lymphogenesis and the greatly prolonged lifetime of the resulting *bcl*-2/J_H+ cells.

In 40% of the *bcl*-2/J_H+ PB DNA samples, two or more closely spaced major product bands were seen. These are likely to be oligoclonal in origin and, thereby, derived from more than one independently arising *bcl*-2/J_H translocation-containing cell. These multiple bands can be distinguished especially well in our study because of the use of a radiolabeled primer to visualize the PCR products. When these samples were analyzed after treatment with nuclease S1, as well as without such treatment, the multiple bands persisted indicating that they were not single-stranded products of PCR. Furthermore, the junction points for the bcl-2/J_H translocations in a 720-bp band and a 880-bp band of one PB sample were sequenced and were shown, as expected, to be non-identical, indicating that they arose from independent translocation events.

The small size differentials, about 50 to 90 bp for most of the bands appearing as doublets, precludes their arising by the use of alternative J_H sequences for annealing to the J_H consensus primer during PCR because the number of residues between J_H sequences 1 through 6 in the human genome is 210, 602, 373, 445, and 605 bp, respectively. Some of the PB DNA samples appear to have up to seven independent bcl-2/J_H translocations. Furthermore, clonally diverse bcl-2/J_H translocations in one DNA sample could coelectrophorese due to the clustering of translocation breakpoints within the approximately 150-bp MBR at the end of the bcl-2 gene, giving an underestimate of oligoclonality. It should be noted that in our study, only MBR-localized bcl-2/J_H translocations were scored and about 40% of bcl-2/J_H translocations in follicular lymphomas do not occur at this region of the bcl-2 gene. Despite these caveats, our data suggest that the prevalence of individuals with multiple types of bcl-2/J_H translocations may be one source of the higher-than-expected percentages of follicular lymphomas that are biclonal in origin.

We compared the level of bcl-2/J_H translocations in PB mononuclear cells of people of different ages (1). The level of these B cell localized translocations from apparently healthy people showed a statistically significant age-dependence. Follicular lymphoma (FL), the cancer most frequently associated with the bcl-2/J_H translocation, usually occurs in people over the age of 50 and only rarely in those less than 40. Despite the a significant steady increase with age (P <0.001) in the percentage of individuals having rather high translocation frequencies, namely, more than 20 translocations per 5×10^6 PB mononuclear cells, three individuals less than age 40 had quite high levels of translocation-bearing PB mononuclear cells (60–900 per 5×10^6 PB mononuclear cells). Because FL and diffuse large cell lymphoma, the only cancers associated with MBR-type bcl-2/J_H translocations, are rare before middle age, the possibility that these individuals already have occult lymphoma and, hence, a high translocation frequency in their PB can be *discounted. Therefore, much higher than normal levels of bcl-2/J_H translocations in PB samples from a small portion of apparently cancer-free individuals are seen in younger adults as well as in older adults despite the general relationship between translocation frequency and age.

In addition, a significant correlation was observed between those donors who had multiple-sized bcl-2/J_H translocation products and age (Spearman rank r = 0.32, P <0.001). That 3 out of 43 people between the ages of 21 and

40 appear to have three or four independently generated bcl-2/J_H+ PB mononuclear cell populations while the majority (58%) of this age group had no detectable translocations (1) suggests that some individuals may be especially prone to generating or accumulating multiple, independently arising bcl-2/J_H translocations. Furthermore, individuals who are more likely than normal to rearrange their bcl-2 and immunoglobulin gene loci might also be more likely to generate cells with illegitimate recombination products involving other proto-oncogenes and so also be at risk for certain types of cancer in addition to those associated with bcl-2/J_H translocations.

The bcl-2/J_H+ PB cells detected in normal individuals should have been in noncancerous cells. We also quantitated the levels of bcl-2/J_H translocations in PB from five FL patients. One case was particularly interesting. By quantitative PCR, this patient carried the highest biochemically detected level of blood-associated bcl-2/J_H translocations, namely, 35 000 translocations per 5 $\times 10^6$ PB mononuclear cells (2). This corresponds to about 5%–7% of his B lymphocytes containing the translocation. With our standard sets of primers, a single specific PCR product band of about 740 bp was seen in this patient's PB sample. We also used a closer set of primer pairs, so that we could examine a 10-year-old archival tumor sample embedded in paraffin, which had partially degraded DNA. With this set of primer pairs, we obtained a smaller product of 250 bp from the PB sample and the same-sized translocation-derived PCR product from the paraffin-embedded tumor. Therefore, the very high frequency of bcl-2/J_H-translocation-bearing cells in the peripheral blood of this patient are most likely to be cancer cells and not noncancerous, but translocation-containing B cells as in PB samples from translocation-positive cancer-free individuals. However, in looking for trace amounts of bcl-2/J_H translocation-positive cells in blood samples from treated cancer patients (testing for minimal residual disease), the possibility must be considered that the positive cells are simply normal cells carrying the mutation, as are found in many healthy adults.

The FL patient with the very high level of MBR-type translocations in his PB was a 62-year-old man initially diagnosed with Stage IV FL in 1986. He received standard chemotherapy treatments several times during the last 10 years because of recurrent skin lesions. At the time of the blood sampling for PCR analysis (May, 1996), his WBC count was unremarkable and his performance status was excellent. However, in June 1996, he was admitted with severe anemia and thrombocytopenia and expired two days after admission to the hospital. Circulating lymphoma cells have been detected in lymphoma patients using flow cytometry and PCR, and a correlation with treatment has been reported. However, the prognostic significance of these findings is unclear. For the FL patient whose clinical course suddenly worsened one month after the PCR analysis, the extraordinarily high bcl-2/J_H translocation level in his blood may have been predictive for an aggressive transformation of his FL into leukemia.

REFERENCES

1. **Ji, W., G.A. Qu, P. Ye, X.-Y Zhang, S. Halabi and M. Ehrlich.** 1995. Frequent detection of bcl-2/JH translocations in human blood samples by a quantitative PCR assay. Cancer Res. *55*:2876-2882.
2. **Zhang, X.Y., H. Safah, R. Mudad, E. Maher, J. Krause, A. Miller and M. Ehrlich.** Frequency of BCL-2/JH translocations in peripheral blood of follicular lymphoma patients. Am. J. Hematol. (In press).

Reverse transcription polymerase chain reaction
(RT-PCR) 3, 7-10, 21-29, 31-32, 58-59, 63, 95,
98-100, 125-126, 129-135, 138-139, 153-
154, 166, 168, 170, 173, 213-214, 229, 245,
247, 250-259, 264, 267-268, 295
RNA copy number 55, 201, 212
RNA extraction 8, 104, 111-112, 115-116,
250-251
*Rsa*I 221

S

Sapphire II 175-176, 178, 180
Scavenger receptor 95, 97, 99
SEAPlaque 155
Semi-nested PCR 297-298, 300-308, 310
Single-tube assay 285-286, 288, 290, 292-294,
296
Spectrophotometric measurement 239
Spectrophotometer 34
Streptavidin-coated paramagnetic beads 287
Streptavidin-coated surface 173

T

t(14;18) 297, 298
T cell receptor (TCR) 33-34, 38-40
TaqMan 19, 117, 171
Terminal deoxynucleotidyl transferase (TdT)
216, 250
2,3,7,8-tetrachlorodibenzo-p-dioxin (TCDD)
247-248, 251, 254-257, 261
Titration 10, 22, 24-27, 29, 44, 58, 72-73,
125-126, 129, 131-133, 135, 192, 196,
221, 223, 229, 240, 282, 291
Translocations 297-298, 302, 304, 307-309
Tyrosine hydroxylase (TH) 31-32

V

V beta (BV) repertoire 33-36, 38-39
v-*erb* B oncogene 5, 69-70
Variability 9, 39, 44, 52, 54, 63, 102, 109,
111-112, 115, 122, 154, 157, 164-165,
167-168, 245, 258, 264
Video image analysis 11, 13, 19, 47, 62,
119-120, 122-124